PORTABLE SYNCHROTRON LIGHT SOURCES AND ADVANCED APPLICATIONS

Related Titles from AIP Conference Proceedings

To learn more about these titles, or the AIP Conference Proceedings Series, please visit the webpage **http://proceedings.aip.org**

PORTABLE SYNCHROTRON LIGHT SOURCES AND ADVANCED APPLICATIONS

International Symposium on Portable Synchrotron
Light Sources and Advanced Applications

Shiga, Japan   13 – 14 January 2004

Editors
Hironari Yamada
Noriko Mochizuki-Oda
Makoto Sasaki

Ritsumeikan University
Shiga, Japan

SPONSORING ORGANIZATIONS
Ritsumeikan University
JSPS (Japan Society of Promotion of Science)

Editors:

Hironari Yamada
Noriko Mochizuki-Oda
Makoto Sasaki

Ritsumeikan University
1-1-1 Nojihigashi, Kusatsu-City
Shiga 525-8577
JAPAN

E-mail: hironari@se.ritsumei.ac.jp
        odan@se.ritsumei.ac.jp
        maks715@se.ritsumei.ac.jp

L.C. Catalog Card No. 2004110101
ISBN 0-7354-0195-0
ISSN 0094-243X

Printed in the United States of America

CONTENTS

ORAL PRESENTATIONS

Light Sources and Instruments

Applications to Materials Science

Applications to Chemistry and Biophysics

Imaging and Medical Applications

POSTER PRESENTATIONS

Preface

This book is devoted to the recent dynamical progress on synchrotron light sources and their advanced applications.

Synchrotron radiation has led to many innovations in material and life sciences. The construction of huge facilities such as SPring-8 was a trend in the 20th century. Recently, big efforts have been made to downsize the light source. It is expected that powerful portable X-ray sources will change the paradigm of the X-ray business and open up completely new research fields that were impossible before.

Significant progress on the laser-plasma and laser-Compton X-ray sources has been made. A unique advancement, made by Yamada at Ritsumeikan University, is the development of the portable synchrotron named MIRRORCLE. In MIRRORCLE, electron energy as low as 6 MeV and electron orbit diameter as small as 15 cmϕ, are achieved using a normal conducting magnet. The quality of the X-ray beam generated by this machine, using a novel method, exceeds that of GeV synchrotrons in some regards. Namely, the X-ray source spot size is of the order of micrometers, which is much smaller; the covered X-ray energy range is from a few keV to 6 MeV, which is much larger; the radiation angle is much wider; and the total X-ray flux 10^9 photons/s is brighter, although the brilliance is lower. This machine provides excellent quality X-ray imaging and brings a new era of medical diagnosis. The hard X-ray microscope is not a dream anymore. X-ray lithography will be soon introduced in factory processes.

The 20 MeV version of MIRRORCLE provides intensity of FIR synchrotron light, by integrating the emission over the whole arc of electron orbit, of the order of Watts on average, utilizing the mirror technique, while the conventional synchrotron provides milli-Watt order. This instrument will make it possible to study the dynamics and behavior of living specimens at each levels such as protein, cell, and organ by exciting directly and selectively a particular state of proteins.

The Ministry of Education and Science of Japan selected the Synchrotron Light Life Science Center (SLLS) of Ritsumeikan University to become a 21st Century COE (Center of Excellence) because of these unique instruments. We organized, for the first time, international symposium on portable synchrotrons and their advanced applications. One hundred and twenty people from Japan, North America, and Europe, working in different scientific fields, such as accelerator, laser, and beam line scientists, material scientists, chemists, biologists, medics, and medical doctors, gathered. Twenty-three oral and invited papers and 28 posters were presented in a two-day symposium. Hot topics in the instrumentations and applications of X-rays as well as

FIR rays were represented.

We are happy to provide information, in this book, on the features of advanced X-ray sources and FIR sources for potential users.

Hironari Yamada

Director Synchrotron Light Life Science Center

Leader Of 21st Century COE program

Symposium Organization

Editors

Hironari Yamada, Noriko Mochizuki-Oda, and Makoto Sasaki, *Ritsumeikan University*

Organizing Committee

Hironari Yamada, *Director SLLS, 21ˢᵗ Century COE leader, Ritsumeikan University (Co-chair)*

Jun-ichi Chikawa, *Director CAST (Co-chair)*

Yoshihiro Taniguchi, *Professor, Ritsumeikan University (Co-chair)*

Takao Nanba, *Professor, Kobe University*

Hiroshi Kihara, *Professor, Kansai Medical University*

Hitoshi Tamiaki, *Professor, Ritsumeikan University*

Shigeru Imai, *Professor, Ritsumeikan University*

Hiroshi Fukami, *Professor, Ritsumeikan University*

Msakazu Kikuchi, *Professor, Ritsumeikan University*

Toshio Matsuda, *Professor, Ritsumeikan University*

Naotake Nakamura, *Professor, Ritsumeikan University*

Symposium Secretariat

Secretary General: Makoto Inoue, *Ritsumeikan University*

Scientific Secretary: Noriko Mochizuki-Oda, *Ritsumeikan University*

Supported by

The Japanese Society for Synchrotron Radiation Research

The Japan Society of Applied Physics

The Laser Society of Japan

Acknowledgement

The symposium was supported by the following companies:

APOLLOMEC CO.,LTD

NEC TOKIN Corporation

NTT Advanced Technology Corporation

Photon Production Laboratory, Lt'd

TOYAMA Co., Ltd

nichicon corporation

Bunkoh-Keiki Co., LTD

Yoshizawa-LA CO., LTD

Canberra Japan KK

Kanno-Giken CO., LTD

Fujikin Incorporated

Welcome to the International Symposium on Portable Synchrotron Light Sources and Advanced Applications

We are happy to meet with people from all over the world gathered to make hot discussion on the frontiers brought by the portable synchrotron light sources.

Synchrotron radiation has led to many innovations in material and life science. The construction of huge facilities such as the Spring-8 was the trend in the 20th century. Recently, big efforts are, however, made on down sizing the light source to open up its wider applications and new research fields. Significant progresses have been made in the laser-plasma and laser-Compton x-ray sources.

At Ritsumeikan University portable synchrotrons with electron energy as low as 6-MeV and an electron orbit diameter as small as 15cm has been developed. The brightness of x-ray generated by novel method is comparable to that of GeV synchrotrons. Intensity of IR synchrotron lights from 20 MeV versions reaches a Watt on average by a mirror technique to integrate emission from the whole electron orbit. These instruments make possible to study the material science as well as dynamics and behavior of the living specimens in each level such as protein, cell, organ and whole body.

Synchrotron Light Life Science Center (SLLS) of Ritsumeikan University established by the 21st Century COE program by MEXT organizes for the first time the international symposium on portable synchrotrons and its advanced applications. Through the symposium, we hope to provide numerous opportunities to learn about new frontiers in portable synchrotron-based experiment, which will impact your research interests. This symposium will make ideal interact among the wide variety of scientists including the light source scientists, biologists, chemists, medicine and material scientists.

Co-Chair

Hironari Yamada   Junichi Chikawa   Yoshihiro Taniguchi

Oral Presentations

Light Sources and Instruments

Ultrabright Multikilovolt Coherent Tunable X-Ray Source at ~ 2.71 – 2.93 Å for Biological Microimaging

Alex B. Borisov, Xiangyang Song, Ping Zhang, Jonas Moses, Jeremy Callner, Maria Vogrinc, Keith Boyer, and Charles K. Rhodes

Laboratory for X-Ray Microimaging and Bioinformatics, Department of Physics, University of Illinois at Chicago, Chicago, IL 60607-7059, USA

Abstract. The recent observation of strong amplification on multikilovolt Xe(L) hollow atom transitions in the ~2.8 Å spectral region can be seen as a consequence of the combination of (1) a new concept for amplification that involves the creation of a highly ordered state combining ionic, plasma, and coherent radiative components and (2) the use of two recently discovered (c. ~1990) forms of radially symmetric energetic matter, namely, hollow atoms and self-trapped plasma channels. This approach enables the demanding power densities necessary for x-ray amplification (~10^{19} W/cm^3) to be reached under conditions for which (α) the effective phase space volume of the interaction is profoundly limited and (β) the energy transfer is radiation dominated. A leading application will be the realization of a new mode of microimaging of living biological matter having a spatial resolution ~1000-fold superior to conventional light microscopy.

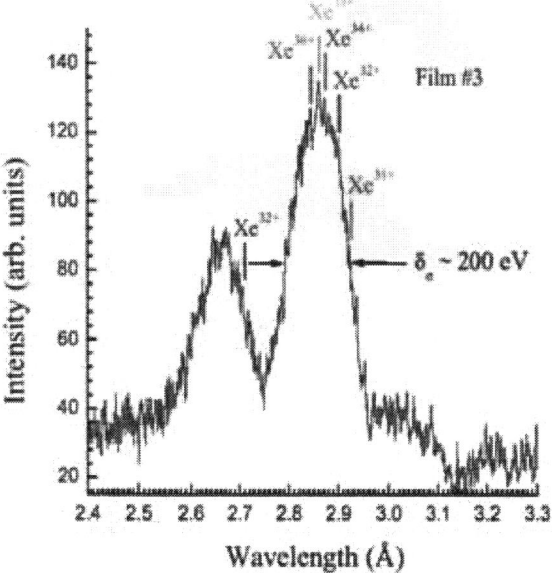

FIGURE 1. Spontaneous emission profile of Xe(L) hollow atom states.

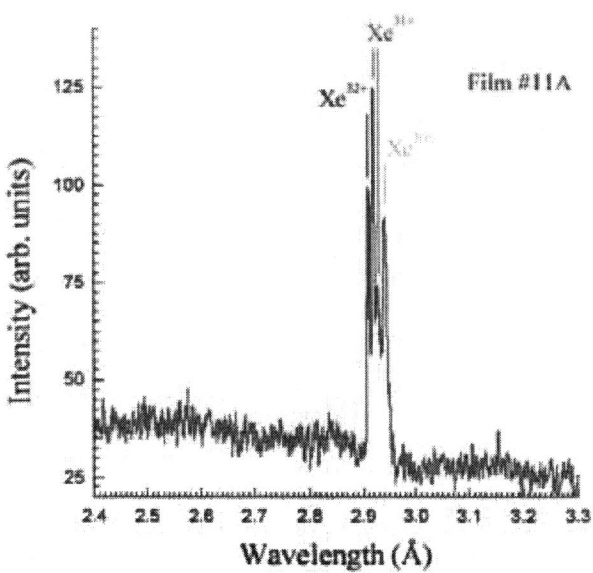

FIGURE 2. Amplified components of the Xe(L) emission on the Xe^{30+}, Xe^{31+}, and Xe^{32+} arrays observed from the plasma channel in the axial direction.

CP716, *Portable Synchrotron Light Sources and Advanced Applications,*
edited by H. Yamada, N. Mochizuki-Oda, and M. Sasaki
© 2004 American Institute of Physics 0-7354-0195-0/04/$22.00

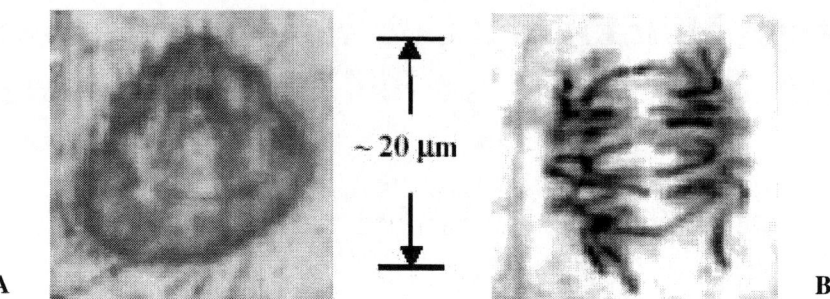

FIGURE 3. (A) Surface damage to the 12.7 μm thick Ti foil used to shield the film pack at an axial distance of ~12.5 cm from the source as viewed under a light microscope with front surface illumination. The damage does not fully penetrate the Ti foil. (B) Photograph of a cell in the early anaphase stage of cell division obtained with a light microscope that illustrates the favourable spatial match of the x-ray beam with the requirements of an illuminator for biological microimaging. This cellular image is taken with permission from Sobotta/Hammersen, Third edition, *Histology, Color Atlas of Microscopic Anatomy* (Urban and Schwarzenberg, Baltimore-Munich, 1985), figure no. 73f on page 45.

Detailed molecular structural information of the living state is of enormous significance to the medical and biological communities. Since hydrated biologically active structures are small delicate complex three-dimensional (3D) entities, it is essential to have molecular scale spatial resolution, high contrast, distortionless, direct 3D modalities of visualization of naturally functioning specimens in order to faithfully reveal their full molecular architectures. An x-ray holographic microscope equipped with an x-ray laser as the illuminator would be uniquely capable of providing these images [1,2,3]. A concordance of physical evidence [4,5], that includes (a) the observation of strong enhancement of selected spectral components of several Xe^{q+} hollow atom transition arrays (q = 31, 32, 34, 35, 36, 37) radiated axially from confined plasma channels, (b) the measurement of line narrowing that is spectrally correlated with the amplified transitions, (c) evidence for spectral hole-burning in the spontaneous emission, a manifestation of saturated amplification, that corresponds spectrally with the amplified lines, and (d) the detection of an intense narrow ($\delta\theta_x \sim 0.2$ mr) directed beam of radiation, (1) experimentally demonstrates in the $\lambda \cong 2.71$–2.93 Å range ($\hbar\omega_x \cong$ 4230–4570 eV) the operation of a new concept capable of producing the ideal conditions for amplification of multikilovolt x-rays and (2) proves the feasibility of a compact x-ray illuminator that can cost-effectively achieve the mission of biological x-ray microholography. The development of this new mode of seeing of living systems will represent the third major advance in three centuries in biological microimaging. The suitability of the x-ray source for the illumination of small biological entities is illustrated in Fig. (3) above by the matching of the beam size (A) with a typical specimen of interest (B). The measurements also (α) establish the property of tunability in the quantum energy over a substantial fraction of the spectral region exhibiting amplification ($\Delta\hbar\omega_x \sim 345$ eV) and (β) demonstrate the coherence of the x-ray output through the observation of a canonical spatial mode pattern. An analysis of the physical scaling revealed by these results indicates that the capability of the x-ray source potentially includes single-molecule microimaging, the key for the *in situ* structural analysis of membrane proteins, a cardinal class of drug targets. An estimate of the peak brightness achieved in these initial experiments gives a value of $\sim 10^{31}$–10^{32} photons $\cdot s^{-1} \cdot mm^{-2} \cdot mr^{-2}$ (0.1% Bandwidth)$^{-1}$, a magnitude that is $\sim 10^7$–10^8-fold higher than presently available synchrotron technology, as shown in Fig. (4).

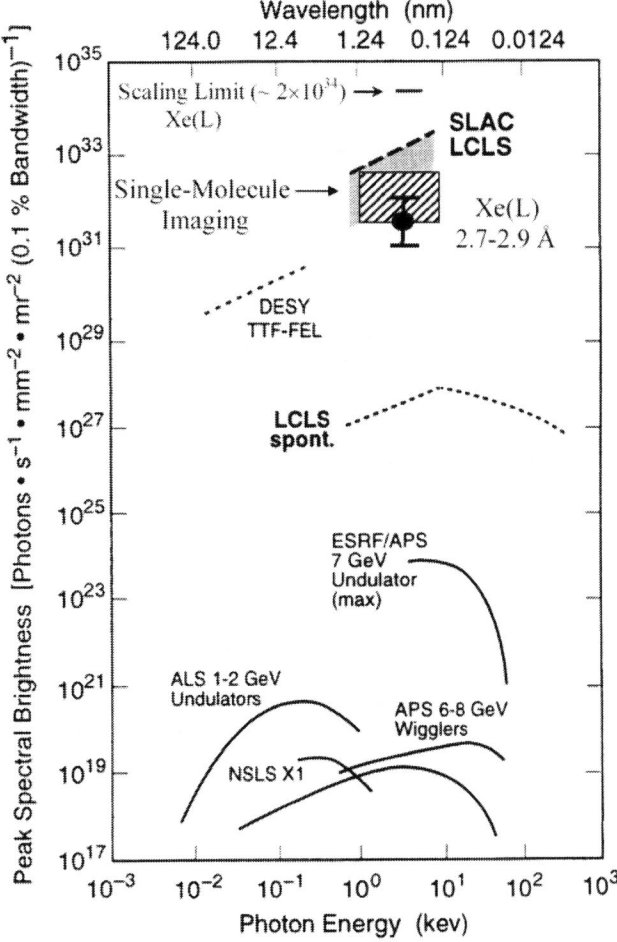

FIGURE 4. Peak spectral brightness comparisons of the present Xe(L) source at 2.7–2.9 Å with existing (solid contours) and projected (dashed contours) facilities. The present performance is given, the estimated requirement for single-molecule imaging is represented by the striped zone, and the scaling limit for a compact laboratory instrument based on the findings of this study is indicated. Figure used with permission and adapted from Tatchyn, R. et al, "X-ray optics design studies for the SLAC 1.5–15 Å Linac coherent light source, *Nucl. Instrum. Methods Phys. Res.* A **429,** 397 (1999).

Sandia is a multiprogram laboratory operated by the Sandia Corporation, a Lockheed Martin Company, for the United States Department of Energy under contract no.DE-ACO4-94AL85000.

REFERENCES

1. Solem, J. C., and Baldwin, G. C., *Science* **218,** 229 (1982).

2. Boyer, K., Solem, J., Longworth, J. W., Borisov, A. B., and Rhodes, C. K., *Nature Med.* **2,** 939 (1996).

3. Netze, R., Wouts R., van der Spoel D., Weckert E., and Hajdu, J., *Nature* **406,** 752 (2000).

4. Borisov, A. B., Song, X., Frigeni, F., Koshman, Y., Dai, Y., Boyer, K., and Rhodes, C. K., *J. Phys. B* **36,** 3433 (2003).

5. Borisov, A. B., Davis, D., Song, X., Koshman, Y., Dai, Y., Boyer, K., and Rhodes, C. K., *J. Phys. B* **36,** L285 (2003).

ACKNOWLEDGMENTS

Support for this research was partially provided under contracts with the Army research Office (DAAG55-97-1-0310), the Naval Research Laboratory, and the Department of Energy at the Sandia National Laboratories (contracts 17733, 11141, and BF 3611).

R&D Results on Laser-Compton Photon Beam Generation *

Junji Urakawa, Masahiro Nomura[1] and Mikio Takano[1]

High Energy Accelerator Research Organization(KEK), 1-1 Oho, Tsukuba-shi, Ibaraki, Japan

[1]National Institute of Radiological Sciences (NIRS), Chiba, Japan

Abstract.

I review recent studies of X-ray generation based on Laser-Compton scattering. Especially, our recent experimental results for the generation of γ-rays and plans for X-ray generation will be explained. The stability of the photon beam source by the Laser-Compton scattering will also be discussed from issues of laser and electron beam stability.

1 INTRODUCTION

We succeeded the establishment of CW laser-wire system for precise transverse emittance measurement within the resolution of 1pm-rad in the electron storage ring. Therefore, I explain the art of the technologies for the laser-wire in next section. As for the measurement of the beam emittances, the laser-wire beam profile monitor has been improved [1]. Two laser-wires were installed to measure both vertical and horizontal beam sizes. The width of the horizontal wire was reduced to $5\mu m$ to be able to measure smaller electron beams. The effective laser power was upgraded about one order of magnitude to make data taking time shorter. The laser-wire profile monitor demonstrated enough position stability which was less than 100nm and made precise collision between the laser and electron beams with 90 degrees crosssing angle during the period of one beam size measurement.

Also, we are generating high brightness polarized γ-rays for the production of polarized positron beam through inverse Compton scattering [2, 3]. In this head-on collision experiment, we generated 1×10^6 γ-rays with a time duration of 30ps in rms, leading to a peak brightness of $7.3 \times 10^{18}/(mrad^2 mm^2 0.1\% bandwidth sec)$ near to the maximum energy of 56MeV. In the section 2, I also describe the essential part of precise head-on collision techniques for generation of high brightness photon beam.

Recent proposals [4, 5] concerning the radiative laser cooling of electron beams for the production of intense beams of quasi-monoenergetic X-rays have suggested the possibility of compact photon beam source based on Laser-Compton scattering. If we assume an enhancement factor of 20000 in the laser power and a 50 μm rms laser-beam size at the beam-laser interaction point (IP), which can be achieved by using a high-finesse optical cavity, the peak power exceeds 1 TW/cm^2 in the case of a 7 ps FWHM pulsed laser. Mode-locked lasers with these parameters are commercially available (for example $Nd : YVO_4$). This power lies well within the linear Compton scattering region. For this proof-of-principle experiments, we have de-

signed an electron storage ring for medical applications. Its circumference is 13.44 m, the beam energy ranges from 40 to 60 MeV, and the energy acceptance amounts to ±3%. In parallel, a 42 cm long optical cavity was constructed which can provide a beam-laser interaction at a 10 degree crossing angle [6].

We also intend to conduct other proof-of-principle experiment for the radiative laser cooling at the KEK-ATF [7]. Assuming the same 10 degree crossing, electrons stored in the KEK-ATF damping ring would quickly be lost due to the repeated interaction with the laser and top-up injection would be necessary. So, we will mainly study the timing jitter and alignment issues at the KEK-ATF. From the experimental results we can then extrapolate to the high average-brilliance γ-ray generation.

In the following sections I describe the laser-wire technologies for inverse Compton scattering which have been developed at the KEK-ATF. In the section 3, status of the proof-of-principle experiment on laser pulse stacking is given shortly. I explain the stability study on Compton scattering at the KEK-ATF, and then the experimental plan for radiative laser cooling itself. Finally, discussion and summary are given.

2 LASER-WIRE TECHNOLOGIES FOR INVERSE COMPTON SCATTERING

We have developed a laser-wire beam profile monitor for measuring the electron-beam emittance at the KEK-ATF. This monitor is based on the inverse Compton scattering with a laser light target. A thin and intense laser target is produced by exciting a Fabry-Perot optical cavity with a CW laser. In the process of this monitor development, the following technologies have been established [8]:

- A relative positioning of the mirrors with a precision of 0.1 nm (rms) was achieved, using a feedback system based on a piezoelectric transducer (PZT). This precision corresponds to a width control of the laser-wire size within 5 μm rms.

- The mode matching technique, required for maintaining a power enhancement factor above 1000, was precisely demonstrated. The ultimate enhancement factor depends on the reflectivity of the mirrors.

- We found that the electron rms beam orbit jitter in the storage ring is less than 1 μm rms.

- The laser-wire position was stabilized to within 0.1 μm rms.

---

* Correponding author: J.Urakawa, email:junji.urakawa@kek.jp

CP716, *Portable Synchrotron Light Sources and Advanced Applications,*
edited by H. Yamada, N. Mochizuki-Oda, and M. Sasaki

In the case of head-on collision, we developed a special e^--laser colliding system with a short focal length. We constructed a Compton chamber in which a pair of off-axis parabolic mirrors with a focal length of 150mm were installed in the e^--beam line. Fig. 1 shows a cross-sectional view of the Compton chamber. At both sides of the chamber, there are two beam-position monitors (BPM) to measure the positions of the e^- beam. Each mirror has a through hole of 5mm diameter at the center to pass the e^- beam and backscattered γ rays. At the collision point, screen monitors can be inserted when the positions of the e^- and laser beams are measured. Near to the exit of the laser device, the laser beam is expanded two and a half times by an expander system consisting of concave and convex lenses. In the chamber, the laser beam is focused by a parabolic mirror and makes a head-on collision with the e^- beam at the collision point. During this experiment,

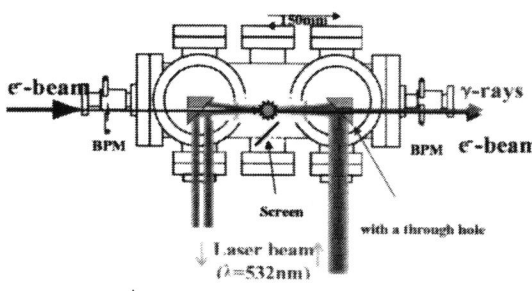

Figure 1: Cross-sectional view of the Compton chamber.

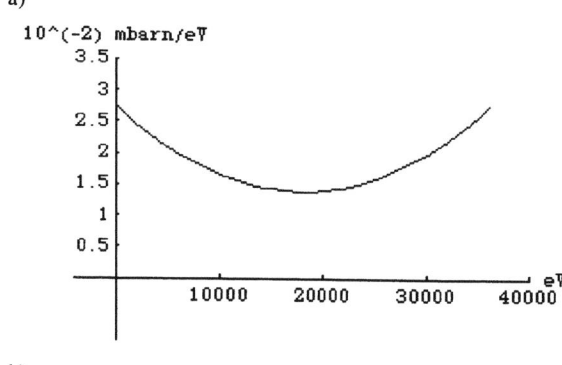

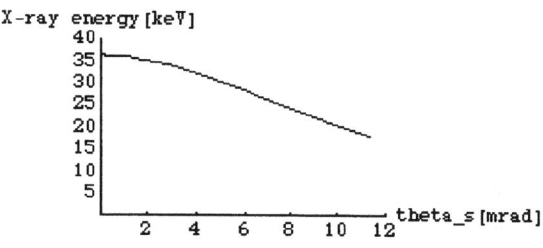

Figure 2: (a) Differential cross section of Compton scattered X-ray, horizontal axis is energy of X-ray. (b) Angular distribution of X-ray energy within $1/\gamma$, horizontal axis is scattering angle of laser photon.

we studied the stability of both beams and achievable beam size. If we use new mode-locked laser and e^- beam with normalized emittance of $1mmmrad$, both beam size can be focused to $5\mu m$. Position jitter of both beam can be reduced until sub-μm level ,which depends on system of laser and electron beam sources. In the case of ideal collision , Fig. 2 shows a differential cross section and an angular distribution of Compton sacttered X-ray, assuming $5\mu m$(rms) beam size of both beams, $3psec$ pulse width of both beams, laser wavelength of 1064nm and electron energy of 45MeV. If we operate the electron linac with repetition rate of 50Hz, 200 bunches/pulse and 0.5nC/bunch, $10mJ$/pulse in the optical cavity with 357MHz mode-locked laser can be converted to $3.54 \times 10^{14}/sec$ X-rays. The total yields near 35keV X-ray which are cut within $0.2/\gamma$ by a collimator is $1.96 \times 10^{13}/sec$, which is enough to do dynamic Intravenous Coronary Arteriography (IVCAG).

3 PROOF-OF-PRINCIPLE EXPERIMENT ON LASER-PULSE STACKING

We have started the proof-of-principle experiment on short pulsed laser stacking in a Fabry-Perot optical cavity. We selected a commercially available mode-locked laser (High Q Laser IC-6000 ps VAN357) and a 42-cm optical cavity for this experiment. For an easy and quick study, we

chose a very simple set-up, consisting of the mode-locked laser, the optical system, 99.75% reflective mirrors and several detectors, as shown in Fig. 3. The frequency of the mode-locker was modulated for measuring the cavity finesse during the pulse stacking. This means that the central frequency of the longitudinal mode from the mode-locked laser oscillator is shifted. If the central frequency is changed by about 500 Hz which corresponds half a wavelength of 1064 nm, the optical cavity for the pulse stacking oscillates between reflection and accumulation of the laser pulse. Table 1 shows the results of a simple measurement, indicating a good agreement with the calculation for the employed 99.75% reflective mirrors. The mirrors are also commercially available and of spherical type (curvature radius 250 mm). The discrepancy of 39% and 11% between calculation and measurement in the transmission arises from the absorption by the mirror material, errors in the surface curvature of the reflective coating, for which a precision at the sub-nanometer level is required, and an insufficient optical mode matching. The table also lists the values calculated for higher-quality mirrors, whose employment would benefit from the other technologies developed for the laser-wire monitor. In the near future, we will add a stepping motor and piezoelectric ac-

Table 1: Proof-of-Principle experiment on laser pulse stacking.

Items	Measured Values	Calculated Values
First experiment with 90% **reflectivity**		
Transmission	0.61	1
Enhancement Factor	6	10 ± 4
Finesse	40 ± 5	30 ± 9
Second experiment with 99.75% **reflectivity**		
Transmission	0.89 ± 0.01	1
Enhancement Factor	~ 300	836 ± 500
Finesse	1040 ± 40	1254 ± 750
Final experiment with 99.99% **reflectivity**		
Transmission	non	1
Enhancement Factor	non	20940 ± 1904
Finesse	non	31414 ± 2856

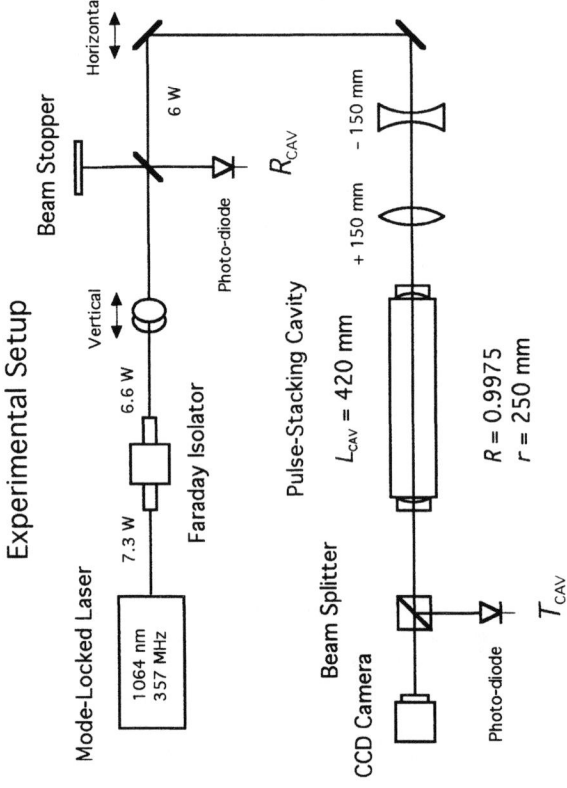

Figure 3: Schematic diagram of the experimental set-up for laser-pulse stacking.

tuators (PZT) to control the length of the optical cavity as is done for the mode-locked laser oscillator [9]. Figure 4 shows a conceptual diagram for the proof-of-principle experiment on radiative laser cooling. The laser oscillator, the pulse-stacking system and the beam of the storage ring are synchronized by a single signal generator with a relative frequency stability of 10^{-13}. This synthesizer and

also mirrors with 99.9%, 99.99% and 99.999% reflectivity are available from various companies. The high-reflection mirrors exhibit an ultralow loss of a only few ppm from scattering and absorption [10].

4 STABILITY STUDY ON COMPTON SCATTERING AT KEK-ATF

The laser-wire beam profile monitor at the KEK-ATF is working well and it measures the beam profile of each bunch in multi-bunch operation. If this monitor is replaced by the optical cavity for the laser-pulse stacking, we can check the stability of the waist size of the pulsed laser beam using the low-emittance electron beam as a probe. The scheme of Fig. 4 promises a precise synchronization within 0.1 psec. However, we have to confirm that this number can be achieved under practical conditions and demonstrate the generation of $10^{13}\gamma$-rays per second.

5 EXPERIMENTAL PLAN OF RADIATIVE LASER COOLING

The storage ring with the radiative laser cooling (LESR) uses an intense laser pulse stored in a high-finesse resonator, which repetitively interacts with the circulating electron beam. The rapid damping caused by the laser-electron interaction counterbalances the intrabeam scattering effect (IBS), thus allowing electron beams of relatively low energy to be cooled or stabilized in the storage ring, down to extremely low transverse emittances.

In nonconservative systems, a steady-state particle distribution can be established due to a balance of cooling and heating processes. In an electron storage ring, the only cooling process is the radiation damping. We here consider an electron beam with natural emittance ϵ_n due to the synchrotron radiation power from bending magnets p_n. If, in addition, a radiation damping by laser-Compton scattering is present, characterized by the partial emittance ϵ_l and the radiation power p_l, the steady-state emittance without the

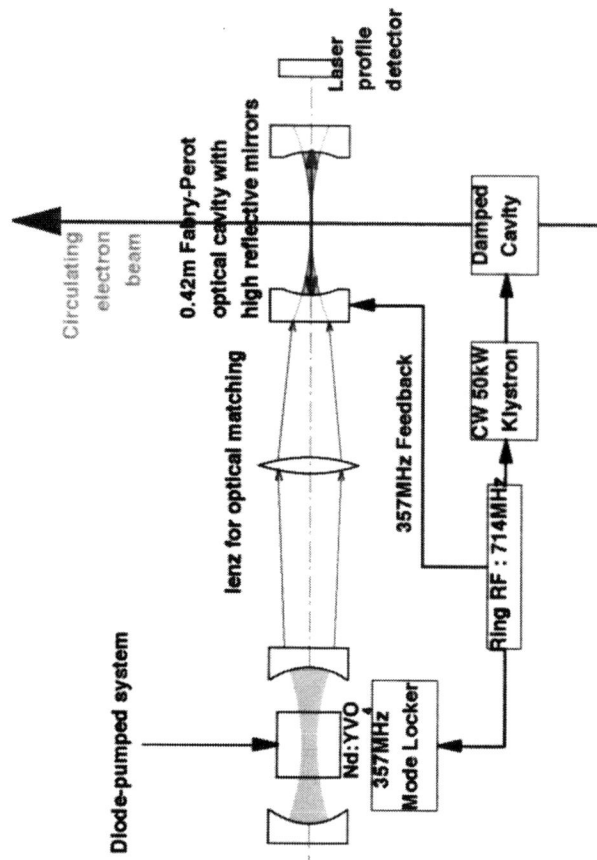

Figure 4: Conceptual diagram for the proof-of-principle experiment on radiative laser cooling.

intra-beam scattering effect becomes

$$\epsilon = \frac{p_n \epsilon_n + p_l \epsilon_l}{p_n + p_l}, \qquad (1)$$

where we treated the bending-magnet synchrotron radiation and the laser-beam interaction as statistically independent.

Due to the Compton interaction, the energy spread increases. The partial Compton energy spread is:

$$\delta^2 = \frac{7}{10} \gamma \frac{E_{las}}{E_0} \frac{1 + \cos \varphi}{2}. \qquad (2)$$

The steady state energy spread due to both synchrotron and Compton fluctuations can be derived from Eq.(1), where emittances should be replaced by the squared energy spreads without IBS. Thus, the total energy spread in the steady state depends on the intensity of the Compton interaction [11].

6 DISCUSSION AND SUMMARY

We are inquiring the effect of the position instability of both beams and the energy spread of the electron beam to produce an applicable high brightness X-ray beam. So, the

construction of photo-cathode RF gun test bench started for the generation of 200 bunches/pulse e^- beam and the study on quality of the high current beam will be scheduled in JFY2004. On the other hand, the compact storage ring for LESR is under design as an international collaboration project. For an electron bunch of 2 nC (about 1.3×10^{10} electrons) and 10% coupling, the steady-state longitudinal emittance approaches 1.0×10^{-4} m, provided the RF accelerating voltage is sufficient (0.4 MV). The storage ring will be equipped with a 714 MHz RF system, which can accommodate a beam of 11.67 mm rms bunch length and 1.67% rms energy spread, as expected to result from the quantum fluctuation of the Compton scattering. Since the longitudinal direction is heated by both the laser-Compton scattering and the intra-beam scattering, a momentum acceptance of 3.0% is necessary to alleviate chromatic effects at the laser-beam interaction point and to ensure an adequate dynamic aperture. In a ring with the target momentum acceptance, we can circulate the electron beam together with an extremely high-power laser pulse, $e.g.$, 5 mJ/pulse, assuming that the ideal theory of Compton scattering applies.

We have roughly estimated the yield of X-rays scattered along the electron trajectory within the solid angle of $1/\gamma$. It is 10^{14} photons per second and applicable for digital differential angiography.

Acknowledgements

I would like to thank all members of the KEK-ATF group and the laser-wire R&D group for their helpful support and discussion. This research was partially supported by the budget for the Advanced Compact Accelerator Project of the National Institute of Radiological Sciences

7 REFERENCES

[1] Y.Honda, K.Kubo et al., Physical Review Letters, Vol.92, No.5, 054602-1 (2004)

[2] M.Fukuda et al., Physical Review Letters, Vol.91, No.16, 164801-1 (2003)

[3] I.Sakai et al., Physical Review Special Topics-Accelerators and Beams, 6, 091001(2003)

[4] Zh. Huang and R.D. Ruth, Phys. Rev. Lett., Vol.80, No.5, 1998, p.976.

[5] V. Telnov, SLAC-PUB-NSF-ITP-96-142, 1996.

[6] J.Urakawa et al., to be published in Nucl. Instr. and Meth.

[7] Edited by F. Hinode et al., KEK Internal 95-4, 1995.

[8] Y. Honda et al., to be submitted to Nucl. Instr. and Meth.

[9] M. Nomura et al., to be submitted to Nucl. Instr. and Meth.

[10] G. Rempe et al., Optics Letters, Vol.17, No.5, 1992, p.363.

[11] P. Gladkikh and E. Bulyak, private communication.

Features of the portable synchrotrons named MIRRORCLE

Hironari Yamada

The Synchrotron Light Life Science Center, Ritsumeikan University
1-1-1 Nojihigashi, Kusatsu-City, Shiga 525-8577, Japan

Abstract. The Synchrotron Light Life Science Center of Ritsumeikan University seeks the next generation life science beyond genomics and proteomics by the unique portable synchrotrons named MIRROCLE. MIRRORCLE-6X generates brilliant hard x-rays in spite of its 6-MeV electron energy. The power of IR/FIR beam from MIRRORCLE-20 is much higher than any synchrotrons by its unique mirror system introduced to collect the synchrotron light from whole arc of the electron orbit. Unique feature of MIRRORCLE is described in this paper.

INTRODUCTION

The Synchrotron Light Life Science Center granted by the MEXT 21st Century COE Program [1] seeks unique Life Science. The overview and the target of our project are drawn in Fig.1. We study the dynamics and behavior of the living specimens in each level such as protein, cell, organ, etc. by developing new research instruments based on portable synchrotrons.

Yamada and collaborators have succeeded in developing two tabletop synchrotron light sources having 15 cm orbit radius made of single peace of normal conducting magnets for brilliant IR radiation named MIRRORCLE-20 as seen in Fig. 2(a) and for hard x-ray productions, named MIRRORCLE-6X as seen in Fig. 2(b).

MIRRORCLE20 provides the IR rays much brighter than any other SR sources. The IR radiation from the entire electron orbit of exact circular is collected, which can reach $10W/mm^2$ in average [2-4]. The IR synchrotron light spectrum is centered at 8 μm wavelength and covers from few to 100μm.

The unique beam line is the dispersive type IR irradiation system. We are able to place specimens along 15 cm long focal plane, where one to 0.5 % band monochromatic IR-rays are available through 30 of 1mm slits by using a grating monochrometer [5]. The powerful CW white beam of SR is in some extent more useful compared with the monochromatic pulse beam of free-electron laser. Detection of IR rays is another task, but we are able to detect chemical

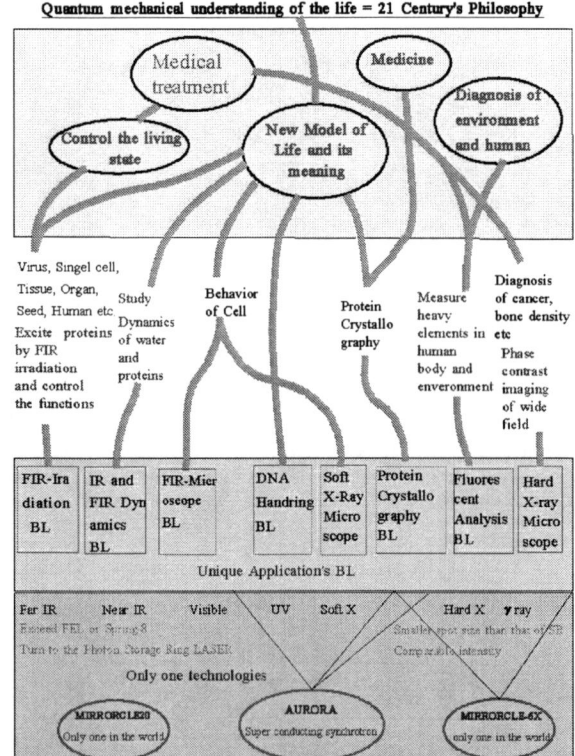

FIGURE 1. Top column shows the destination of the project. The second column describes proper targets of the project. The third column is devoted for specific beam lines to be prepared. The bottom line indicates unique synchrotron light sources dedicated for IR, soft x-ray, and hard x-ray.

CP716, *Portable Synchrotron Light Sources and Advanced Applications,*
edited by H. Yamada, N. Mochizuki-Oda, and M. Sasaki
© 2004 American Institute of Physics 0-7354-0195-0/04/$22.00

FIGURE 2(a). MIRRORCLE-20, the brilliance of Far- and Mid-IR radiation is much brighter than any other SR sources.

FIGURE 2(b). MIRRORCLE-6X is made of normal conducting magnet, but generates hard X-rays up to 6MeV as bright as SR sources.

products due to the specific wavelength IR irradiation onto the specific proteins or DNA in water. We will find switches, which trigger the protein's functions [5].

Changes in the characteristics of water by the specific IR irradiation are one of the important research subjects to understand the meaning of heat in human body.

Characteristics of cancer tumor or cholesterol under IR-rays irradiation are another subject to be studied. We will find specific wavelength, which heat up cancer tumor selectively, or resolve cholesterol. If these wavelengths are not absorbed by the skin or water, we may open up new medical treatment opportunity.

MIRRORCLE-6X [8] has been completed in November 13, 2003, and is operated for studying the characteristics of machine since December. The obtained x-ray brightness is the kind that 20cm wide X-ray image of chicken can be taken in 0.5 sec. This implies that the total X-ray flux is higher than conventional SR sources, since the radiation angle is wide. The brilliance is the kind of 10^9 photons/mrad^2, mm^2, 0.1 %band at for instance 30keV X-ray energy, while we are able to increase another two orders by increasing the accelerator cavity voltage.

The quality of the X-ray images is so wonderful as seen in Fig. 3, which shows not only all organs but also feather and brad cells of chicken of normal size without special imaging aids.

The X-ray microscope is one of the important application subjects for MIRRORCLE-6X, since the X-ray emitting point can be an order of nano-scale, which enables the hard X-ray microscope without optical elements [9,10].

The 8-GeV synchrotron has been the only facility, which accepts the X-ray fluorescent analysis of heavy elements before [11.12]. The routine operation becomes possible by MIRRORCLE-6X, since it provides well-collimated abandant high energy X-ray flux. The X-ray fluorescent spectrum of heavy elements taken by MIRRORCLE-20 is shown in Fig. 4.

FEATURES OF MIRRORCLE-6X

X-ray emission mechanism

Yamada first gave the mechanism of brilliant hard X-ray source by the normal conducting magnet theoretically in 1996 [13]. He proposed the use of collision of relativistic electrons and target in the synchrotron orbit. We could easily expect the X-ray characteristics so as in Table 1.

TABLE 1. The emission mechanism and the X-ray Characteristics.

We can easily expect the following
1. We use atomic force instead of magnetic 2. Source size is determined by the target 3. Radiation angle is determined by 1/γ, since photon is emitted from the relativistic electrons 4. Electrons is re circulating   →   1. X-ray energy go up to the electron energy. 2. Promise the high brilliance. Hard X-Ray microscope without optical elements. 3. Wide radiation field is wonderful 4. Regardless the target size and materials the total flux is same. which realize the less than one micron source size.
Brilliance=Photon number/mrad^2,mm^2

The source size is the most important facts in the quality of the X-ray beam. This is the reason why most of SR sources are made in the large scale to minimize the beam emittance. Strong magnetic field and low energy electron beam increase the emittance. In the present scheme no matter how the beam size is large, the target size determines the X-ray source point. Those electrons, which pass through the target, are re-circulating; so all electron energy is converted to X-rays [14].

FIGURE 3. X-ray image of regular size chicken taken by MIRRORCLE-6X. Feather, bone edge as well as all the details of inside are seen clearly. The 25-micron wide target is used.

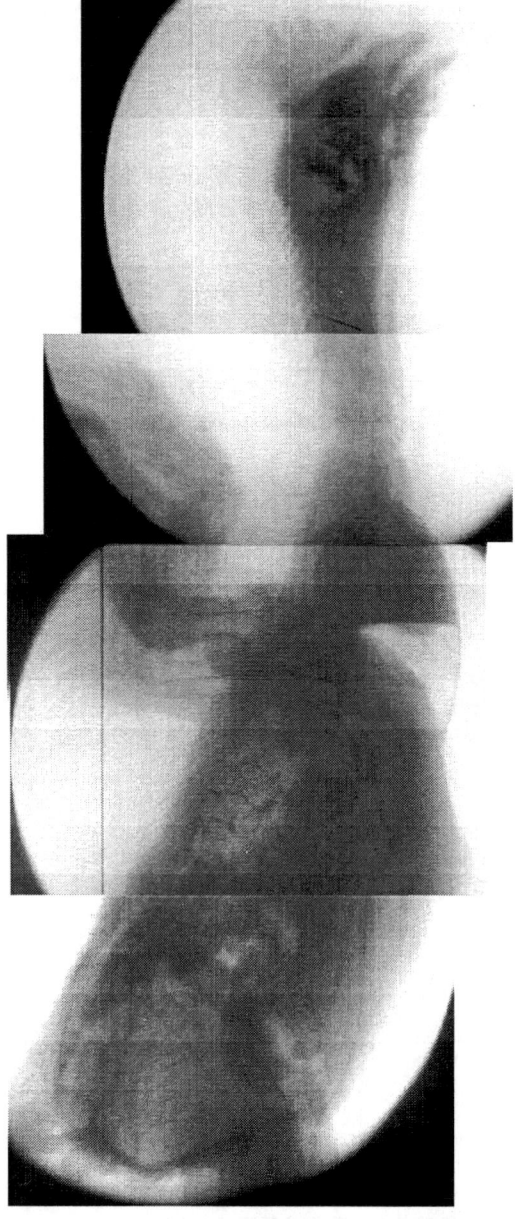

Indeed we have observed higher intensity with smaller target as seen in TABLE 2 by using MIRRORCLE-20 [9,15]. With wider targets the X-ray beam spread more widely, thus X-ray intensity drops with wider targets. This phenomenon is predicted by the theory [13] as the calculation is shown in TABLE 3. These are the reasons why we could observe the extremely fine resolution X-ray image.

TABLE 2. The target size dependence of the X-ray intensity is measured.

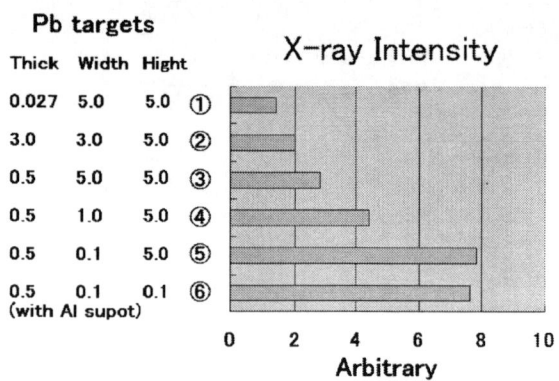

TABLE 3. The theoretical calculations of X-ray intensity from MIRRORCLE-6X

Machine parameters	Case 1	Case 2	Case 3	Case 4
e-Energy[MeV]	6	6	6	6
Gass presure or target shape	7.60E+00	C-foil	C-wire	W-wire
Particle density[/m3]	2.51E+23	1.13E+29	1.13E+29	6.25E+28
Photon energy to see[keV]	30	30	30	30
Atomic number	7	6	6	74
Cross section[m2]	1.297E-29	9.7272E-30	9.7272E-30	1.2354E-27
Photon spectrum band width	1.00E-03	1.00E-03	1.00E-03	1.00E-03
Target thickness/turn[m]	0.942	1.00E-05	1.00E-04	1.00E-04
Target width	-	-	1.00E-05	1.00E-05
Interaction area	1.51E-07	1.51E-07	4.00E-10	4.00E-10
Photon number/turn,electron	2.58E-08	9.23E-08	9.23E-07	6.76E-05
Number of electron/bunch,A	1.81E+09	1.81E+09	1.81E+09	1.81E+09
Total photon flux/A,0.1%band	1.61E+11	5.77E+11	5.77E+12	4.22E+14
Photon flux/mrad.A,0.1%band	2.57E+07	7.59E+10	7.59E+11	5.56E+13
Radiation loss/electron[eV]	1.91E+01	6.83E+01	6.83E+02	5.00E+04
Target A/Beam A ratio	1.00E+00	1.00E+00	2.65E-03	2.65E-03
Beam loss rate by Brem.[/sec]	2.11E+03	7.64E+03	2.03E+02	1.25E+04
Beam loss rate by nuclear scat.[/sec]	5.15E+05	1.84E+06	4.89E+04	3.58E+06
Injection rate(Hz)	400.00	400.00	400.00	400.00
Peak current(A)	0.10	0.10	0.10	0.10
Efficiency*Effective pulse width	6.00E-08	6.00E-08	6.00E-08	6.00E-08
Max beam current(A)	2.00E-03	5.59E-04	2.11E-02	2.88E-04
Brilliance	1.43E+04	1.18E+07	4.45E+09	4.45E+09
Total photon flux/A,0.1%band	3.23E+08	3.23E+08	1.22E+11	1.22E+11

The X-ray energy spectrum is shown in Fig. 4, which was taken by 3"x3" NaI scintillator at 4 m distance from the source point. To avoid saturation and pile up of X-rays the measurement was carried out 20 msec after the beam injection; so the small amount of beam is circulating. The spectrum continues up to the electron energy as known as the bremsstrahlung spectrum. Not big difference is seen between 5μm Al and 100μm Pb targets, which are predicted theoretically. We see here that MIRRORCLE provides the highest energy X-ray beam among any X-ray sources.

FIGURE 4. X-ray spectrum from MIRRORCLE-6X taken by NaI scintillator detector.

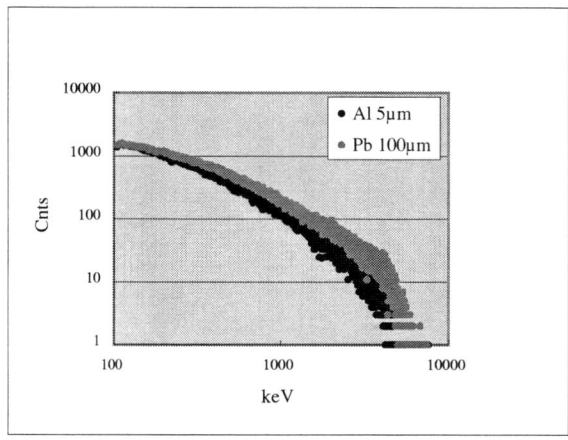

Specification of MIRRORCLE-6X

In Table 4, we show the specification of MIRRORCLE-6X. The integrated DOSE is measured by the calibrated ion chamber. The brilliance and the total flux are the theoretical values but consistent with the observation made by NaI with the correction of the lifetime. We keep injection contentiously at 400Hz with 100mA injector peak current. Our injection scheme called the resonance injection enables the injection without disturbing the circulating beam [16,17,18]. We can increase the X-ray intensity as much as user requested by increasing the beam injector current and the repletion. So far the heat up problem of the target is not observed at this beam current. We assume that the target is thin enough for the secondary electrons and soft x-rays escape from the target as predicted by Yamada.

The radiation angle, which is defined by the kinematics of electron and the target atom, is say 82 m radian. We observe very uniform distribution, which is excellent for imaging.

We conclude that MIRRORCLE-6X is a valuable X-ray source; in some extent more valuable than SR sources in the field of imaging, microscope, and cancer therapy. No other machine such as CT and MRI can

provide both functions of fine diagnosis and cancer treatment by the same machine [19,20].

TABLE 4. Specifications of MIRRORCLE-6X

	Storage Ring type , Electron E=6MeV Injector=100mA , Repetition=400Hz
Radiation schme	Collision with target nucleus
Radiation Angle	83 mrad
Spectrum	White 1keV ~ 6MeV
Time structure	Pulse width 100ns ~ 10ms Repetition : 2.45GHz Current : 50A/pulse
Intensity	0.4 (0.006) Gy / PULSE (200ns)
Imaging time	1 (100) pulse (200ns (250ms)) / flame(576cm²)
Average	160 (2.4) Gy / s
Brilliance	2.5E+11(9) photons/s/mrad²/mm²/0.1% λ at 30keV
Total Photons	5.5E+11(9)photons/s/0.1% λ at 30keV

FEATURES OF MIRRORCLE-20

FIR-ray emission scheme

MIRRORCLE-20, which has 15cm orbit radius, is under operation at the injection of 40 Hz repetitions with 20 mA peaks current. Since the circumference is extremely small, 600 mA beam current is stored by one shot of injection. We found that the injection efficiency is nearly 100%. Last two years MIRRORCLE-20 has been operated for x-ray production to perform the x-ray imaging [9], the irradiation experiments of cancer cells, the x-ray fluorescent analysis of heavy elements and so on. Since the activities of the x-ray experiments are shifted to MIRRORCLE-6X, MIRRORCLE-20 serves presently as a dedicated FIR generator.

Key issue is the special mirror system. The designed beam optics is somewhat different from usual SR beam line optics used for the SR sources [2,3,4]. The complex mirror system collects all radiations emitted from the whole arc of the exact circular 15 cm diameter electron orbit. An exact circular mirror and a quasi-ellipsoidal mirror are used as seen in Fig. 4. The quasi-ellipsoidal mirror focuses the Far infrared SR, and parabolic mirror shape the FIR-beam to the parallel beam of 1x3 cm² cross section.

When one ampere of beam current is stored in the MIRRROCLE-20, the 20 MeV synchrotrons, the FIR power is less than 10^{-6} W/mrad, 1%band at the characteristic wavelength 8 μm, which is essentially the same as that of GeV synchrotrons. When the whole radiation is corrected, the extracted IR-SR power reaches 1mW/1%band, which is denoted as 2π

[14]. When the interference appears by the SR radiations emitted from the short bunches and stored by the exact circular mirror, we will see coherent radiation name Photo Storage Ring. The fundamental wavelength can be adjusted by changing the electron orbit radius. The lasing mechanism of Photon Storage Ring has been proposed by Yamada in 1989 [21-24].

FIGURE 5. Inside of MIRRORCLE-20 vacuum chamber. An exact circular mirror and a quasi-ellipsoidal mirror are installed.

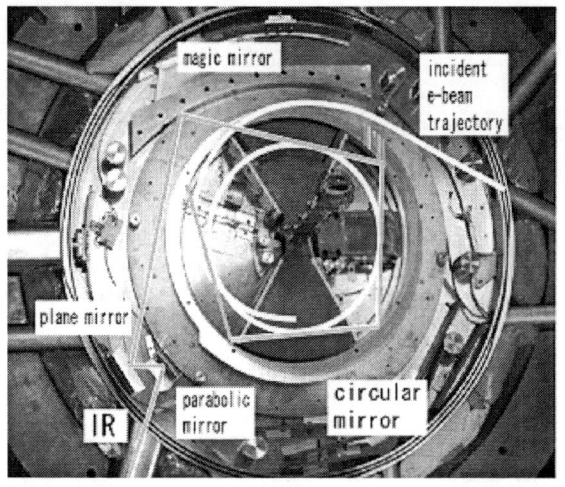

FIGURE 6. IR-ray spectrum from MIRRORCLE-20.

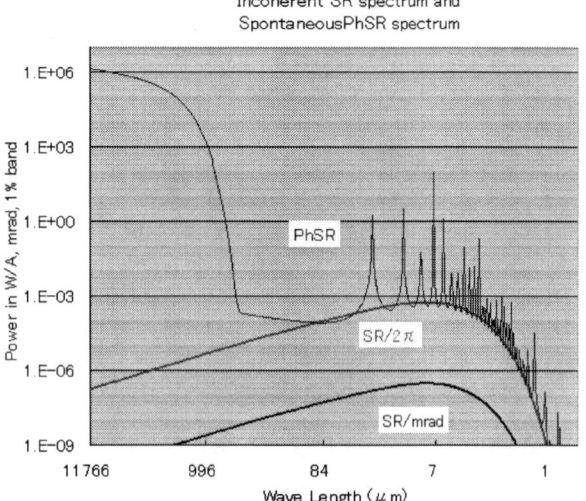

FIR-ray characteristics

MIRRORCLE-20 has a potential to provide the most powerful white FIR and Mid-IR rays among any

IR sources and coherent FIR radiation by the photon storage ring scheme. The power of MIRRORCLE-20 is, however, presently limited by the lifetime of electron beam, which is namely affected by the degree of vacuum. The measured total Mid-IR power is 270 mV, which correspond to about 1 W peak power in total, at the beam port consisting of 3cmϕ ZnSe window, and 200mV at 1 m distance from the exit port as shown in Fig. 7. The extracted IR beam is parallel with spot size of 10x30 mm^2 wide. In 10 msec beam stops suddenly, which could be caused by the microwave beam instability. The 600 mA beam current is extremely large in the usual sense for the low energy synchrotron; so many kind of instability could be caused to this sudden stop.

We are planning to increase the repetition rate to 100Hz, vacuum pressure to 10-7 Pa, and beam current to more than 20 mA. By all these means we are able to proceed the planned experiment from April, 2004.

FIGURE 7. IR-ray time spectrum from MIRRORCLE-20.

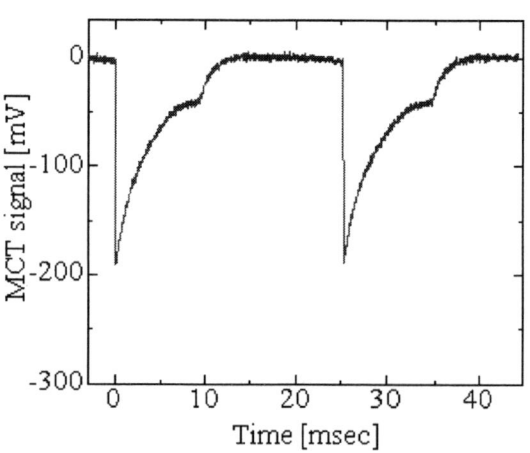

ACKNOWLEDGMENTS

Author express many thanks to colleagues, Daisuke Hasegawa, Taichi Hayashi, Makoto Inoue, Yasuji Kitazawa, Andrey I. Kleev, Nobuhiro Miura, Yuko Miyoshi, Ahsa Moon, Noriko Oda, Yoshiko Okazaki, Eitoku Ro, Hideo Saisho, Makoto Sasaki, Norio Toyosugi, Isao Tohyama, Takanori Yamada, Reiko Yamada, to the Doctor course students; Tohru Hirai, and Takeshi Kikuzawa, and to the Master course students; Hirano, Maki, Narazaki, Sugihara, Suetsugu, Takashima, Tohma for their helps and discussions.

REFERENCES

1. H. Yamada, The Synchrotron Light Life Science Center Granted by the MEXT 21st Century COE Program, Proc. Int. Conf. on Synchrotron Radiation Instrumentation, San Francisco, 2003, Aug. 24-29: Accepted for publication in Review of Scientific Instruments.

2. H. Yamada, T. Toma, and Y. Nakamura, "Present status of MIRRORCLE-20 and the observed brightest FIR synchrotron radiation", Proc. IRMMW2003 Inter brightest FIR synchrotron radiation, Proc. IRMMW2003 Inter. Conf. October 3, 2003, Otsu

3. H. Yamada, A. Moon, Y. Nakamura, T. Kikuzawa, N. Oda, and N. Miura, "Brilliant FIR or IR BL for Life Science Research", Proc. IRMMW2003 Inter. Conf. October 3, 2003, Otsu.

4. A. Moon, Y. Nakamura, T. Toma, H. Yamada, FIR beam line for MIRRORCLE-20, contribution in this conference.

5. H. Yamada, Current Status and Biological Research Ramifications of Photon Storage Ring as a Noble Infrared Laser Source, Advances in Colloid and Interface Sci. 71-72, (1997) pp. 371-392.

6. T. Kikuzawa, H. Yamada, H. Mochizuki-Oda, N. Miura, A. Moon, Protein Dynamics Research Station on IR Beam Line of MIRRORCLE-20, contribution in this conference.

7. Nobuhiro Miura*, Hironari Yamada*#, Ahsa Moon#, Kishi Nishikawa, "Dynamical Study of Water Structure by Infrared Synchrotron Light", contribution in this proceedings.

8. D. Hasegawa, H. Yamada, A.I. Kleev, N. Toyosugi, T. Hayashi, T. Yamada, I. Tohyama, and Y.D. Ro, The tabletop synchrotron MIRRORCLE-6X, Proc. 14th Sympo. Accelerator Science and Technology, Tsukuba, Nov. 2003, p.111.

9. H. Yamada, "Novel X-ray Source based on a tabletop synchrotron and its unique features", Nuclear Instruments and Methods in Physics Research B199, 2003 pp.509-516.

10. T. Hirai, H. Yamada, Y.Sonoda, S. Maki, T. Takashige, T. Takashima, D. Hasegawa, N. Toyosugi, "Novel Edge-Enhanced X-ray Imagining by MIRRORCLE", Proc. Int. Conf. on Synchrotron Radiation Instrumentation, San Francisco, 2003, Aug. 24-29: Accepted for publication in Review of Scientific Instruments.

11. H. Yamada, H. Saisho, T. Hirai, and J. Hirano, X-ray fluorescent analysis if heavy elements with the portable synchrotron MIRRORCLE, Spectrochemica Acta B, in press.

12. Saisho, et al. contribution in this conference.

13. H. Yamada, Super photon generator using collisions of circulating relativistic electrons and wire targets, Jpn. J. Appl. Phys., Vol. 35 (1996) pp. L182-L185.

14. H. Yamada, The smallest Electron Storage Ring for High Intensity Far-Infrared and Hard X-ray Generation, Journal of Synchrotron Radiation, (1998) pp.1326-1331.

15. H. Yamada, et al., Development of the hard x-ray source based on a tabletop synchrotron, Nucl. Instrum. Methods. Phys. Res.A467-468 (2001) 122-125.

16. H. Yamada, Present status of AURORA#1-Potential of compact SR-ring as a hard X-ray source, (Synchrotron Radiation Facilities in ASIA, IONICS PUBLISHING)(1994).

17. Yamada, Commissioning of AURORA: the smallest synchrotron light source, J. Vac. Sci. Tech. B8(6) pp.1628 (1990)

18. T. Takayama, Nucl. Instr. and Meth. B 24-25 (1987) 420.

19. Y. Suetsugu, H. Yamada, D. Hasegawa, T. Hirai, Y. Sugihara, H. Yamada, Problem of Radiation Safety in the Diagnosis using MIRRORCLE, contribution in this conference.

20. T. Tesima, T. Ogata, A. Kawaguchi, Y. Suzumoto, D. Hasegawa, N.M.-Oda, H. Yamada, The biological effects on cancer cells by synchrotron radiation generated from MIRRORCLE-6X, contribution in this conference.

21. H. Yamada, "Photon storage ring", Japan J. Appl. Phys. 28 pp.L1665-L1668) (1989)

22. H. Yamada, "Novel free-electron laser named photon storage ring", Nucl. Instrum. Methods A304 pp.700-702, (1990)

23. Mode structure and amplification in the photon storage ring (K. Mima, K. Shimoda, H. Yamada)(IEEE J.Quantum Electronics 27 pp.2572-2579)(1991)

24. A.I. Kleev and H. Yamada, IEEE Journal of Quantum Electronics 39(6), 2003, pp. 1-9.

The X-Ray Microscopy And Micro-Spectroscopy Facility At The ESRF

J. Susini, A. Somogyi, R. Barrett, M. Salomé, S. Bohic, B. Fayard, D. Eichert, O. Dhez, P. Bleuet, G. Martinez-Criado, and R. Tucoulou

X-ray Imaging Group, Experiments Division,
European Synchrotron Radiation Facility, B.P. 220, 38043 Grenoble, France

Abstract. Among the 40 beamlines in operation at the European Synchrotron Radiation Facility, three beamlines are fully dedicated to X-ray microscopy and micro-spectroscopy techniques in the multi-keV range. Offering a unique combination of non destructive analytical techniques which aim to satisfy the growing demand from experimental research fields such as medicine, geology, archaeology, earth, planetary and environmental sciences. Following a brief discussion on the strengths and weaknesses of X-ray microscopy and spectro-microscopy techniques in the 1-20keV range, characteristics of the beamlines are briefly described. Examples of applications are given in the reference list.

INTRODUCTION

The unique properties of X-rays make them a powerful probing radiation for use in microscopy techniques. For example, soft X-ray microscopes have established capabilities in absorption contrast imaging of thick hydrated biological material in near-native environments at spatial resolutions well beyond those achievable with visible light microscopy. In the past decade there has been a strong tendency in X-ray microscopy to develop alternative contrast mechanisms and spectroscopic methods, which can provide both valuable complementary information on the sample nature and/or reduce the necessary radiation doses. Simultaneously, the development of high brilliance high energy X-ray sources coupled with advances in manufacturing technologies of focusing optics has led to significant improvements in sub-micrometer probes for spectroscopy, diffraction and imaging applications in the multi-keV 'hard' X-ray range. Both by extrapolation of the experience gained in the soft X-ray regime and by the development of new techniques, 'hard' X-ray microscopes now offer a complementary analytical tool which can contribute to a wide range of existing and new applications of X-ray microscopy.

Among the 40 beamlines in operation at the European Synchrotron Radiation Facility, three beamlines are fully dedicated to X-ray microscopy and micro-spectroscopy techniques in the multi-keV range. ID21, ID22 and ID18F are method-oriented beamlines offering a unique panel of non destructive analytical techniques. The main fields of applications are driven by the unique attributes of X-ray microscopy in the multi-keV energy range: i) access to K-absorption edges and fluorescence emission lines of medium-light elements and L,M - edges of heavy materials for micro-spectroscopy, chemical or trace element mapping; ii) higher penetration depths compared to soft X-rays allowing imaging of thicker samples; iii) favorable wavelengths for diffraction studies and iv) generally large focal lengths and depths of focus which are advantageous for the use of specific sample environments (in-situ, high pressure, controlled temperatures....). Typical experiments can be broadly divided into two categories: i) morphological studies which require high spatial resolution and are therefore well adapted to 2D or 3D transmission full-field microscopy. ii) studies dealing with co-localization and/or speciation of trace elements in heterogeneous matrices at the micron scale. Scanning X-ray microscopy, in transmission and/or fluorescence modes, tends to be better suited for the latter cases, which often require both low detection limits and spectroscopic analysis capabilities. An increasing number of experiments require a complete analysis of the same sample by screening light and heavy elements. Furthermore, the need for combined morphological and chemical information is a clear

CP716, *Portable Synchrotron Light Sources and Advanced Applications,*
edited by H. Yamada, N. Mochizuki-Oda, and M. Sasaki

trend, which is pushing our strategy toward a coupled access to full-field sub-micron imaging and micro-probe based spectroscopy. The three beamlines share a common concept in combining several detection modes and imaging techniques (fluorescence, absorption or phase contrast, 2D and 3D imaging) (see Table 1).

THE X-RAY MICROSCOPY BEAMLINE ID21

ID21 is a beamline dedicated to X-ray imaging and spectro-microscopy [1]. The beamline is installed on a low-beta straight section, which is equipped with three different undulators and provides beam for two independent end-stations on two separate branch-lines: The Scanning X-ray Microscope (SXM) served by the "direct" branch-line, is equipped with a fixed-exit silicon double crystal monochromator operating in the 2-8 keV energy range, and the full-field imaging transmission X-ray microscope (TXM), is served by the side-branch and optimized for imaging techniques at 4 keV. Both microscopes use zone-plates as focusing lenses. The beamline covers several disciplines over a broad energy range. Therefore various types of zone plates are currently needed for each kind of application. i) Fluorescence or diffraction applications in the multi-keV energy range require large diameter and medium-resolution (thus long focal length) zone-plates, offering high flux and a better access to the sample. ii) Full-field X-ray microscopy necessitates high-resolution zone-plates, to be used as objective lenses and large diameter zone-plates for condensers. Several collaborations were therefore initiated with different laboratories with particular emphasis upon high efficiency zone-plates for the 2-8 keV range on one hand [2,3] and high-resolution zone plates for the Ca K-edge region (~ 4 keV) on the other [4]. The high spatial resolution focusing combined with spectroscopic methods in the few keV domain has demonstrated unique capabilities in a number of experiments from diverse fields. Sulphur [5,6,7], Calcium [8], Vanadium [9], Chromium [10], and Iron [6] K-edges are the most requested edges/lines. The reduction of air absorption by using the scanning microscope in secondary vacuum is a clear asset for energies below 4keV. Indeed, the capability of performing micro-XANES at the sulphur K-edge with a spatial resolution of about 0.1μm is a unique feature. A wide range of the applied sciences uses the X-ray fluorescence spectrometry [11], morphological imaging [12], and alternative contrast modes [13] available on the ID21 SXM.

Multi-keV X-ray microscopy often suffers from a lack of absorption resulting in low contrast images. Furthermore, the use of absorption contrast can subject the specimen to high radiation doses leading to possible structural changes. Even for radiation hard materials, many of the samples imaged using fluorescence yield are insufficiently absorbing to provide high contrast images in transmission mode. Therefore accurate morphological localization of trace elements is difficult or even impossible. The development of phase contrast methods fully compatible with detection in fluorescence yield is therefore essential. Those general considerations led us to focus our R&D program on the investigation of various possible optical schemes for phase contrast imaging. Three original strategies are currently developed on ID21: Differential Phase Contrast (DPC) using configured detectors on the SXM [14]; Differential Interferential Contrast (DIC) with a configured zone-plate on both SXM and TXM [15, 16] and finally, Zernike X-ray microscopy on the TXM [17,18] were successfully developed and are now routinely used.

ID22/ID18F: A MULTI-TECHNIQUE BEAMLINE

ID22 is a high-energy multi-technique X-ray microprobe beamline served by 2 undulators. The double crystal monochromator equipped with both Si(111) and Si(311) offers a wide energy range (6-70 keV) for optimal excitation conditions of a broad range of elements. Several routinely available micro-analytical techniques, such as micro-X-ray fluorescence analysis (μ-XRF), micro-X-ray absorption spectrometry (μ-XAS), micro-diffraction (μ-XRD) and absorption/phase contrast imaging and tomography, provide the possibility of the complex analysis of the sample. The microprobe is formed by direct imaging of the X-ray source with a distance from the source to focusing optics of approximately 55m. Generally focusing is performed using a Kirkpatrick Baez mirror system. For some specific applications refractive lenses and Fresnel zone-plates can be used. Corresponding fluxes and spot sizes are given in Table 1.

The ID18F end-station [19] is a dedicated microprobe instrument occupying an experimental hutch on the ID18 Nuclear Resonance beamline with which it shares extremely stable and high throughput beamline optics (slits, monochromator). Three 32mm period undulators are installed on the ID18 high beta straight section providing an exceptionally high brilliance beam in the 6-29keV energy range. Originating as collaboration between the ESRF and the Micro and Trace Analysis Center (MITAC) of the University of Antwerp, (Belgium), and the ID18F microprobe was first designed with an emphasis placed on X-ray fluorescence microprobe measurements [20]. In a second step, the instrument has evolved into a multi-technique station offering simultaneous focused-probe fluorescence and wide or small angle X-ray scattering measurements [21] and the possibility for fluorescence tomography and absorption tomography acquisitions.

The multi-technique capability of ID22/ID18F attracts a wide variety of research fields and both end stations find widespread applications in various disciplines spanning from the medical and biological sciences through to geology [22, 23]. More precisely, non-destructive investigation of the spatial distribution, concentration and speciation of trace elements in single cells or tissues can give useful information about normal or pathogen biological functioning of living organisms or about the biological effects and mechanism of different high-Z labeled drugs [24,25]. In order to estimate the possible environmental impact of different pollutants (e.g. heavy metals) originating from micrometer sized particles (e.g. radioactive fuel particles [26], fly ash [27]), the complex investigation of the elemental composition, speciation, morphology and crystalline structure of the particles are needed. The non-destructivity of the method is also a crucial requirement for the study of unique samples such as extraterrestrial grains from space collections [28] or samples of high archaeological value [29]. The increasing interest and demand of several research fields also motivate the development of new methods at ID22 such as quantitative μXRF tomography [30].

ACKNOWLEDGMENTS

The authors wish to thank R. Baker, S. Labouré, E. Gagliardini, G. Rostaing, G. Berruyer, A. Homs, H. Witsch and F. Thurel for their invaluable involvements in the X-ray microscopy projects at the ESRF, B. Kaulich and U. Neuhaeusler for their major roles in the design, construction and commissioning of the full-field X-ray microscope on ID21. We are indebted to external collaborators for technical and scientific developments: G. Schneider, M. Panitz, C. David, T. Weitkamp, E. Di Fabrizio, S. Cabrini, P. Charalambous, G. Morrison and T. Wilhein.

TABLE 1. Main characteristics of the micro-spectroscopy and micro-imaging end-stations at the ESRF.

	ID21 beamline	ID22 Beamline	ID18F end-station
Source	Low beta	High beta	Low beta
Undulators (period, pole, K_{max})	Linear U42 (42, 38, 2.1) Linear W80 (80, 20, 5.9) Helical H52 (52, 59, 1.5)	Linear U42 (42, 38, 2.1) Linear U23 (23,86, 1.45)	Linear U32 (42, 50, 1.9) Linear U32 (42, 50, 1.9) Linear U32 (42, 50, 1.9)
Energy range (keV)	2-8	7-60	6-29
Monochromator	Si 111 and Si 220	Si 111 and Si 311	Si 111
Harmonics rejection	2-bounce mirror horizontally deflecting	1 mirror horizontally deflecting	----
Imaging modes	Scanning 2D Full-field 2D	Scanning 2D/3D Full-field 2D/3D	Scanning 2D/3D Full-field 2D/3D
Contrast modes	Fluorescence Absorption Phase	Fluorescence Absorption Phase	Fluorescence Absorption Phase
μ-spectroscopy	XANES	XANES - EXAFS	----
μ-diffraction	----	SAXS/WAXS	SAXS/WAXS
Detection modes	Fluorescence Transmission	Fluorescence Transmission	Fluorescence Transmission
Focusing optics	Fresnel zone-plates	Kirkpatrick Baez (KB) Refractive lenses (CRL) Fresnel zone-plates (FZP)	Refractive lenses
μ-probe size (vert x horz.)	$0.2 \times 0.2 \mu m^2$ (scanning) $0.05 \times 0.05 \mu m^2$ (full-field)	$1 \times 3 \mu m^2$ (KB) $1 \times 12 \mu m^2$ (CRL) $1 \times 12 \mu m^2$ (FZP)	$1.5 \times 12 \mu m^2$
Flux (ph/s/Si111bw) in the spot	10^9	10^{11} (KB) $10^9 - 10^{10}$ (CRL) $10^9 - 10^{10}$ (FZP)	10^{10}

REFERENCES

1. J. Susini, M. Salomé, B. Fayard, R. Ortega and B. Kaulich, *Surface Review and Letters*, 9(1), 203 (2002).

2. E. Di Fabrizio, F. Romanato, M. Gentili, S. Cabrini, B. Kaulich, J. Susini, R. Barrett, *Nature*, **401**, 89 (1999).

3. C. David, B. Kaulich, R. Barrett, M. Salomé, and J. Susini, *Appl. Phys. Letters*, **77(23)**, 3851 (2000).

4. M. Panitz, G. Schneider, M. Peuker, D. Hambach, B. Kaulich, S. Oestreich, J. Susini, and G.Schmahl, in *6th International Conference on X-ray microscopy*, edited by W.Meyer-Ilse et al., AIP Conference Proceedings **507**, Berkeley 1999, 676 (2000).

5. J.P. Cuif, Y. Dauphin, J. Doucet, M. Salomé and J. Susini, *Geochimica & Cosmochimica Acta*, 67(1), 75 (2003).

6. N. Métrich, M. Bonnin-Mosbah, J. Susini, M. Salomé, *Geophysical Research Letters*, **29(11)**, 33 (2002).

7. J. Prietzel, J. Thieme, U. Neuhäusler, J. Susini & I. Kögel-Knabner, *European Journal of Soil Science*, **54(3)**, 423 (2003).

8. C. Merigoux, F. Briki, F. Sarrot-Reynauld, J. Doucet, B. Fayard, J. Susini, M. Salomé, *Biochimica & Biophysica Acta*, **1619**, 53 (2003).

9. T. Ueki, K. Takemoto, B. Fayard, M. Salome, A.Yamamoto, H. Kihara, J. Susini, S. Scippa, T. Uyama, H. Michibata, *Zoological Science*, **19(I)**, 27 (2002).

10. R. Ortega, G. Deves, M. Bonnin-Mosbah, M. Salomé, L. Anderson, S.K. Kasprzak, *Nuclear Instruments and Methods A*, **10**, 215 (2001).

11. L. Di Gaspare, A. Notargiacomo, F. Evangelisti, E. Palange, S. Pascarelli and J. Susini, *Solid State Communications*, **122**, 359 (2002).

12. I. Bihannic, L. J. Michot, B.S. Lartige, D. Vantelon, J. Labille, F. Thomas, J. Susini, M. Salomé, B. Fayard *Langmuir*, **17**, 4144 (2001).

13. P.Bergonzo, R.Barrett, O.Hainaut, D. Tromson, C. Mer, B. Guizard, *Diamond and Relat. Mater.* ,**11**, 418 (2002).

14. G. R. Morrison, W. Eaton, R. Barrett, P.S. Charalambous, in Proceedings of the 7th International Conference on X-ray Microscopy edited by J. Susini et al., *J. Phys. IV*, **104**, 547 (2003).

15. T. Wilhein, B. Kaulich, E.Di Fabrizio, F. Romanato, S. Cabrini and J. Susini, , *Appl. Phys. Letters*, **78(14)**, 2082 (2001).

16. B. Kaulich, T. Wilhein, E. Di Fabrizio, F. Romanato, M. Altissimo, S. Cabrini, B. Fayard, and J. Susini, *JOSA A*, **19 (4)**, 797 (2002).

17. Schneider G., Hambach D., Niemann B., Kaulich B., Susini J., Hoffmann N. and Hasse W., *Appl. Phys. Letters*, **78(13)**, 1936 (2001).

18. U. Neuhäusler, G. Schneider, W. Ludwig, M.A. Meyer, E. Zschech, D. Hambach, *J. Phys. D: Appl. Phys.*, **36**, 79 (2003).

19. A. Somogyi, M. Drakopoulos, L. Vincze, B. Vekemans, C. Camerani, K. Janssens, A. Snigirev, F.Adams, *X-Ray Spectrom.*, **30**, 242 (2001).

20. L. Kempenaers, K. Janssens, L. Vincze, B. Vekemans, A. Somogyi, M. Drakopoulos, A. Simionovici, F. Adams, *Analytical Chemistry*, **74**, 5017 (2002).

21. T. Wess, I. Alberts, J. Hiller, M. Drakopoulos, A.T. Chamberlain, M. Collins, *Calcified Tissue Int.*, **70**, 103 (2002).

22. B. Ménez, P. Philippot, M. Bonnin-Mosbah, A. Simionovici, F. Gibert, *Geochimica et Cosmochimica Acta*, **66**, 561 (2002).

23. M. Bonnin-Mosbah, A.S. Simionovici, N. Metrich, J.P. Duraud, D. Massare, P. Dillman, *J. of Non-crystalline Solids*, **288**, 103 (2001).

24. S. Bohic, A. Simionovici, A. Snigirev, R. Ortega, G. Deves, D. Heymann, C.G. Schroer, *Appl. Phys. Letters*, **78**, 3544 (2001).

25. R. Ortega, G. Devès, S. Bohic, A. Simionovici, B. Ménez and M. Bonnin-Mosbah, *Nucl. Instrum. and Meth. B*, **181**, 480 (2001).

26. B. Salbu, T. Kregling, O.C. Lind, D.H. Oughton, M. Drakopoulos, A. Simionovici, A. Snigirev, I. Snigireva, T. Weitkamp, F. Adams, K. Janssens, V.A. Kashparov, *Nucl. Instrum. and Meth. A*, **467**, 1249 (2001).

27. M. C. Camerani Pinzani, A. Somogyi, A. Simionovici, S. Ansell, B.M. Steenari, O. Lindquist, *Environmental Science and Technology*, **36**, 3165 (2002).

28. J. Borg, E. Quirico, A. Simionovici, P. I. Raynal, P. Chevallier and Y. Langevin, *Planetary and Space Science*, **50**, 939 (2002).

29. P. Martinetto, M. Anne, E. Dooryhée, M. Drakopoulos, M. Dubus, J. Salomon, A. Simionovici, P. Walter, *Nucl. Instrum. Meth. B*, **181**, 744 (2001).

30. B. Golosio, A. Somogyi, A. Simionovici, P. Bleuet, L. Lemelle, J. Susini, accepted for publication in *Appl. Physics Letters*.

Large-Area Phase-Contrast X-ray Imaging System
Based on a Two-Crystal X-ray Interferometer

A. Yoneyama[1], T. Takeda[2], Y. Tsuchiya[2], J. Wu[2], T.T. Lwin[2], and K. Hyodo[3]

[1]Advanced Research Laboratory, Hitachi Ltd., Hatoyama, Saitama 350-0395, Japan; [2]Institute of Clinical Medicine, University of Tsukuba, Ibaraki 305-8575, Japan; and [3]Institute of Materials Science, High Energy Accelerator Research Organization, Tsukuba, Ibaraki 305-0801, Japan.

Abstract. With the aim of applying the phase-contrast X-ray imaging technique to biomedical imaging, we have been developing a large-area imaging system based on a skew-symmetric two-crystal X-ray interferometer. This latest of our imaging systems has a 60×30-mm field of view and has been used to observe fine two- and three-dimensional images of several biological samples by using a 17.7-keV synchrotron X-ray beam at the Photon Factory. In this paper, we give the outline of the imaging system and recent observational results of large biological samples.

INTRODUCTION

Phase-contrast X-ray imaging, which detects phase shifts within a sample, has a great potential for biomedical imaging. The phase-shift cross-sections for light elements are about 1000 times bigger than absorption cross section [1,2], so the phase-contrast imaging technique provides a method for the fine observation of biomedical samples without the need for contrast agents and harmful X-ray exposure. This high-sensitivity imaging technique was first implemented in a monolithic triple Laue-case (LLL) X-ray interferometer [3] (Fig. 1(a)), and used in the radiographic observation of small pieces of rat cerebellum and human metastasized liver tumor tissue [1,2]. In addition, this imaging technique has been

combined with computed tomography for three-dimensional observation (phase-contrast CT [4]), which has been used to observe small columnar pieces of tissue from various organs [5-7].

To apply this imaging technique for wider biomedical application, larger fields of view and suppression of the thermal disturbance caused by radiated heat from the sample are two key requirements. However, the monolithic X-ray interferometer cannot meet these requirements; therefore it is necessary to divide the interferometer into two crystal blocks. To operate a two-crystal X-ray interferometer, the relative positions between the two blocks must be controlled to an accuracy in the order of X-ray wavelengths. The skew-symmetric two-

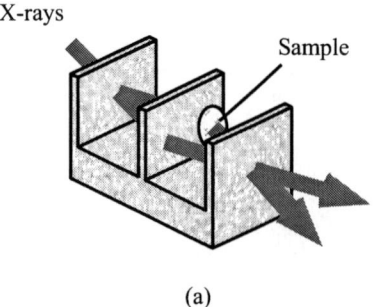

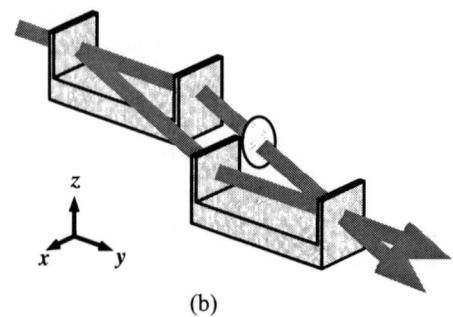

(a)   (b)

FIGURE 1. (a) LLL X-ray interferometer and (b) skew-symmetric two-crystal X-ray interferometer

CP716, *Portable Synchrotron Light Sources and Advanced Applications,*
edited by H. Yamada, N. Mochizuki-Oda, and M. Sasaki
© 2004 American Institute of Physics 0-7354-0195-0/04/$22.00

crystal X-ray interferometer (STXI, Fig. 1(b)) [8] has the advantage that only two rotational axes need to be controlled (θ and ρ); therefore, its operation is simple in comparison with other possible forms of two-crystal interferometer, which require more complex control [8]. With this advantage in mind, we have been developing a large-area phase-contrast X-ray imaging system based on the STXI [9-13]. The latest imaging system (namely, our fourth one) has a 60×30-mm field of view, and has been used to observe phase-maps (spatial distributions of the phase shift caused by samples) and three-dimensional CT images of several biological samples [12,13].

IMAGING SYSTEM WITH 60×30-MM FIELD OF VIEW

Figure 2 shows a schematic view of the fourth-imaging system. The incident synchrotron X-ray beam is horizontally expanded by the asymmetric crystal, and then input to the STXI. The STXI generates two interference beams, one for CCD-based imaging detector 1 [14] used for measuring the phase-maps and the other for image detector 2 used for the operation of the feedback positioning system [13].

The STXI was positioned by a mechanical positioning system composed of three tables: S1 for adjusting the STXI under the Bragg condition against the incident X-rays; and S2 and tilt tables for tuning the θ and ρ rotation between the divided crystal blocks,

respectively. To attain the required sub-nanoradian mechanical positioning and stability around the θ axis for the operation of the STXI, the composition of the positioning system was simplified as far as possible, the pivots direction of tables S1 and S2 were oriented vertically to suppress the mechanical vibration transmitted from the floor, and a sleeve bearing was used to improved the rigidity of table S2. In addition, a fine positioning mechanism driven by a laminated piezoelectric actuator (PZT) was adopted to table S2. These measures together achieved a minimum incremental rotation (step size) of 0.03 nrad for table S2 [12].

To accurately stabilize the phase fluctuation of the interference beams caused by the drift rotation around the θ axis, a feedback positioning system was used to control table S2. The drift rotation can be detected as the motion of the interference pattern, so a compensatory voltage is applied to the PZT; that is, the PZT expands by just enough to cancel out the motion of the interference pattern [13].

RESULTS AND DISCUSSION

The observations of biological samples were carried out with this imaging system at the beam-line BL-14C1 of the Photon Factory in Tsukuba, Japan. A vertically fan-shaped beam emitted from a vertical wiggler was monochromated to 17.7 keV by a Si(220) double-crystal monochromator and then input into the

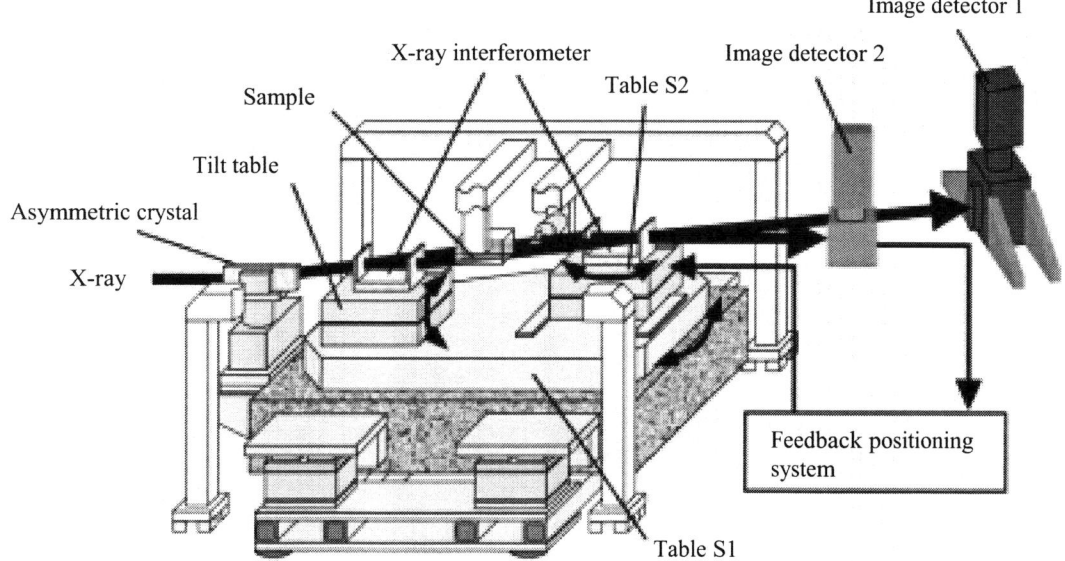

FIGURE 2. Schematic view of the imaging system based on a two-crystal X-ray interferometer.

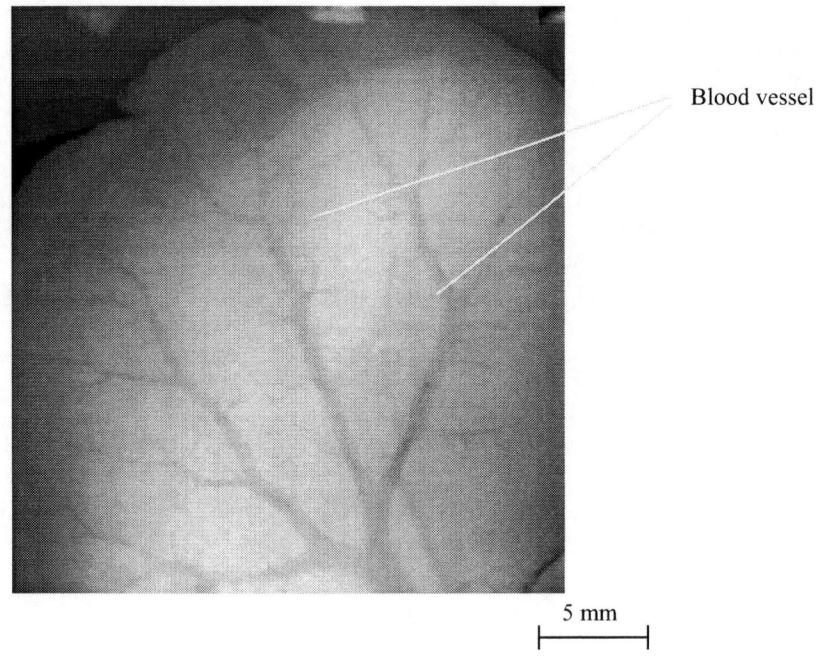

Blood vessel

FIGURE 3. Phase map of the liver from a rabbit. The blood vessels are clearly depicted.

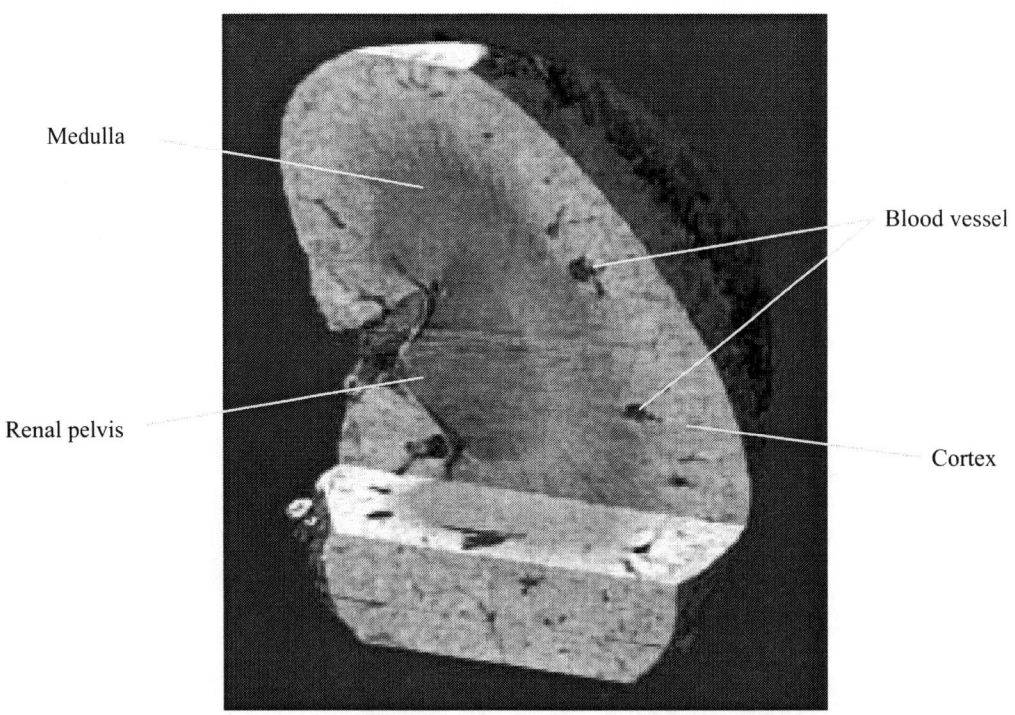

Medulla

Blood vessel

Renal pelvis

Cortex

FIGURE 4. Thee-dimensional image of the kidney of a hamster.

imaging system.

Figure 3 shows a phase map of the liver from a rat obtained by the fringe-scanning method with subtraction of the background phase [12]. The number of fringe scans was set to five with a 10-s exposure to obtain each of the interference patterns. The image size is 25 mm wide by 30 mm high. The feedback positioning system was operated during the measurement, and the phase fluctuation was stabilized within $\pi/15$ rad. Since the phase shift from the physiological saline in the blood vessels and that from the surrounding liver are different, blood vessels with a diameter of about 50 μm can be clearly depicted.

Figure 4 shows a three-dimensional image of a kidney from a hamster. To obtain this image, the sample was placed in a water-filled sample cell and rotated in 0.72-degree steps (250 projections). A phase map for each projection was obtained by the same process as was used to obtain the image in Fig. 3, except that the number of fringe scan was three and the exposure time for the detection of each interference pattern was 5 s. To reconstruct the tomograms, filtered back-projection with a Shepp-Logan filter was used. The sample was about 10 mm in diameter and 15-mm high. The cortex and the medulla are clearly differentiated, and tubular structures (blood vessels and renal pelvis) are depicted.

Note that these two experiments were approved by the Medical Committee for the Use of Animals in Research of the University of Tsukuba, and it conformed to the guidelines of the American Physiological Society.

CONCLUSION

Observations of fine two and three-dimensional images of several biological samples were performed by using our newly developed large-area phase-contrast X-ray imaging system based on a two-crystal X-ray interferometer. Various structures such as blood vessels in the samples were depicted clearly.

In the next step of our work, we are increasing the energy of the X-ray beam. The usage of high-energy X-rays gives a way to shorten the measurement period, reduce X-ray dose, and improve the spatial resolution. In addition, it enables thick samples with a larger difference in densities to be observed. The observation of a 3-cm diameter sample, such as a small animal's heart, with a 35-keV X-ray beam is our first target with this system.

ACKNOWLEDGMENTS

This study was carried out with the support through Special Coordination Funds of the Ministry of Education, Culture, Sports, Science and Technology of the Japanese Government and under Proposal No. 2002S2-001 approved by the High Energy Accelerator Research Organization.

REFERENCES

1. Momose, A., and Fukuda, J., Med. Phys. 22, 375-380 (1995).

2. Takeda, T., Momose, A., Itai, Y., Wu, I., and Hirano, K., Acad Radiol. 2, 799-803 (1995).

3. Bonse, U., and Hart, M., Appl. Phys. Lett. 6, 155-156 (1965).

4. Momose, A., Nucl. Instrum. Meth. A352, 622-628(1995).

5. Momose, A., Takeda, T., Itai, Y., and Hirano, K., Nature Med. 2, 473-475 (1996).

6. Beckmann, F., Bonse, U., Busch, F., and Gunnewig, O., J. Comp. Assist. Tomogr. 21, 539-553 (1997).

7. Takeda, T., Momose, A., Hirano, K., Haraoka, S., Watanabe, T., and Itai, Y., Radiology 214, 298-301 (2000).

8. Becker, P., and Bonse, U., J. Appl. Cryst. 7, 593-598 (1974).

9. Momose, A., Yoneyama, A., and Hirano, K., J. Synchrotron Rad. 4, 311-312 (1997).

10. Yoneyama, A., Momose, A., Seya, E., Hirano, K., Takeda, T., and Itai, Y., Rev. Sci. Instrum. 70, 4582-4586 (1999).

11. Yoneyama, A., Momose, A., Koyama, I., Seya, E., Takeda, T., Itai, Y., Hirano, K., and Hyodo, K., J. Synchrotron. Rad. 9, 277-281 (2002)

12. Yoneyama, A., Takeda, T., Tsuchiya, Y., Wu, J., Lwin, T. T., Koizumi, A., Hyodo, K., and Itai, Y., Nucl. Instrum. Meth. (in press).

13. Yoneyama, A., Takeda, T., Tsuchiya, Y., Wu, J., Lwin, T. T., and Hyodo, K., Y., AIP Conf. Proc. (in press).

14. Momose, A., Takeda, T., Yoneyama, A., Koyama, I., and Itai, Y., Nucl. Instrum. Meth. A 467, 917-920 (2001).

Application of infrared synchrotron radiation to various fields of science

Takao Nanba

Graduate School of Science and Technology, Kobe University, Kobe 657-8501, Japan

Abstract. Up to now, infrared synchrotron radiation (IRSR) has been recognized as a powerful light source because of its many excellent properties such as high brilliance, a small divergence angle of the light beam in the direction perpendicular to the plane of the ring, a continuous spectral distribution over the entire energy range from x-ray to far-infrared regions in addition to a well defined pulsed structure, and so on. There exist two different types of storage ring among the rings in the world. One is a compact ring such as UVSOR (Okazaki,Japan) with a rather small bending radius (2.2 m) of a diple magnet and the other a large scale of ring such as SPring-8 (Harima,Japan) with a large radius (39.6 m). Both types of rings will open a new opportunities in the various fields of science because of its very high brilliance over conventional light sources.

Introduction

Infrared synchrotron (IRSR) is a powerful light source because of its many excellent properties such as high brilliance, a small divergence angle of the light beam in the direction perpendicular to the plane of the ring, a continuous spectral distribution over the entire energy range from x-ray to far-infrared regions in addition to a well defined pulsed structure, and so on. For example, on the polarization, IRSR is naturally a polarized light beam with a quite different polarization features compared to black body radiation sources. Its electric vector contains mainly the component parallel to the plane of the ring and a small perpendicular component out of the plane. So we can get a linear polarization in the plane of the ring and a circular polarization by extracting the radiation out of the plane of the ring without any polarizer. Such linearly and circularly polarized SR can make possible respectively an infrared linear and circular dichroism experiment, for example, on magnetic materials in the entire infrared region These prominent characteristics of IRSR have been proved by several groups at different SR facilities [1-8].

The first opening of the utilization of the IRSR light to the common users in the field of solid state physics has been done at UVSOR and the proof of the advantage of the IRSR over black body sources has given a trigger to prompt the construction of the beamline (BL) dedicated to the IR spectroscopy at following SR facilities. SPring-8 at Harima in Japan is the third generation x-ray light source and a largest storage ring for synchrotron radiation in the world. The construction of infrared (IR) beamline BL43IR at SPring-8 has been completed in the fiscal year of 2000 for IR spectroscopic study and has been utilized by many scientists. The beamline covers a very wide wavelength region from 500 nm (20,000 cm^{-1}, 2.5eV) to 0.1mm (100cm^{-1}, 12.5 meV) and offers very high brilliant IR photon beam to the experimental end stations [9].

In this report, we will describe the present stage and a perspective of the actual application of IRSR to the various fields of science.

CP716, *Portable Synchrotron Light Sources and Advanced Applications*,
edited by H. Yamada, N. Mochizuki-Oda, and M. Sasaki
© 2004 American Institute of Physics 0-7354-0195-0/04/$22.00

Extraction of IRSR radiation

Fig.1(*a*) shows the computed beam spread of the IRSR at the wavelength of 10 μm emitted into a vertical angle (ψ) to the ring with a unit horizontal angle (θ =1 mrad) with different bending radii of ρ = 40 m (a) with the acceleration energy *E* of 8GeV and 2.2 m (b) with *E* = 0.7 GeV, respectively, and (*b*) the schematic drawing of the emitting solid angle of the radiation. The figure shows clearly that more intense IRSR from a larger orbit of electrons is emitted into a narrower vertical angle. In addition, the intensity of the parallel component of the electric vector of the photon beam concentrates more close to the ψ = 0 plane. This concentration effect in addition to the increase in the intensity with the factor $\rho^{\frac{1}{3}}$ suggests that the brilliance of the IRSR increases roughly with the factor of $\rho^{\frac{1}{2}}$. Here, ρ is the radius of the electron orbit given by the bending radius of the dipole magnet.

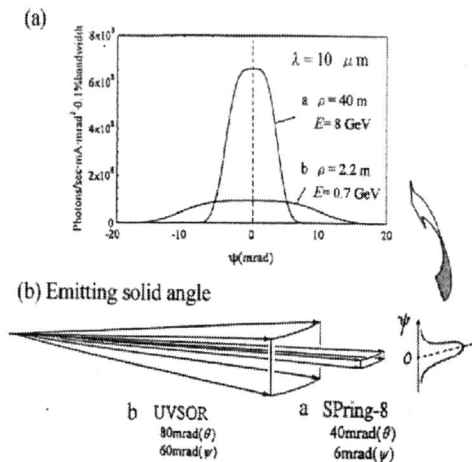

(a)

(b) Emitting solid angle

Figure 1. (*a*) The computed beam spread of IRSR at 10 μm emitted into a vertical angle (ψ) to the ring with a unit horizontal angle (θ =1 mrad), a bending radii of ρ =40 m (a) with the acceleration energy *E* of 8 GeV and 2.2 m (b) with *E* = 0.7 GeV, respectively, and (*b*) the schematic drawing of their emitting solid angle.

This means that a compact ring such as UVSOR emits the IRSR photon beam into a larger solid angle in comparison to a large ring. Then a compact ring requires an installation of a large collection mirror for the incident photon beam into a large solid angle. On the other hand, an extraction of IRSR from a large ring

composed of a dipole magnet with a large radius leads to a long emission length. For example, SPring-8 has the extremely long emission length of 1.43 m because of the radius of the electron beam orbit of 39.3 m and the horizontal acceptance angle of 36.5 mrad. Consequently, for both types of ring, the specially designed mirror is required to install in order to realize a well focused beam at the focal position without any astigmatism. We propose the installation of a specially designed mirror, a so-called magic mirror shown in Fig.2 ((*a*) for the schematic shape and (*b*) for the geometrical configuration of the set up). A magic mirror possesses a vertical curvature with a gradual change along the horizontal axis from one side to the other side of the mirror and then can realize a perfect focus for the circular orbit[10].

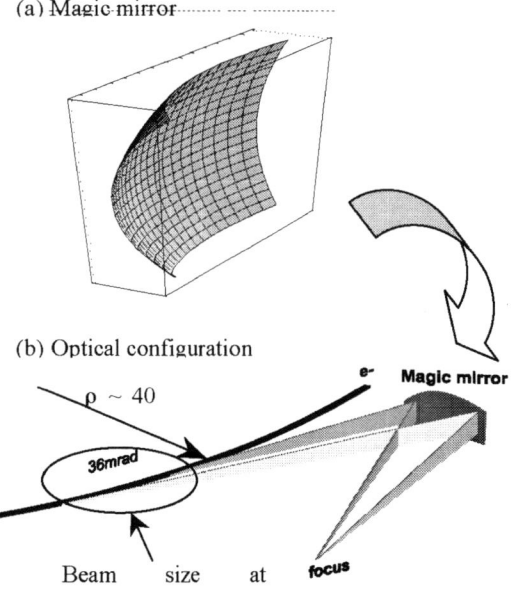

(a) Magic mirror

(b) Optical configuration

Figure 2. A schematic drawing of the so-called magic mirror at BL43IR, (*a*) the shape and (*b*) the geometrical configuration of the set up.

Experimental

Because of its high brilliance over conventional light sources, the utilization of IRSR can make following experiments possible, for example, (1) IR spectroscopy under extreme experimental conditions such as high pressure, and high magnetic field, (2) IR microscopic analysis on very small domains in synthetic specimens, (3) the 2D imaging spectroscopy on specimens with material inhomogeniety such as an off-stoichiometry and impurity distribution. These experiments (1)-(3) play a very important role in studies of not only solid state physics but also various fields of science including chemistry, earth science and biology.

At BL43IR of SPring-8, we have proved that the spectromicroscopy station can provides with a very narrow photon beam with a special resolution smaller than 15 μm in the FWHM[11]. Using this well focused photon beam with a small size at the end-station, the far-infrared reflection measurement on very small solid samples under high pressure up to 16 GPa has been performed by using a diamond anvil cell (DAC). The observation of the change in the electronic state very close to the Fermi level under pressure becomes possible by this kind of experiment. Recently the change in the electronic state of YbS due to the pressure induced insulator-metal transition was successfully observed for the first time[12]. The second example at the spectromicroscopy station is the observation of the spectral change in the IR transmission spectra of single crystal of natural brucite using a DAC under the conditions at high temperature up to 360 K and high pressure up to 17 GPa [13]. Brucite is a prototype of hydrous magnesium silicate minerals and the behavior of hydroxyls (OH-) under high pressure and high temperature is a very interesting subject from stand point of earth science. The experiment made clear the existence of two kinds of dipole moment of hydroxyls (OH-) under high pressure and high temperature.

On the other hand, at the compact ring such as UVSOR and Miracle (Ritsumeikan Univ., Kusatsu, Japan) it is possible to install a large collection mirror with a large solid angle very close to an electron orbit for collection of an intense photon beam with a large stored current up to 1 A. Then it.is considered to become a very powerful infrared light source. One of the target of the application seems to be a study on an irradiation effect to biological specimens by an intense IRSR.

Conclusions

Both a compact and a large ring provide with a brilliant IRSR over conventional light sources. The utilization of IRSR can make possible experiments in the various fields of science from solid state physics to life science.

Acknowledgment

The synchrotron radiation experiments were performed both at the UVSOR and the SPring-8 facilities. We are grateful to all staff members of the UVSOR and the SPring-8 for their kind cooperation in the experiments.

References

1. Duncan W.D. and Jarwood J., *Daresbury laboratory Technical Memorandum* No.DL/SCI/ TM32E (1982).
2. Schweizer E.*et al.,Nucl. Instr. and Meth.* **A239** 630 (1985).
3. Yarwood J.*et al.*, *Nature* **312** 742-744 (1984).
4, Nanba T. *et al.*, *Int. J.Infrared and Millimeter Waves* 7 1769-1776 (1986).
5. Nanba T., *Rev.Sci.Instrum.* **60** 1680–1685 (1989).
6. Williams G.D. *et al*, *Phys. Rev. Lett.* **62** 261(1989).
7. Nelander B., *Vibrational Spectroscopy* **9** 29 (1995).
8. Roy P.*et al*, *Nucl. Instru. Meth.***A 325** 893 (2001).
9. Kimura H *et al.*, *Nucl. Instru. Meth.***A 467-468** 441-444 (2001).
10. Kimura S. *et al.*, *Nucl. Instru. Meth.***A 467-468** 437-440 (2001).
11. Kimura S., Nanba T. *et al.* , *Nucl. Instru. Meth.***A 467-468** 893-896 (2001).
12. Matsunami M.,Chen L. Nanba T. et al, *Acta physica Polonia* **B34** 1011-1014 (2003).
13. Shinoda K. et al., Phys. Chem. Minerals **29** 396-402 (2002).

Applications to Materials Science

Application of Synchrotron Radiation in the Study of the Structure and Dynamics of Amorphous Ices and Clathrate Hydrates

John S. Tse

Steacie Institute for Molecular Sciences, National Research Council of Canada
Ottawa, Ontario, Cancda K1A 0R6

Abstract

Major advantages of a synchrotron is the energy tunability and very high brightness. These unique properties permit different types of spectroscopy which probe different physical and electronic properties to be performed. In this presentation, various experiments, from infrared (~0.01 eV) to Compton scattering (125 keV), on ice and clathrate hydrates will be discussed. Using low-temperature infrared spectroscopy, we have studied the mechanism for the novel iso-structural phase transition of high pressure ice-VIII. High resolution powder diffraction using high energy radiation was used to provide very accurate atom radial distribution functions for the characterization of the structural changes associated with the HDA to LDA ice transformation. Very high energy x-ray was used to study the interactions between simple rare gas atoms and methane with the ice framework in clathrate hydrates. The "resonant scattering" of the guests with the ice vibrations was identified using x-ray scattering and the specific role of the guest, in particular Kr in Kr hydrate was studied with the nuclear inelastic scattering technique.

CP716, *Portable Synchrotron Light Sources and Advanced Applications,*
edited by H. Yamada, N. Mochizuki-Oda, and M. Sasaki
© 2004 American Institute of Physics 0-7354-0195-0/04/$22.00

UV Photo-Oxidation of Silicon: A Novel Growth Method of Ultra-Thin SiO$_2$ Films

Hiroyuki Oyanagi and Atsuyuki Fukano

National Institute of Advanced Industrial Science and Technology, 1-1-1 Umezono, Tsukuba, Ibaraki 305-8568, Japan

Abstract. Superior insulating performance is found for dense silicon dioxide ultra-thin films (~3 nm) grown by UV photo-oxidation of silicon. Density profile obtained by glazing incidence x-ray reflectivity shows that the high density (2.32 g/cm^3) SiO$_2$ is formed on Si(100) surface at much lower temperature (<450 °C) than thermal oxidation, using 126 nm photons. The sharp and flat interface within 1-2 monolayers is revealed by high resolution transmission electron microscopy. The film density is strongly dependent on wavelength around 172-126 nm, suggesting that the specific excited species of oxygen is involved in the growth mechanism. Unique properties of photo-oxidized silicon dioxide are related to the modified Si-O network structure and ring statistics.

INTRODUCTION

Tracking "Moore's law" is getting more and more difficult as transistor gate dielectric scaling proceeds and gate insulators using silicon dioxide (SiO$_2$) approach the limit of keeping dielectric property of defect-free SiO$_2$ (*e.g.*, 1.2 nm for 90 nm design rule, Intel). Limitations for SiO$_2$ as a gate insulator are mainly due to the presence of "imperfect" SiO$_2$-Si interface which leads to electrical breakdown at lower voltage and greater leakage current as the film thickness approaches ~1.5 nm, which is a typical thickness of "suboxide" SiO$_x$ layer in conventional thermally oxidized SiO$_2$. Two opposing requirements, *i.e.*, scaling and power consumption, introduced a shift to "high-k" materials, expected to take place in mid 2000's.[1] In Fig. 1, the structure of CMOS device is schematically illustrated where the SiO$_2$ thin film is used as a gate insulator. To improve the SiO$_2$ dielectric limit, various photo-assisted growth techniques were proposed and tested.[2-4] Here, we report that the UV oxidation technique of silicon has superior insulating properties of ultra-thin (~3 nm) SiO$_2$ films[3] and discuss on future prospects realized by application of synchrotron radiation as a tunable light source in the 100 nm wavelength range. As the insulating performance, *i.e.*, breakdown voltage and leakage current, depends on the quality of SiO$_2$-Si interface determined by the density of oxygen defects (vacancy), numerous efforts have been directed toward the search of a new growth technique which minimizes oxygen insufficiency.

As the UV photo-oxidation method requires a simple setup and easy procedure, it can be readily incorporated into one of the processing steps in MOS device manufacturing. In contrast to thermal oxidation technique which needs high temperature growth (>900 °C), the UV photo-oxidation can be performed at much lower temperature (<400 °C). Strain-induced defect formation is thus suppressed and surface damage can be avoided, in contrast to RF plasma[5] or implantation of oxygen ions.[6,7] Inspite of the low substrate temperature during exposure, however, the SiO$_2$ films grown by this technique showed higher density than conventional films, *i.e.*, 2.27-2.32 g/cm^3 *vs.* 2.20-2.24 g/cm^3 for conventional SiO$_2$.[3]

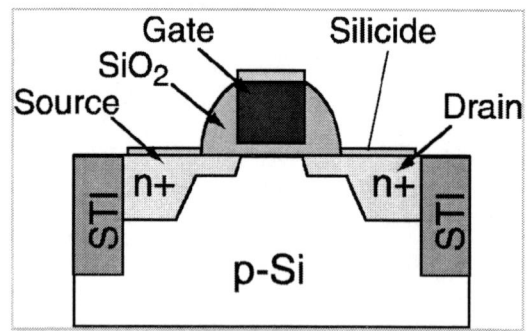

FIGURE 1. Schematic structure of CMOS device. Thin SiO2 layer is used as a gate dielectric which reaches 1.2-1.5 nm in 65 nm scale devices in 2004.

Densification of SiO$_2$ by UV irradiation was previously reported.[8] In the present case, modified Si-O network structure by densification was successfully related to improved dielectric properties, *i.e.*, the

CP716, *Portable Synchrotron Light Sources and Advanced Applications*,
edited by H. Yamada, N. Mochizuki-Oda, and M. Sasaki
© 2004 American Institute of Physics 0-7354-0195-0/04/$22.00

rearrangement of Si-O bonds could reduce oxygen vacancy at the SiO_2-Si interface which increased the dielectric property of ultra-thin SiO_2 films.

EXPERIMENTAL

The growth chamber of SiO_2 ultra-thin films is schematically illustrated in Fig. 2. As a light source for excitation, conventional excimer lamps filled with various kinds of gas (Ar_2, Xe_2 and KrCl) were attached to the reaction chamber and the substrate was irradiated by UV photons through a MgF_2 viewport. The wavelengths used were 172 nm, 222 nm and 126 nm. The substrate, silicon wafer (*p*-type, ρ=14-22 Ω/cm), was placed on a heater and gently heated during the growth. The substrate temperature was kept below 450 °C throughout the growth. After evacuation by turbo molecular pump, pure oxygen gas (1 atm) is introduced and flown continuously with a typical flow rate of 30 cc/min. The film thickness and growth rate were determined by transmission electron microscopy (TEM). The interface structure was studied by high-resolution TEM (HRTEM) and high-resolution Rutherford back scattering (HRRBS).

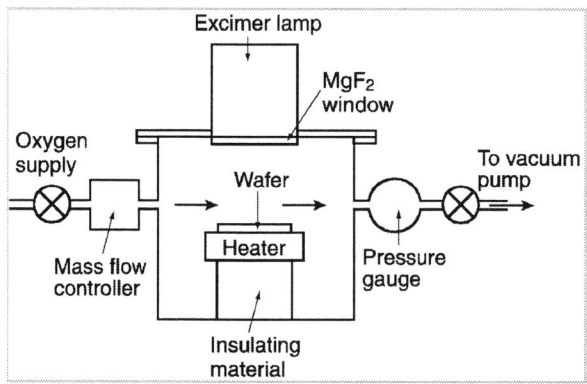

FIGURE 2. Schematic of growth chamber.

The dielectric property was evaluated by a two-point probe method. The metal (aluminum) insulator semiconductor (MIS) was fabricated using 3 nm thick SiO_2 prepared by 126 nm photon irradiation. The metal-SiO_2 and SiO_2-Si interfaces were evaluated by HRTEM. The composition of the SiO_x layer was analyzed by XPS Si 2p core peak profiles.

RESULTS AND DISCUSSION

The SiO_2 growth rate in UV photo-oxidation is dependent on the wavelength but for all wavelength photons used in this experiment, the growth rate slowed down around 4 nm in thickness and saturated at ~5 nm. The insulating property of SiO_2 gate insulator strongly depends on the quality of Si-SiO_2 interface, or the defect density in the oxygen-poor "suboxide" layer. We found that UV photo-oxidized SiO_2 ultra thin films (<5 nm) have maximum 5% higher density (2.32 g/cm³) than that of SiO_2 formed by thermal oxidation techniques.[3] A striking difference between the photo-oxidized and thermally oxidized SiO_2 films is the superior insulating performance of the former. The SiO_2 films grown by photo-oxidation of silicon using an excimer lamp having various wavelength values (λ =126-222 nm) on Si(100) substrates were compared. The density profile was obtained from x-ray reflectivity measurements which showed that the UV-oxidized SiO_2 films have almost the same density with silicon or mineral cristobalite (polymorph of quartz), having a constant density distribution along growth direction. Chemical state of silicon and Si-O vibrational property were investigated by x-ray photoelectron spectroscopy (XPS) and Fourier transform infrared (FTIR) spectroscopy. The results show that the suboxide layer is negligible in thickness whereas the Si-O stretching vibration frequency red-shifted in agreement with the trend of densification through the decreased O-Si-O bond angle and change in ring statistics.

The high-resolution transmission electron microscopy (HRTEM) image (Fig. 3) shows a smooth interface having flatness about 1-2 monolayers.

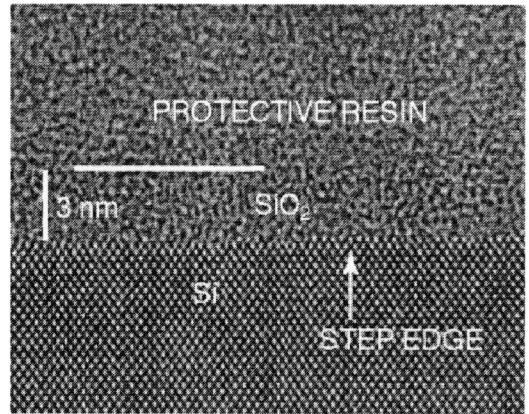

FIGURE 3. HRTEM image of UV photo-oxidized SiO_2 (3 nm in thickness).

The HRTEM image was measured by H-9000NAR (Hitachi) with the electron beam voltage of 200 kV. The image shown in Fig. 3 was taken with the electron beam along the <110> direction and the magnification of 10, 000, 000. The film thickness calibrated by the atomic spacing of silicon was 3.0 nm. The sharp image of ordered silicon atoms along the interface shows that the growth of SiO_2 does not induce structural disorder. The fact that atomically smooth SiO_2-Si interface was preserved suggests that the photo-assisted growth mechanism involves diffusion of oxygen atoms and not silicon atoms, *i.e.*, the growth does not alter the original silicon network. The formation of such a sharp interface may be related to the saturation of oxide film thickness around 4-5 nm. This saturation may correspond the lifetime of mobile oxygen atoms or depth of photo-excitation.

The density profile can be obtained from x-ray reflectivity profile measured over a wide range in incidence angle. Figure 4a shows the reflectivity as a function of incidence angle measured for the SiO_2 film (3.5 nm in thickness) by UV photo-oxidation (λ=172 nm), where open circle and solid line indicate the experimental data and calculation, respectively. Fitting the theoretical curve, the density profile is obtained as a function of depth from surface. The obtained density function is plotted in Fig. 4b. The density of SiO_2 (172 nm), 2.27 g/cm^3 is slightly greater than the average density of thermally oxidized SiO_2, 2.20-2.24 g/cm^3.

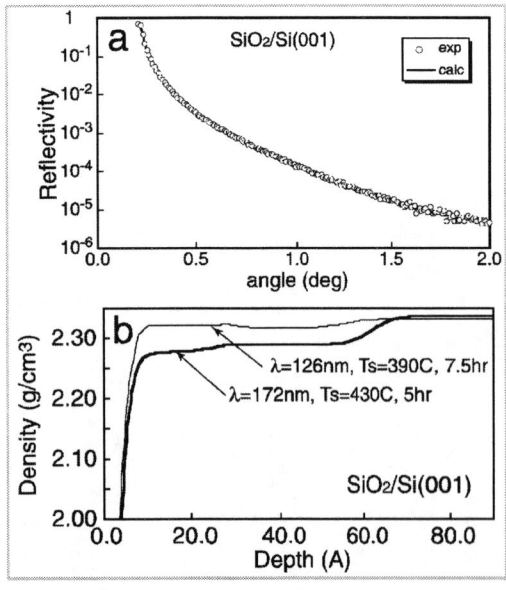

FIGURE 4. a., X-ray reflectivity of UV photo-oxidized SiO_2 (3 nm in thickness), b., density profile for SiO_2 films with 172 nm and 126 nm irradiation.

The highest density (2.32 g/cm^3) was observed for the film prepared under 126 nm photon irradiation. Moreover, the density of SiO_2 layer for 126 nm growth is remarkably flat indicating homogeneous structure along the growth direction. The SiO_x layer is generally distinguished as an intermediate density region which smoothly connects with the SiO_2 layer as seen in case of 172 nm irradiation. However, in case of high density SiO_2 film grown by 126 nm irradiation, it is difficult to find such a low density layer between the SiO_2 and Si. XPS Si 2p levels showed that the amount of Si species with incomplete bonding with oxygen is negligible considering the silicon atoms partly bonded to Si at the interface (ideally abrupt interface model). So far some of the SiO_2 films grown by other techniques with flat density profile were reported.[9] To our knowledge, however, the observed density (2.32 g/cm^3) is the highest value, almost equal to those of silicon and one of the polymorphs, cristobalite (2.33 g/cm^3).

The density values measured for SiO_2 films grown by UV photo-oxidation are summarized in Table 1 together with those of conventional SiO_2 and polymorphs of quartz. In SiO_2, each silicon atom is bonded with four oxygen atoms, forming a SiO_4 tetrahedron, which is linked with each other sharing two-fold coordinated oxygen atom. In the Si-O network, the Si-O-Si angle θ_0 is related to the density and ring statistics. In a high density form of SiO_2 glass, θ_0 decreases whereas in low density forms, it increases. Because of increasing repulsive interaction of bonding charge as θ_0 decreases, the structure becomes unstable under compression and transforms into high-pressure phases. However, in case of silicate glasses, high density forms are quenched as a metastable state after removing pressure. The density increase (5%) in UV photo-oxidized SiO_2 is not surprising as high density forms (2.2-3.0 g/cm^3) are formed by UV irradiation.[8]

TABLE 1. Density of SiO_2 and polymorphs of quartz.

Material	Density (g/cm^3)	
SiO_2 (UV)		
126 nm	2.32	Ref. 3
172 nm	2.27	Ref. 3
222 nm	2.27	Ref. 3
Dry	2.24	
Wet	2.20	
Cristobalite	2.32	
Si (bulk)	2.33	

Superior dielectric property of SiO_2 film (3.0 nm) prepared by 172 nm photo-oxidation is shown in Fig 5. The current density of thermally grown SiO_2 film (3.3

nm) gradually increases as the dielectric voltage is increased,[10] whereas that of UV photo-oxidized SiO_2 shows smaller leakage current until it breaks down at the maximum dielectric voltage of 18 MV/cm. Further, it was found that the breakdown voltage of $\lambda = 126$ nm photo-oxidized SiO_2 film was about 45 MV/cm with leakage current less than 10^{-4} A/cm^2. This demonstrates that both breakdown voltage and leakage current are improved by a factor of four and six orders of magnitude, respectively. This is presently the highest value of ever reported gate insulating performance of SiO_2 films.

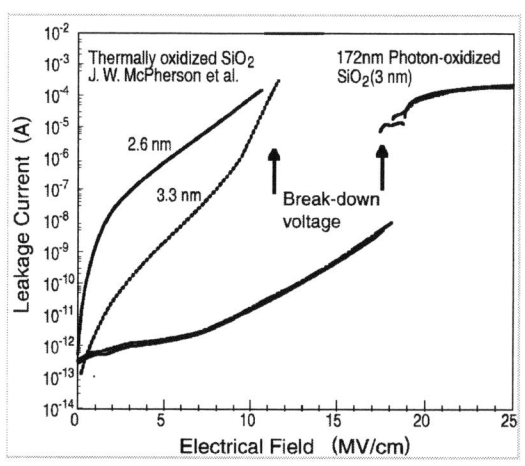

FIGURE 5. J-V characteristic for thermally grown SiO_2 (3.3 nm) and 172 nm photo-oxidized SiO_2 (3.0 nm).

In this report, we describe the growth of high-density SiO_2 ultra-thin films by UV photo-oxidation technique and their physical properties. The growth mechanism of dense silicon dioxide is still unclear although photo-induced densification and increase of refractive index are well-known.[8] UV photons excite valence electrons in SiO_2 into conduction bands and create paramagnetic oxygen vacancy centers. In Ge-doped SiO_2 glass, two-fold oxygen coordinated paramagnetic germanium atoms (Ge E' centers) are formed by UV irradiation.[11] Such divalent states would be distinguished in XPS Si 2p core level peak. Our XPS data found a small amount of Si^{2+} states which could be ascribed to the photo-induced paramagnetic centers. In SiO_2 glass, the ring statistics is related to density whereas reversible compression without modifying the ring statistics reduces the O-Si-O bond angle. In photo-induced densification, the ring statistics can be modified as oxygen radicals (O^*) migrate, break the Si-Si bonds and finally form the Si-

O-Si bonds. Oxygen radicals created by dissociation of O_2 molecules diffuse into Si interstitial sites. The two processes, *i.e.*, formation of paramagnetic defect centers and Si-O bonds compete each other. In order to obtain detailed information on kinetics, the substrate temperature dependence is to be studied. The saturation of growth indicates that either the migration or the formation of Si-O bonds are strongly photo-assisted. Information on symmetric and asymmetric Si-O vibration is important to discuss the ring statistics. Vibrational properties of Si-O bonds are under investigation by FTIR spectroscopy. Because of strong wavelength dependence, synchrotron radiation can be a powerful tunable light source to investigate the optimum wavelength and mechanism of photo-assisted growth and densification.

ACKNOWLEDGMENTS

The authors express their thanks to H. Hashimoto for x-ray reflectivity measurements and A. Takada and N. Hata for FTIR measurements. They also thank A. Sakai for technical assistance in photo-oxidation experiments and .

REFERENCES

1. Moore, G. E., *Electronics* 38, 114-117 (1965).

2. Kazor, A. and Boyd, I. W., *J. of Appl. Phys.* 75, 227-231 (1994).

3. Fukano, A. and Oyanagi, H., *J. of Appl. Phys.* 94, 3345-3349 (2003).

4. Kazor, A. and Boyd, I. W., *Appl. Phys. Lett.* 63, 2517-2519 (1993).

5. Zhang, J. F. , *Semicond. Sci. Technol.* 5, 824-830 (1990).

6. Afanas'ev, V.V., Stesmans, A. and Twigg, M.E., *Phys. Rev. Lett.* 77, 4206-4209 (1996).

7. Fujimaki, M., Nishihara, Y., Ohki, Y., Brebner, J. L., and Roorda, S., *J. Appl. Phys.* 88, 5534-5537 (2000).

8. Kato, H., Fujimaki, M., Noma, T. and Ohki, Y., *J. Appl. Phys.* 91, 6350-6353 (2002).

9. Hashimoto, H., private communication.

10. McPherson, J.W. and Khamankar, R.B., *Semicond. Sci. Technol.* 15, 462-470 (2000).

11. Uchino, T., Takahashi, M. and Yoko, T., *Phys. Rev. Lett.* 84, 1475-1478 (20—00).

Hard X-Ray Spectro Microprobe Analysis of Inhomogeneous Solids: A Case Study. Element Distribution and Speciation in Selected Iron Meteorites

Ronald G. Cavell[1,2,*], Renfei Feng[2], Elspeth M. Barnes[2], Patricia A. Cavell[3], Alistair J. McCready[2] and M. Adam Webb[2]

[1]Dept. of Chemistry, University of Alberta, Edmonton, Alberta, Canada, T6G 2G2. [2]The Alberta Synchrotron Institute,University of Alberta Research Transition Facility, 8308 -114 Street, Edmonton, Alberta Canada T6G 2E1, [3]Dept of Earth and Atmospheric Sciences, University of Alberta, Edmonton, AB, Canada T6G 2E3.

Abstract. The hard X-ray microprobe provides an effective methodology for the non-destructive analysis of inhomogeneous materials. Application of X-ray absorption/fluorescence spectroscopy techniques (XANES and EXAFS) permits the speciation of the elements and yields information about the local structural environment. Microfocussed, monochromatic, tunable X-rays allows examination of small areas of micrometer dimensions with spectroscopic procedures. Typically the materials which are presented are thick and cannot be altered for the experiment. This condition introduces difficulties which may compromise the results. Herein we discuss those difficulties and show that the system can yield reliable results in spite of the compromises. Some results are presented on the two iron meteorites we have examined. These specimens are representative of highly inhomogeneous materials and illustrate the difficulties encountered with compositional variations which may occur at sub-millimeter dimensions and also illustrate the difficulties presented by the need to analyze components present at ppm concentration levels in a concentrated matrix. In these particular samples the major constituent is Fe which ranges from 90% to 70%, balanced by Ni which ranges from 10% to 30%. The critical diagnostic trace elements Ga and Ge which must also be analyzed are present at the 80 and 340 ppm level respectively. These diagnostic elements have been shown by EXAFS to be substitutionally placed in the matrix of the major element species in these meteorite samples.

INTRODUCTION

Many important materials are highly inhomogeneous and the analysis of such materials often poses problems. We wish to know what elements are present, the distribution of these elements within the material, often to micrometer or smaller dimensions, and the nature of the chemical valence and/or surroundings of these elements. Many different microscopic probes have been developed for such measurements. Among the most powerful are the hard X-ray microprobes now being developed at third generation synchrotron sources. Micrometer areas can be illuminated with monochromatic X-rays and the X-rays can be energetically scanned to provide high resolution spectroscopic measurements. Application of these principal X-ray absorption/fluroescence spectroscopy techniques (XANES and EXAFS) permits the speciation of the elements and yields information about the local structural environment of the species in the microscopic area under examination. We shall illustrate this technique with studies of very inhomogeneous materials - the iron meteorites [1]. The X-ray techniques briefly described here are, however, equally applicable to many kinds of inhomogeneous samples; zoned minerals, metal alloys and composite materials present situations which require the determination of elemental distribution as well as elemental chemical speciation and local structural environments, of micron-scale areas.

Meteorites are fascinating objects which provide clues to the evolution of the universe [2]. Full knowledge of their character and hence their origin

CP716, *Portable Synchrotron Light Sources and Advanced Applications,*
edited by H. Yamada, N. Mochizuki-Oda, and M. Sasaki

requires detailed study of the chemical state, concentrations and distribution of the elements within. Variations occur over spatial dimensions on the order of micrometers. Iron meteorites (based on Fe-Ni alloys) are subclassified into 13 groups on the basis of structure and their Ga, Ge, and Ni content [2, 3]. The system is representative of inhomogeneous materials in the presentation of significant analytical challenges. The first goal of such analysis is to determine the element distribution across the sample. The problem is often intensified by the inhomogeneity - in this case there are very large variations in the Fe/Ni concentration ratios. These materials are composed primarily of bulk iron metal (>70%). Regions of high nickel concentrations (*c.a.* 30% Ni) are named taenite,

those with low nickel concentrations (*c.a.* 7% Ni) are named kamacite. The structure type is different in each of these phases (or "minerals"). The matrix is relatively dense. Also present are inclusions or intergrowths of iron or iron/nickel carbide, sulphide and phosphide "minerals". Significant trace elements also occur, some in the high ppm to percentage concentration range. Of these, Ga and Ge are the most important because they are used to classifiy the meteorites into "types" which are thought to be indicative of their detailed origin [2, 3]. In general also, the meteorite samples are rare and valuable materials so non-destructive techniques are preferred. Special sample preparations are also difficult to make because the material is hard and brittle.

ANALYSIS OF THICK SAMPLES

Constraints Presented by Thick Samples, Self Absorption Effects

The rules for obtaining good XAS spectra are that (a) concentrated samples should be thin, (b) samples should be uniform, and (c) a good contrast of absorption should be seen at the edge (a pronounced "jump" is desirable). Dilute samples may be made thicker for higher signal intensity [4]. Materials which we wish to examine (and many related inhomogeneous materials) however do not satisfy these criteria. Because the samples are infinitely thick, all our measurements must be performed with the fluorescence detection mode. In these thick sample environments, self absorption in the sample distorts the spectrum. We must be concerned with extent of penetration of the X-rays, the absorption of the fluorescent X-rays and other such effects. Most of this can be calculated so that spectra can be adjusted [5, 6] however this aspect of the analysis is at present incomplete. We also need to be concerned (in the case of the polished mounts used herein) with the 3-dimensional profile of the sample – that is, do the structural features that we see on the surface descend into the material in parallel with our exciting beam (so that we are sampling the same material to depth or are they tilted so that our X-rays sample different materials as they penetrate the sample.

Systems with wide limits of concentration also provide constraints on the choice and efficacy of detectors. We must measure small (ppm) concentrations of materials in the presence of high concentrations of the element(s) which form the matrix. Solid state detectors (especially the multi element type) are commonly used because they possess high

sensitivity allowing the measurement of elements in low concentration but such detectors are often swamped by the signals from the concentrated species. Attenuating the strong signals with filters also affects the weak signals. The advantage offered by such detectors is the measurement of the complete fluorescent spectrum with one sampling interval. The energy resolution is however not as good as may be required by particular cases and overlaps between signals often inhibits or prevents the differentiation of the components. This is presently a major problem when the elements of concern are close neighbors in the periodic table.

It can be shown that if the absorption of a sample is high then the ratio of the emergent fluorescent signal to the intensity of the exciting radiation (I_f/I_o) will involve serious distortions of the absorption coefficient of interest, $\mu_i(E)$ [5, 6]. The most accurate determination of the absorption coefficient is required because this provides the chemical and structural information which is sought. To ascertain whether reliable results can be obtained in practice, we examined a relatively thick foil (of iron) which, although thin enough to allow transmission measurements to be made, was also sufficiently thick to show significant self absorption attenuation effects. We measured transmission, fluorescence and total electron yield simultaneously and the spectra are shown in Figure 1. The sample was mounted at a 45° angle with respect to the beam (as is generally required to perform fluorescent measurements) so the effective thickness of the foil was greater than is applicable when foils are used in the usual absorption only mode.

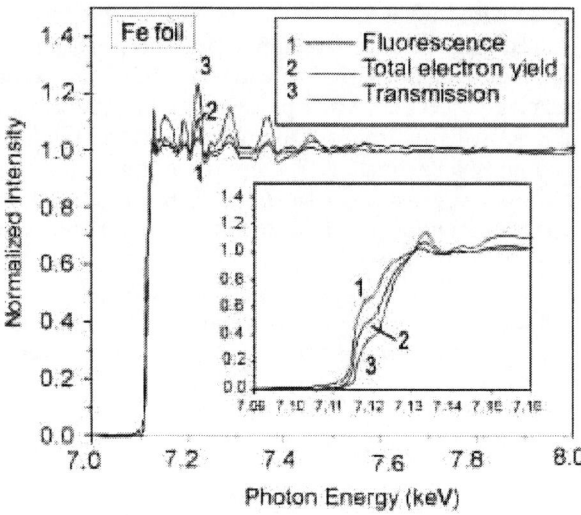

FIGURE 1. EXAFS spectra of Fe foil obtained with different modes of detection.

At the experimental angle the fluorescent absorption coefficient, $\mu_{Fe}(E_f)$, will be much lower than $\mu_{Fe}(E)$ above the edge. Therefore the edge in the measured spectrum (displayed as I_f/I_o) will be suppressed as we see in the spectrum (Figure 1 insert). We note that there is considerable difference in the spectral shape provided by the different detection methods. In particular the edges are highly distorted. The analysis of this data is shown in Figures 2-4. We note that despite the differences in magnitude, the EXAFS oscillation shapes are quite similar for all. The spatial information revealed by all final transforms, when normalized, is equivalent. Thus, even if the fluorescent data are compromised, the EXAFS analysis results can be expected to be quite reasonable especially with regard to the determination of the major features.

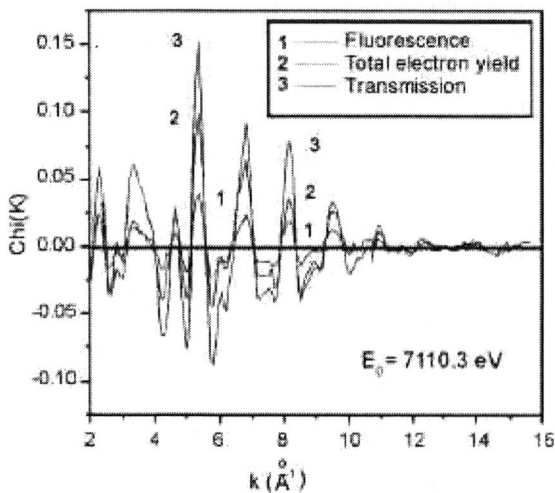

FIGURE 2. EXAFS analysis of Fe foil data.

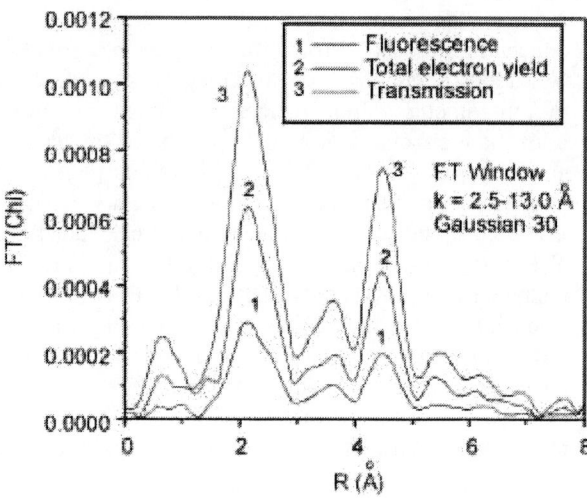

FIGURE 3. R space Fourier Analysis of Fe foil. Amplitudes are those determined by the raw data.

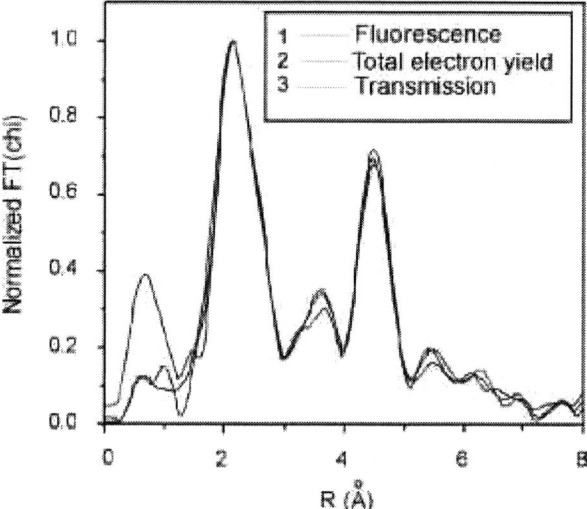

FIGURE 4. R space Fourier analysis of Fe foil. Amplitudes are normalized to each other.

RESULTS AND DISCUSSION

The first step in studies of this type is to map the elemental distributions in the sample by successively irradiating each small area and collecting the emitted X-ray fluorescence signal with solid state multi-element detectors. Typical results are shown in Figure 5 for the Canyon Diablo Meteorite sample. The XAFS data for Fe, and Ga in kamacite and Fe in the foil, were obtained at the APS during a single session.

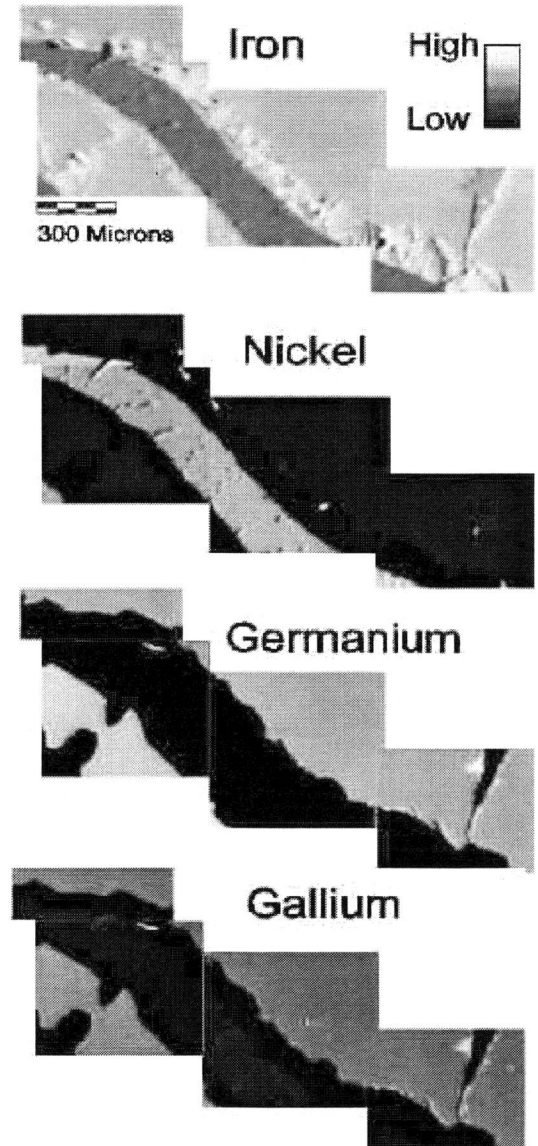

FIGURE 5. Element Distribution maps for a region of the Canyon Diablo meteorite specimen. Concentration scale values are arbitrary and independent in each case .

Mapping the elemental distribution in the Canyon Diablo specimen reveals a feature in center area of the material which has reduced iron and enhanced nickel content; this is an iron/nickel phosphide inclusion. The highest (relative) Ge and Ga concentrations occur along with the highest iron concentration. Close examination of the concentration profiles shows that the Fe/Ni phosphide inclusion also carries a small proportion of Ga but not Ge. EXAFS analysis of the minor elements (Ga and Ge) in the major matrix area (the large featureless outer area in the map) revealed that these elements are present as true solid solutions imbedded in the body centered cubic lattice of the kamacite. Extensive analysis of other areas of the Canyon Diablo specimen are reported elsewhere [1].

More interesting, perhaps, is the rich inhomogeneous structure exhibited by the Mayerthorpe meteorite. The electron backscattered image (Figure 6) obtained in a separate experiment

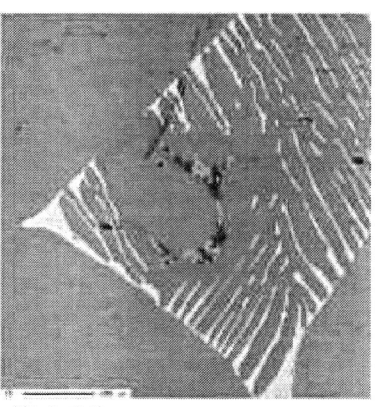

FIGURE 6. Inhomogeneity patterns in an area of the Mayerthorpe specimen as revealed by the electron backscattered image. A part of the lower right hand corner area is mapped with the X-ray microprobe in Figure 7

with the electron microprobe showed an extensive pattern of differentiated material. Mapping a portion of this larger area with the X-ray microprobe (Figure 7) showed that the thin filaments are areas of high nickel concentration infiltrated into the area of lower nickel concentration. This surrounding matrix is the mineral kamacite (approx 7% Ni in Fe) and the filaments are the mineral taenite (approx 30% Ni in Fe). Ge and Ga distribution appears to follow the nickel.

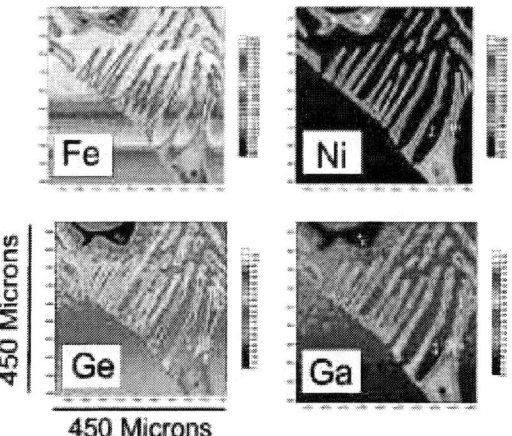

FIGURE 7. Element distribution maps for a segment of the Mayerthorpe meteorite specimen shown in Figure 6. Concentration scales are arbitrary and independent. Light areas indicate high concentration.

We plot the relative concentrations in a profile across the mapped area as illustrated in Figure 8; the increase in Ni (and Ga and Ge) is paralleled by a decrease in the Fe level as expected.

Analysis of the EXAFS spectrum (Figure 10) in the high nickel area of the sample (Spot B in Figure 8) reveals that the nickel structure is face centered cubic (as is known for taenite). Additional analysis of the Ga and Ge EXAFS likewise indicates that these elements are emplaced in the fcc lattice structure of the taenite.

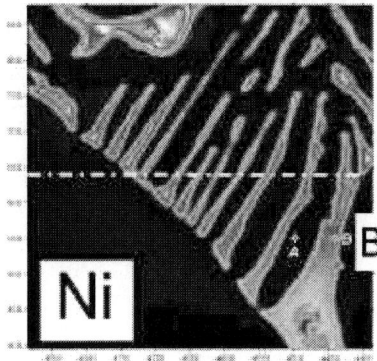

FIGURE 8. The Nickel concentration Map and the Profile track. The concentration profile across the sample is shown by the dotted line and the profile is graphed in Figure 9.

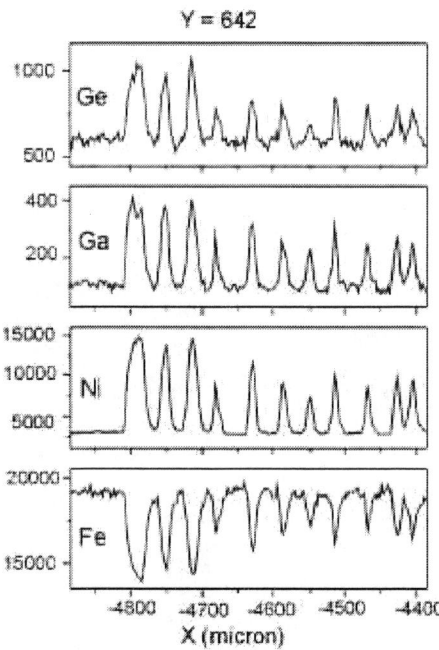

FIGURE 9. Relative concentration profiles in one track across the Mayerthorpe mapped area. Note that the bands range in width from 10 to 50 microns. The intensity scale in each case is an arbitrary count value.

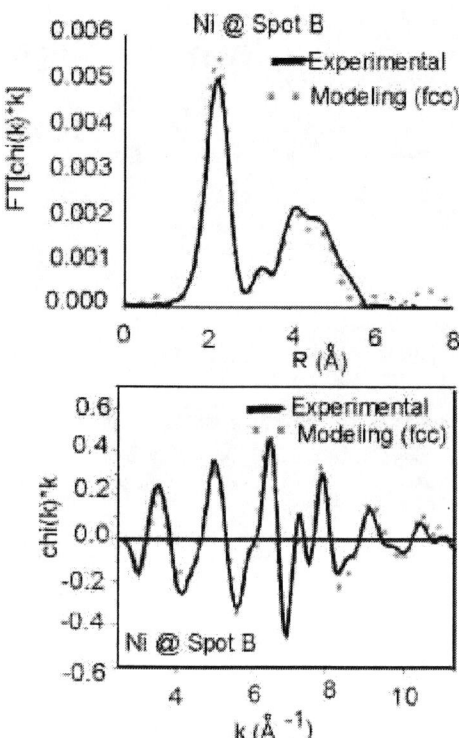

FIGURE 10. EXAFS analysis of the Nickel spectrum at Spot B (taenite, high Ni concentration) located in Figure 9.

SUMMARY

Elemental distribution and element speciation in complex inhomogeneous can be determined with microscopic X-ray fluorescence spectroscopy. EXAFS can reveal the structure of the major phases (here, Fe and Ni) without crystallography and it is also possible to evaluate the structural characteristics of species present in low concentrations. We show that self absorption effects, although important, do not overly compromise the analysis. Concentration profiles across phase boundaries can be established. At the present time, quantification of the analysis is not fully established. More work on calibrations and corrections to achieve quantitative analysis is needed.

EXPERIMENTAL

The meteorite samples were provided as mounted specimens in an epoxy resin block. The iron foil was mounted on a simple frame. XAFS microprobe data for Fe, Ni, Ga and Ge in the meteorite sample and the iron foil data were obtained at the Pacific Northwest Consortium - Collaborative Access Team (PNC-CAT) undulator beamline (Sector 20) at the Advanced Photon Source (APS), Argonne National Laboratory [7]. X-rays from a Si (111) double crystal monochromator (75% tune at 7500 eV) were incident at an angle of 45 degrees to the sample which was mounted on a high precision x-y translation stage. The monochromator had an energy resolution of ? E/E ~1 × 10^{-4} eV and was calibrated using Cu foil and amorphous Ge against the reference edge values (E_o) of 8980 eV [8] and 11103 eV [9]) respectively. A Kirkpatrick-Baez mirror pair [10] focussed the beam to approximately 5µm x 5µm for both the element distribution mapping and the XAFS data collection. Fluorescent X-rays were collected with a 13-element Ge(Li) solid state detector (Princeton Gamma-Technology) mounted at 90 degrees to the beam, with several layers of aluminum foil (150 µm total) being used to attenuate the strong Ni and Fe signals. A Lytle detector was used for measuring the high intensity Fe signal during XAFS data collection. Typical counting intervals for the mapping were on the order of 0.3 to 1.0 seconds/pixel point, and between 1.0 and 4.0 seconds/point for the XAFS. The Fe, Ni, Ge and Ga X-ray absorption spectra were signal averaged and processed with WinXAS [11] and IFFEFIT [12] software. The EXAFS interference functions $\chi(k)$ were extracted from the $\mu(E)$ data of individual scans using a first order polynomial background subtraction and normalization to the edge jump. Fourier transforms were taken with k^1 weighting, and a 10% Gaussian window over the k-space range using the nearest zero crossings to 2.5 and 9.5 Å^{-1}.

ACKNOWLEDGEMENTS

Synchrotron work was done at the Pacific Northwest Consortium (PNC) at the Advanced Photon Source (APS), Argonne National Laboratory near Chicago, Illinois, USA and (P K edge) at the Canadian Synchrotron Radiation Facility (CSRF), SRC (Aladdin), Madison, WI. ANL is operated by the University of Chicago under contract with the U.S. Department of Energy (DoE) Office of Basic Energy Sciences. The APS (Contract W-31-109-Eng-38) and PNC-CAT (Contract DE-FG03-97ER45628) are supported by the U.S. DoE, Office of Basic Energy Sciences.The Natural Sciences and Engineering Research Council of Canada (NSERC) supports PNC-CAT with capital funding and both PNC and CSRF with Major Facilities Access grants. RGC, EMB, AW, and RF thank the University of Alberta and the Alberta Synchrotron Institute for support and financial assistance. RGC also thanks NSERC for support. We thank the Earth and Atmospheric Sciences Department of the University of Alberta for the loan of meteorite specimens.

REFERENCES

1. Cavell, R. G., Barnes, E. M., Arboleda, P. A. H., Cavell, P. A., Feng, R., Gordon, R. A., and Webb, M. A., *American Mineralogist* **89**, 519-526, (2004).

2. Buchwald, V. F. *Handbook of Iron Meteorites*. Vols. 1 & 2. University of California Press, Berkeley, CA, (1975).

3. Wasson, J. T. and Kallemeyn, G. W., *Geochimica et Cosmochimica Acta* **66**, 2445-2473 (2002.), Wasson, J. T. and Richardson, J. W., *Geochimica et Cosmochimica Acta* **65**, 735-744 (2001).

4. Koningsberger, D. C. and Prins, R., *X-Ray Absorption: Principles, Applications, Techniques of EXAFS, SEXAFS and XANES*. Wiley, New York (1988).

5. Booth, C. H. and Bridges, F., *Proceedings of XAFS 12*, Malmo, Sweden, June 22-27 2003.

6. Cavell, R. G. and Feng, R. unpublished data.

7. Heald, S. M., Stern, E. A., Brewe, D. E., Gordon, R. A., Crozier, E. D., Jiang, D. T., and Cross, J. O., *Journal of Synchrotron Radiation* **8**, 342-344 (2001).

8. Kraft, S., Stümpel, J., Becker, P., and Kuetgens, U., *Review of Scientific Instruments* , **67**, 681-687 (1996).

9. Bearden, J. A. and Burr, A. F., *Reviews of Modern Physics* **39**, 1, 125-142 (1967),

10. Eng, P. J., Rivers, M. L., Yang, B. X. and Schildkamp, W., *Proceedings of SPIE*, Volume: **2516**, 41-51 (1995).

11. Ressler, T., *Journal of Physics IV*, (C2) 269-270 (1997).

12. Newville, M., 2001. *Journal of Synchrotron Radiation* **8**, 322-324 (2001).

Applications to Chemistry and Biophysics

Raman H-Bond Energies and Pair Volumes for Water

George E. Walrafen

Department of Chemistry, University of Kansas
Lawrence, Kansas

Abstract. Maximum values of the H-bond energies and H-bond pair volumes were determined from the OD-stretching Raman contour of HDO in water using dispersion curves obtained from the temperature and pressure dependence of the intensities. The maximum H-bond ΔE value is $5,100 \pm 500$ cal/mol, and the maximum H-bond pair ΔV is 1.4 cm^3/mol. A minimum in the volume dispersion curve indicates that some H-bonds are strongly bent as the volume is decreased by pressure rise to ≈ 10 kbar at 28 °C. Similar effects of H-bond bending are observed from dispersion curves obtained from NaCl and NaBF$_4$ in water. The main effect on liquid water of isothermal pressure rise to 10 kbar is to form H-bonds, both linear and bent, thus explaining freezing to ice VI.

INTRODUCTION

The dispersion of the H-bond enthalpy (approximately equal to the H-bond energy at low pressures), and of the H-bond pair volume, may be determined over the Raman OH- and OD-stretching contours from water using the definitions: $H_i = -R[\partial \ln I_i / \partial(1/T)]_P \approx E_i$, and $V_i = -RT[\partial \ln I_i / \partial P]_T$. I_i is the contour height, above the baseline, at the Raman frequency $\omega = i$. These definitions measure the temperature or pressure dependence of the Gibbs free energy as a function of Raman frequency over the OH- and OD-stretching contours, and as such provide more detailed information than shape changes observed in the individual spectra.

METHOD

The enthalpy and volume dispersion curves over the OH- and OD-stretching Raman contours are obtained as follows. A series of Raman OH- and OD-stretching contours is obtained isobarically and under absolute intensity conditions for a wide range of temperatures, or isothermally over a wide pressure range. Baselines, usually slightly nonlinear, are obtained under the Raman stretching contours, ideally by polynomial least squares fitting using baseline shapes outside the range of the contours. Contour heights are then measured at a series of Raman frequencies, i.e., at ω values in cm^{-1}, for the complete range of temperatures or pressures examined. Least squares slopes are then obtained from the derivatives $[\partial \ln I_i / \partial(1/T)]_P$ or $(\partial \ln I_i / \partial P)_T$. Five to ten well-spaced temperatures or pressures are generally sufficient, and 15 to 35 least squares fits over the main region of the stretching contours are usually adequate to develop the main features of the energy or volume dispersions.

The ends of the energy or volume dispersion curves are easily determined. Difference spectra involving widely different temperatures, for example, indicate that the temperature dependence is zero at the upper and lower ω limits of the OD- and OH-stretching contours. This means that two $E_i = 0$ points may be determined simply by measuring the Raman frequencies at which the intensities become zero. Also, the presence of an isosbestic point, that is, a frequency at which I_i is independent of temperature (or pressure), yields another $E_i = 0$ point. The two $E_i = 0$ end points are valuable because they provide accurate limits to the dispersion curve in regions where the temperature or pressure dependence of the intensity is virtually impossible to measure accurately.

CP716, *Portable Synchrotron Light Sources and Advanced Applications,*
edited by H. Yamada, N. Mochizuki-Oda, and M. Sasaki
© 2004 American Institute of Physics 0-7354-0195-0/04/$22.00

RESULTS

A typical H-bond enthalpy or energy dispersion curve, obtained from absolute Raman intensity measurements,[1] is shown in Figure 1. This curve was obtained from the OD-stretching contour from HDO in water. Three $E_i = 0$ points were obtained from the isosbestic frequency[1] at 2570 cm^{-1} and from the two contour end points near 2200 cm^{-1} and 2800 cm^{-1}. The other 16 experimental points were determined from least square fits.

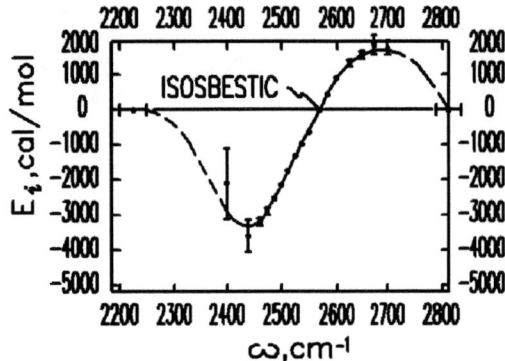

FIGURE 1. Dispersion of Raman H-Bond Energy, E_i across the OD-stretch contour of HDO in H_2O. Reprinted with permission from G.E. WALRAFEN, J. CHEM. PHYS., 15 FEB 04. Copyright 2003, American Institute of Physics.

The <u>minimum</u> of Figure 1 at 2440 cm^{-1} occurs at the same frequency as that of the peak from amorphous HDO-H_2O ice[2] at 4 K. The <u>maximum</u> of Figure 1 occurs at $\approx$ 2650 cm^{-1} which is the peak frequency of HDO in water at 400°C and 4,000 bar.[3] These frequency coincidences mean that the H-bonds in water are similar with regard to O-O distance and H-bond angle to those in amorphous ice and supercritical water, but they do not mean that water is a mixture of amorphous ice and supercritical water.

The difference in energy between the maximum and the minimum of Figure 1, namely, $E_{2650}(MAX) - E_{2440}(min) = 5,100 \pm 500$ cal/mol. This value is in excellent agreement with Pauling's limiting value[4] of about 4.5 kcal/mol.

The minimum in Figure 1 means that OD-oscillators at 2440 cm^{-1} (similar to those in amorphous ice) decrease in intensity/concentration at the greatest rate with temperature rise. The maximum in Figure 1 means that OD-oscillators at 2650 cm^{-1} (similar to those in supercritical water) increase in intensity/concentration at the greatest rate with temperature rise.

In many cases it is extremely difficult to obtain absolute Raman intensity data. Absolute Raman intensities are those resulting from precisely constant excitation intensity and detection efficiency. This difficulty is especially acute in high pressure Raman measurements where sapphire window rupture, pressure induced birefringence, cell corrosion, etc., create problems. Fortunately, one may resort to other definitions to obtain the dispersion curves, namely, $\Delta H = -R[\partial \ln(I_i/I_{REF})/\partial(1/T)]_P$, and $\Delta V = -RT[\partial \ln(I_i/_{REF})/\partial P]_T$. Here, I_{REF} may refer to any suitable ω for which the signal-to-noise ratio is large. However, if the isosbestic frequency is known, the use of $I_{REF} = I_{ISOSBESTIC\ FREQUENCY}$ gives the same result as that obtained from absolute Raman intensity measurements. This is so because a point of constant intensity when used as a standard is equivalent to constant excitation intensity.

The dispersion of the volume was obtained from Raman intensity measurements of HDO in water to $\approx$ 10,000 bar at 28°C. An intense broad minimum was observed at $\approx$ 2400 cm^{-1}, and a maximum at $\approx$2650 cm^{-1}. The difference, $\Delta V_{2650}(MAX) - \Delta V_{2400}(MIN) = 1.4$ cm^3/mol. This volume change refers to the O-D$\cdots$O pair volume in water. This is the maximum volume change because the states refer approximately to amorphous ice and supercritical water.

The above-described minimum in the volume dispersion refers to a <u>rise</u> of I_{2400} with increasing pressure, and the maximum to a <u>decrease</u> in I_{2650} with rising pressure. This means that isothermal rise of pressure to $\approx$ 10 kbar increases the amount of H-bonding, at least near room temperature. The fact that the amount of H-bonding may be increased by high pressure has been little understood, and widely unappreciated. It explains, via the Raman data described above, why water freezes to ice VI near 10,000 bar at room temperature.

A second sizeable minimum was also observed in the volume dispersion curve near 2600-2620 cm^{-1}. The corresponding OH-stretching frequency position may be estimated from 2600-2620 $\times$ 1.353 = 3518-3545 cm^{-1}. This is the corresponding OH-stretching value for HDO in heavy water.

The joint frequency-angle probability distribution function of Lawrence and Skinner[5] for HDO indicates that the O-H···O angle corresponding to this OH-stretching frequency is in the vicinity of 150 degrees.

Bending of H-bonds is a convenient way for water to decrease its molal volume in response to high pressures. Another way involves the formation of hydrogen bonds, because their O-O distances are smaller than those of broken hydrogen bonds. H-bonding bending, however, does not lower the energy of the system so much as linear H-bond formation.

It is known that ice VI contains H-bonds which are strongly bent.[6] The above-described 150 degree bending in liquid water produced by high pressures presages the bent H-bonds in ice VI.

It is possible to estimate the isobaric temperature derivative, $(\partial \Delta \Delta V / \partial T)_P$, from data of this work, where $\Delta \Delta V$ is the maximum H-bond pair volume described above, namely, $\Delta \Delta V = \Delta V_{2650}(MAX) - \Delta V_{2400}(MIN)$. This estimate of the above derivative involves use of the thermodynamic equation of state: $(\partial H / \partial P)_T = V - T(\partial V / \partial T)_P$.

Enthalpy dispersion curves were obtained from pure water at 1 bar and at 1,500 bar. The H-bond ΔH so obtained was found to be nearly independent of temperature over 1,500 bar. Hence, $(\partial \Delta \Delta V / \partial T)_P \approx \Delta \Delta V / T$. The average temperature, T, was ≈ 305 K, which yields a value of $\approx 5 \times 10^{-3}$ cm^3/deg-mol. Other considerations, however, suggest a more conservative value of ≈ 2 to 5×10^{-3} cm^3/deg-mol for the isobaric temperature coefficient of the maximum H-bond volume. The positive sign of this derivative means that the H-bond pair volume should increase <u>isobarically</u> with temperature rise.

The limiting ΔH or ΔE for HDO in water is $5,100 \pm 500$ cal/mol. If one takes the heat capacity of water to be constant and to have a value of about 18 cal/deg-mol, it is suggested from the above ΔH that the states involved in the enthalpy change are separated by about 280 ± 30 K. This roughly 300 K range is certainly sufficient to disrupt most H-bonds at low pressures. Of course, it should be possible, in principle, to determine the dispersion of the heat capacity across the OD-stretching contour from HDO in water. Thus far, however, curvature in the $\ln I_i$ versus $1/T$ plots, if it exists at all, is within the experimental uncertainties of the absolute Raman intensity data. Hence, a direct experimental verification of the states involved in the H-bond ΔH by means of the heat capacities is not currently possible.

Energy and volume dispersion curves may also be obtained for aqueous solutions, e.g., NaCl, NaBF$_4$, or t-butanol in water, as well as for ternary solutions involving a single electrolyte or single nonelectrolyte in HDO/water mixtures.

The Raman OH-stretching energy dispersion curve for NaCl in water gives indications of bent O-H···Cl$^-$ H-bonds.[1] The Raman OH-stretching energy dispersion curve for NaBF$_4$ in water is the most unusual of the curves obtained thus far.[7] This curve displays two large minima, separated by a maximum. The maximum occurs near 3,530 cm^{-1}, which is experimentally at the same position as the minimum described in the H-bond volume dispersion curve mentioned above. This suggests that the BF$_4^-$ ion may engage in bent, relatively weak H-bonds. It should be noted here that the Raman spectrum from NaBF$_4$ in water or in HDO/water mixtures is visually indistinguishable from the corresponding Raman spectrum of NaClO$_4$ in water.[8]

The interesting feature of the NaBF$_4$ energy dispersion curve is that this electrolyte interacts with all H-bonds of the solvent, the very strong as well as the very weak, to form its own probably bent, weak H-bond interaction.

The use of enthalpy or energy and volume dispersion curves resulting from the OD- and OH-stretching contours of aqueous solutions is in a very early stage of development. The expanded application of such dispersion methods to a wide range of aqueous solutions is expected to provide a new type of very detailed information about solute-water interactions in liquid water.

REFERENCES

1. Walrafen, G. E., *J. Chem. Phys.*, accepted and to be published 15 Feb 2004, or later; G. E. Walrafen, *J. Chem. Phys.* **48**, 244 (1968).

2. Iogansen, A. V. and Rozenberg, M. Sh., *Opt. and Spectry.* (USSR) **44**, 49 (1978).

3. Franck, E. U. and Lindner, H., doctoral dissertation of the latter, University of Karlsruhe, 1970.

4. Pauling, L., "The Nature of the Chemical Bond," Cornell University Press, 1948. See page 304.

5. Lawrence, C. P., and Skinner, J. L., *J. Chem. Phys.* **118**, 264 (2003).

6. Kamb, B., *Science* **150**, 205 (1965).

7. Walrafen, G. E., recent unpublished work.

8. Walrafen, G. E., *J. Chem. Phys.* **55**, 768 (1971).

High Pressure FTIR/Raman Studies on Molecular Conformation of Proteins and Model Peptides

Yoshihiro Taniguchi

Department of Applied Chemistry, College of Science and Engineering, Ritsumeikan University, 1-1-1, Noji-higashi, Kusatsu, Shiga 525-8577, Japan
E-mail : taniguti@se.ritsumei.ac.jp

Abstract

Molecular conformational equilibrium is the most central concept in the chemistry and biochemistry of chain molecules. This equilibrium is occasionally affected by environmental conditions such as temperature, pressure and solvents. Such structural flexibility of molecules plays an important role in chemical and biological phenomena in liquid phase. In particular, the equilibrium in water is of vital importance for biological systems.

The pressure effect on the equilibria in liquids can give information of volume differences between trans/gauche of rotational isomers or native/unfolded states of proteins. The volume properties are important to understand the intermolecular interaction between the solute and solvent molecules and the molecular mechanism. It is accepted that the volume changes for trans/gauche equilibrium of rotational isomers in non-polar solvents are less than -5 cm^3/mol [1] and for native /unfolded states of proteins in aqueous media less than -100 cm^3/mol [2].

Recent development of high pressure FTIR spectroscopy combined with resolution enhancement techniques and Raman spectroscopy is able to detect the signal of each rotational isomer in dilute aqueous solution and the secondary structure of the pressure induced structure changes of proteins in water. In this study, the effect of pressure on the conformational equilibrium between rotational isomers of haloacetone [3,4], alanine dipeptide [5], and proteins [6,7] in aqueous media has been studied by the FTIR and Raman spectroscopes. On the base of both observed volume changes of simple chain molecules and proteins, the molecular mechanism on the pressure induced conformational changes will be discussed.

References
1. Y. Taniguchi, *J. Mol. Structure* , **126**, 241 (1985).
2. Y. Taniguchi, K. Suzuki, *J. Phys. Chem.*, **87**, 5185 (1983).
3. M. Kato, Y. Nanba, Y. Taniguchi, *Chem. Phys. Letter*, **289**, 30 (1998).
4. Y. Shiratori, Y. Nanba, M. Kato, Y. Taniguchi, *Bull. Chem. Soc. Jpn.* **76***, 501* (2003)
5. T. Takekiyo, T. Imai, M. Kato, Y. Taniguchi, *Biopolymers*, **73** (2003) in press.
6. N. Takeda, M. Kato Y. Taniguchi, *Biochemistry*, **34**, 5980 (1995).

CP716, *Portable Synchrotron Light Sources and Advanced Applications,*
edited by H. Yamada, N. Mochizuki-Oda, and M. Sasaki

Molecular Mechanism Of Light-Driven Proton Pump Of Bacteriorhodopsin

Mikio KATAOKA

Graduate School of Materials Science, Nara Institute of Science and Technology (NAIST)
Ikoma, Nara 630-0192, Japan

Abstract. Synchrotron radiation X-ray diffraction has been intensively applied to bacteriorhodopsin photointermediates to understand the molecular mechanism of light-driven proton pump. Time-resolved X-ray diffraction with time resolution of 244msec revealed that the two types of the structures appear during a photocycle. In order to realize the time resolution of 5μsec, we developed a new snap shot diffraction technique. With the method, we revealed that the structural change is triggered by the deprotonation of Schiff base. Based on the structures, we proposed a conformation-controlled conformational change model. The apparent discrepancy between crystal structures and low resolution structures is also discussed.

INTRODUCTION

Bacteriorhodopsin (BR) is a sole protein found in the purple membrane of *Halobacterium salinarum*. BR transport protons from the cytoplasmic to the extracellular medium by using absorbed photon energy (a light-driven proton pump). BR is composed of 7 transmembrane α-helices, named A to G, and the chromophore, retinal. Upon absorption of light, BR undergoes a photoreaction cycle composed of J, K, L, M, N and O intermediates [1]. The most important steps for the proton pump are deprotonation and reprotonation of the Schiff base of the chromophore, which are the event at the M and N intermediates. A simple idea of the ion pumping mechanism is that the ion-binding states of the ion pump are linked to protein conformations to switch the ion transport pathway from one membrane surface to another surface [2]. Thus, the structure of the M intermediate of BR has been intensively investigated to reveal the substantial conformational changes [3-8]. It was shown that this structural change is closely related to the deprotonation of the Schiff base; in the original structure (conformation E) the proton channel is open to the extracellular side, and when an M-type conformation is assumed (conformation C) it is open to the cytoplasmic side [9]. Further, we revealed that the structure of the N intermediate is somewhat different from that of the M intermediate, and that the structural transition from the M to the N intermediate is hydration-dependent

[10]. The characteristic change of the M intermediate is appeared near helix B, while that of the N intermediate is near helix F. Based on these findings, we proposed the conformation-controlled conformation change model [11].

Although it is now accepted the substantial conformational change during photoreaction, there are serious and essential discrepancies among the structures of the intermediates. We believe that two distinct structures for the M and the N intermediates are essential [7, 11], while another group insisted that the one structure is sufficient for the pump activity [12]. The high resolution X-ray crystallography has been successfully applied to the structural studies of photointermediates to reveal the role of internal water molecules and the local structural changes [13-17]. However, the large conformational change observed with the low-resolution techniques is not necessarily confirmed with the crystallography. The observed shift of α-helix is about 2Å with low resolution technique [15, 18], while it is less than 1Å by crystallography [13-17]. Further, the crystal structures of the intermediates are not necessarily identical among the researchers [13-17].

It is essential to understand the origin of the discrepancies for the better understanding of the molecular mechanism of proton pump. For the purpose, we carried out time-resolved X-ray diffraction experiment for the purple membrane [19, 20]. We developed the new technique to realize the 5μsec time

CP716, *Portable Synchrotron Light Sources and Advanced Applications,*
edited by H. Yamada, N. Mochizuki-Oda, and M. Sasaki
© 2004 American Institute of Physics 0-7354-0195-0/04/$22.00

resolution. The relationship between the crystal structures and the low-resolution structures is also discussed in this report

TIME-RESOLVED X-RAY DIFFRACTION OF BACTERIORHODOPSIN

Time-resolved X-ray diffraction experiments were carried out at the beamline XU45 in SPring-8 [21]. X-ray diffraction patterns up to 7 Å were recorded by a CCD camera (Hamamatsu Photonics C4880-82) coupled with a 6-inch X-ray image intensifier (Hamamatsu Photonics) [21]. Two hundred frames were recorded with 244 msec time resolution. Temperature was kept at 10 °C. The samples were excited by a xenon flash lamp with Toshiba Y50 and C50D filters (500 < λ < 630 nm) at 2.44 sec after the start of the diffraction measurement, which corresponds to the beginning of the 11th time frame. A series of sequential 2-dimensional ring diffraction patterns were averaged circularly to reduce them into a set of sequential 1-dimensional patterns, and 10 different data sets were averaged. The sequential 1-dimensional diffraction patterns are equivalent to an m × n matrix, whose element is the diffraction intensity at the m-th point of S (=$2\sin\theta/\lambda$) and the n-th time point. The data were analyzed by the singular value decomposition (SVD) method.

The diffraction profile just after the flash illumination showed the characteristic diffraction changes for the formation of the M intermediate, namely the intensity increases of (20) and (32), and the intensity decreases of (11) and (40) reflections. These changes recovered to the original intensities within 40sec. We could resolve three independent significant components as the result of the SVD analysis. The first component is corresponding to the unphotolyzed BR. The other two components showed characteristic time courses. The result clearly shows the existence of two intermediate structures. The diffraction profiles corresponding to the component with slow decay time and that with fast decay time were reconstructed by global fitting with double exponential functions. The differential Fourier maps were synthesized with the resultant profiles. The maps of the slow and fast components have characteristic features of N-type and M-type structure, respectively. The time-resolved diffraction experiment clearly reveals that the two different intermediate structures really exist in the BR photocycle, and that the M type structure appears faster than the N type structure.

The time-resolved diffraction work clarified that the M to N transition accompanies the conformational change, and that the M intermediate possesses the distinct conformation from the unphotolized state. The next question is to reveal the relationship between the deprotonation of Schiff base and the conformational change. For the purpose, we need to measure the L to M transition, the time resolution of which should be less than 50μsec. The present system for the time resolved study is insufficient to do this, because the best time resolution is 36msec for the CCD detector. We developed a new snap shot diffraction method to measure the structural transition with 5μsec time resolution. We developed a new snap shot diffraction method with a fast shutter system. The shutter system generates pulsed X-ray with 6μsec width. The light trigger is synchronous to the fast shutter through a delay. The delay time determines the time resolution. Another merit of this technique is that we can reduce the radiation damage. The accumulation of 20 data sets gave a sufficient diffraction profile to discriminate the structure. By changing the time delay from 5μsec to 0.5sec, we observed a series of diffraction data, which is equivalent to the time-resolved X-ray diffraction with the 5μsec time resolution. Two independent components were separated with the SVD analysis, indicating the single intermediate. The difference Fourier map of the reconstituted diffraction profile clearly shows the characteristic M-type structure. The time course of the rise of the M-type structure, however, shows the significant delay from the spectral change. The decay is almost identical. The result indicates that the deprotonation of the Schiff base is faster than the structural change. This is the first evidence for the structural M1 to M2 transition. The result confirms the proposed conformation controlled conformational change mode.

CONFORMATION-CONTROLLED CONFORMATIONAL CHANGE MODEL

The results of the structural studies on the photointermediates can be summarized as follows.
1. Global conformational changes occur during the photoreaction cycle.
2. The conformational change occurs in the cytoplasmic half of the molecule. The change brings the opening of the proton channel in the cytoplasmic side.
3. The conformational change is triggered by the deprotonation of Schiff base.

4. Subtle but meaningful structural change accompanies with the M2 to N transition.

Based on the results, we proposed the conformation controlled conformation change model [11]. We require the following assumptions.

1. BR can take two conformations. We call them the E conformation and the C conformation each other. In the E conformation, the proton channel is open to the extracellular side, while in the C conformations, the channel is open to the cytoplasmic side. There are two sub-states (M2 and N) in the C conformation.
2. Deprotonation of Schiff base brings the E to C conformational change, while reprotonation of Schiff base brings the C to E conformational change.

The proton pump mechanism is explained as follows [11]. The light absorption induces the isomerization of retinal from all-*trans* to 13-*cis* form. The isomerization induces the proton transfer from the Schiff base to Asp85 (the deprotonation). The deprotonation triggers the conformational change from E to C (assumption 2). During the conformational change, the connection between Asp96 and the Schiff base is established (the M type structure, assumption 1). The further conformational change brings the channel opening of the cytoplasmic side (assumption 1). The opening makes the cytoplasmic half of channel hydrophilic, which lowers the pKa of Asp96. The proton is released to Schiff base through the established connection (reprotonation of the Schiff base). The connection is also necessary for the inhibition of the proton release to the cytoplasm. The reprotonated Schiff base induces the C to E conformational change (assumption 2). The closure of the cytoplasmic channel makes the microenvironment of Asp96 hydrophobic, which causes the reprotonation of Asp96 from the cytoplasm. The original state is recovered because the E conformation is established again (assumption 1).

The essence of the model is that the local chemical reaction and the global conformational change are closely inter-related. In the other words, localized chemical reaction affects protein conformation and the protein conformation regulates local reaction. We consider that such interrelationship guarantees the switch mechanism as the essence of the vectorial proton transportation of BR [11]. The model is

confirmed and is refined to the more elegant model based on the crystal structures [22].

COMPARISON OF CRYSTAL STRUCTURES WITH LOW-RESOLUTION STRUCTURES

The crystal structures of the intermediate with the deprotonated Schiff base revealed the role of internal water molecules and the local structural changes such as the rearrangements of hydrogen bond networks near the Schiff base and in the extracellular channel [13-17]. However the large movement in helix level revealed so far is not necessarily observed. The magnitude of the helix shift was limited to be less than 1 Å, if any, which is significantly smaller than the shift observed by low-resolution diffraction [15, 18]. The apparent discrepancy between the crystal structure and the low-resolution structure is discussed as the effect of the crystalline force constraint [15]. Further, the clarified structural changes are somewhat different among the investigators. Table 1 shows the available PDB codes for the intermediates with the corresponding dark states. Although Takeda *et al.* have not published their results, the coordinates are available through PDB web site. In order to clarify the origin of the discrepancies mentioned above, we calculated the low resolution difference electron density maps with these reported coordinates of the intermediates except for the D96N reported by Sass *et al.* [13]. The atomic coordinates of cytoplasmic side are not included in 1C8S because of the fluctuation.

The calculated difference density maps were categorized into three groups: 1) the map with the characteristic change near F helix; 2) the map with the characteristic change near B helix; 3) the map with no substantial changes. The degree of the calculated diffraction intensity changes is estimated as R-factor, $\sum |\Delta I(h,k)| / \sum I(h,k)$. The values are widely distributed from 0.02 to 0.1. The classification with the R-factor shows good correlation with the categorization with the map. The map with the change near F helix shows large electron density changes with R-factor, 0.1 (Sass *et al.* (2000) [14]), the map with change near B shows small electron density changes with R-factor, 0.05

TABLE 1. The list of deposited PDB codes for the M intermediate.

Author	Protein	Dark	Light	State Reported	State Assigned
Luecke *et al.* (1999)	D96N	1C8R	1C8S	late-M	-
Sass *et al.* (2000)	Wild type	1CWQ	1CWQ	late-M	MN
Luecke *et al.* (2000)	E204Q	1F50	1F4Z	early-M	M1
Faccioti *et al.* (2001)	Wild type	1KG9	1KG8	early-M	M2
Takeda *et al.* (2000)	Wild type	1QM8	1DZE	M	M1

(Facciotti *et al.* (2001) [16]), and the map with little electron density changes has the R-factor of 0.02 (Sass *et al.* (2000) [14], Takeda *et al.* (2000)(unpublished)). In the case of low-resolution studies reported so far, the R-factor is about 0.06 for M-type and about 0.1 for N-type structure, which are comparable to the former two [7, 10].

From these observations, we can conclude that the differences in the crystal structures among the investigators are originated from the fact that the observed intermediate states are different each other. They characterized the intermediate only with the absorption spectrum to confirm that their observed state has the deprotonated Schiff base. However, the sophisticated FTIR spectroscopy revealed the existence of the MN intermediate, which has the deprotonated Schiff base with the N type protein conformation [23]. We also confirmed the MN intermediate [10]. Further, as stated above, there are two M-states, namely the deprotonated Schiff base with no protein conformational change (M1) and that with small conformational change (M2). From the comparison of the calculated maps with the low resolution structures reported so far, we can assign the observed intermediate states for the crystal structures, which are summarized in Table 1.

In order to confirm the conclusion, we need to measure IR spectral change upon illumination for the crystalline sample. Since the available BR crystal is rather small (100μm in diameter), the measurements with the conventional FTIR spectrometer is essentially impossible. We carried out FTIR measurement for the BR crystal with the IR beam line at SPring-8 with a microscope. Since the changes in IR spectrum upon illumination are distinct but very small, we need the sufficient base line stability to discriminate the changes. At the moment, we cannot observe difference IR spectrum between the photointermediate and the unphotolyzed state, because the baseline of the spectrum is seriously unstable. The noise level is almost identical with the expected changes. The improvement of the measurement system is now under way.

ACKNOWLEDGMENT

This work was partly supported by the Grant-in-Aid from the Ministry of Education, Science, Culture, Sports and Technology. The author expresses his sincere thanks to Drs. Toshihiko Oka, Hironari Kamikubo, Janos K. Lanyi and Richard Henderson for their valuable discussions and supports offered throughout this study.

REFERENCES

1. Lozier, R. H., Bogomolni, R. A., and Stockenius, W., *Biophys. J.* **15**, 955-962 (1975).
2. Jardetzky, O., *Nature* **211**, 969-970(1966).
3. Dencher, N. A., Dresselhaus, D., Zaccai, G., and Büldt, G., *Proc. Natl. Acad. Sci. U. S. A.* **86**, 7876-7879 (1989).
4. Koch, M. H. J., Dencher, N. A., Oesterhelt, D., Plöhn, H.-J., Rapp, G., and Büldt, G., *EMBO J.* **10**, 521-526 (1991).
5. Nakasako, M., Kataoka, M., Amemiya, Y., and Tokunaga, F., *FEBS. Lett.* **29**, 273-275 (1991).
6. Subramaniam, S., Gerstein, M., Oesterhelt, D., and Henderson, R., *EMBO J.* **12**, 1-8 (1993).
7. Kamikubo, H., Kataoka, M., Váró, G., Oka, T., Tokunaga, F., Needleman, R., and Lanyi, J. K., *Proc. Natl. Acad. Sci. U. S. A.* **93**, 1386-1390 (1996).
8. Vonck, J., *Biochemistry* **35**, 5870-5878 (1996).
9. Kataoka, M., Kamikubo, H., Tokunaga, F., Brown, L. S., Yamazaki, Y., Maeda, A., Sheves, M., Needleman, R., and Lanyi, J. K., *J. Mol. Biol.* **243**, 621-638 (1994).
10. Kamikubo, H., Oka, T., Imamoto, Y., Tokunaga, F., Lanyi, J. K., and Kataoka, M., *Biochemistry* **36**, 12282-12287 (1997).
11. Kataoka, M., and Kamikubo, H., *Biochim. Biophys. Acta* **1460**, 166-176 (2000).
12. Subramaniam, S., Lindahl, M., Bullough, P., Faruqi, A. R., Tittor, J., Oesterhelt, D., Brown, L., Lanyi, J., and Henderson, R., *J. Mol. Biol.* **287**, 145-161 (1999).
13. Luecke, H., Schobert, B., Richter, H. T., Cartailler, J. P., and Lanyi, J. K., *Science* **286**, 255-261 (1999).
14. Sass, H. J., Büldt, G., Gessenich, R., Hehn, D., Neff, D., Schlesinger, R., Berendzen, J., and Ormos, P., *Nature* **406**, 649-653 (2000).
15. Subramaniam, S., and Henderson, R., *Nature* **406**, 653-657 (2000).
16. Facciotti, M. T., Rouhani, S., Burkard, F. T., Betancourt, F. M., Downing, K. H., Rose, R. B., McDermott, G., and Glaeser, R. M., *Biophys. J.* **81**, 3442-3455 (2001).
17. Luecke, H., *Biochim. Biophys. Acta* **1460**, 133-156 (2000).
18. Oka, T., Kamikubo, H., Tokunaga, F., Lanyi, J. K., Needleman, R., and Kataoka, M., *Biophys. J.* **76**, 1018-1023 (1999).
19. Oka, T., Yagi, N., Fujisawa, T., Kamikubo, H., Tokunaga, F., Kataoka, M., *Proc. Natl. Acad. Sci. U. S. A.* **97**, 14278-14282 (2000).
20. Oka, T., Yagi, N., Tokunaga, F., and Kataoka, M., *Biophys. J.* **82**, 2610-2616 (2002).
21. Fujisawa, T., Inoue, K., Oka, T., Iwamoto, H., Uruga, T., Kumasaka, T., Inoko, Y., Yagi, N., Yamamoto, M., and Ueki, T., *J. Appl. Cryst.* **33**, 797-800 (2000).
22. Lanyi, J. K., and Shobert, B., *J. Mol. Biol.* **328**, 439-450 (2003)
23. Sasaki, J., Shichida, Y., Lanyi, J. K., and Maeda, A., *J. Biol. Chem.* **267**, 20782-20786 (1992).

Protein Crystallography : A "Must" Technology for Drug Design

Takao Matsuzaki

Mitsubishi Chemical Corporation and Zoegene Corporation

Abstract. The history of drug-related protein crystallography and drug design is reviewed to show that "Lead Generation" is high-lighted in the pharmaceutical industry nowadays. A new drug design method has been developed. The method gave very high success rate; 10-60 % gave < 100 µM, 90 % gave < 10 mM. The crystal structures of drug-protein complexes have become even more important to give solid experimental bases for e.g. 1,000 designed structures and to find the new mechanisms of drug action.

INTRODUCTION

Protein crystallography has been utilized in drug design field since early 1980s. Major contribution was that the method visualized drug action by determining drug-protein complex structures. Combined with genetic engineering and synchrotron radiation, the method steadily established its position in a new field, "SBDD (Structure-Based Drug Design)". On the other hand, drug design method stayed in "Docking" for a long time and there still seems to be a bottleneck to create promising compounds.

TABLE 1. History of Drug-related Protein Crystallography & Drug Design

Year	Drug-related Protein Crystallog	Drug Design
-1970s	3D Structures of Proteins	QSAR (Structure Activity Rel.)
1980s	**3D Struc. Of Complexes ≡ SBDD** ↓	**Docking** ↓
1990s	Protein Prod. by **Genetic Eng.** **Synchrotron** MAD (Multi-wavelength Anom. Diff.) ↓	Combinatorial Chemistry HTS (High Throughput Screen) Insilico Screening ↓
2000s	Structural Genomics HTX (High Throughput Cryst.) ↓	NMR Screening **Stagnation with Docking** ↓
2002	**Collapse of HTX Business Model**	→ **Strong Need for** **Lead Generation Method**
	Complex Struc. are more important	←

CP716, *Portable Synchrotron Light Sources and Advanced Applications,*
edited by H. Yamada, N. Mochizuki-Oda, and M. Sasaki

DRUG-RELATED PROTEIN CRYSTALLOGRAPHY

We have determined 7 protein structures and 28 protein-ligand complex structures in 1998-2000. Two examples are given.

Tau Protein Kinase I (TPKI)

The first example is an Alzheimer disease-related protein, TPKI. TPKI inhibitors will prevent the formation of neurofibrilla which is observed in patient's cortex. We determined the 3 dimensional structure of TPKI by overcoming the difficulty of the protein production. Fifty litter E.coli culture broth yielded only 1 mg protein for crystallization. Synchrotron X-ray measurement was indispensable with this small amount of sample.

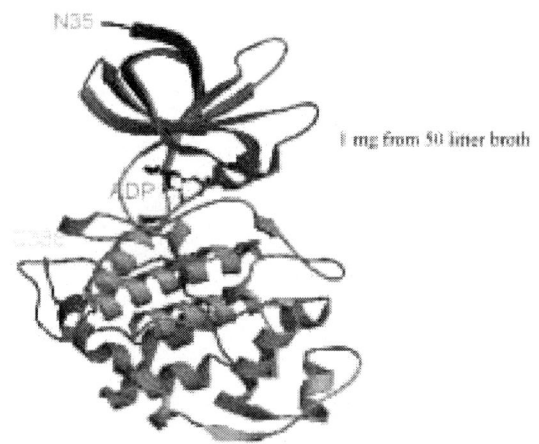

FIGURE 1. X-ray structure of TPKI.

Translin

The second example is an DNA/RNA binding protein, Translin, which may be related to leukemia. We determined the 3D structure by MAD (Multi Wavelength Anomalous Diffraction) method, mutating 2 Cysteins with Methionines for Se-Methionine introduction. Genetic engineering technique is quite important to prepare such samples.

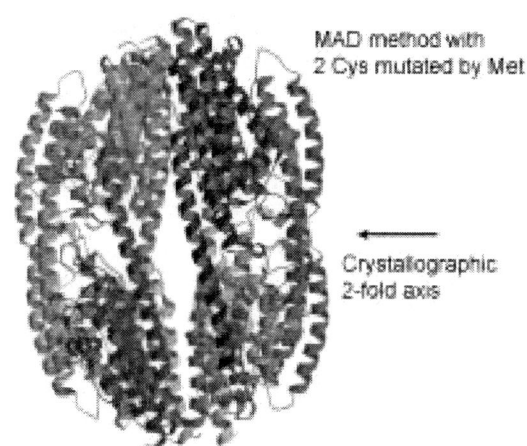

FIGURE 2. X-ray structure of Translin.

Protein Production is a crucial step.

The most crucial step used to be protein crystallization for a long time. But, we now have genome sequence information for protein production and our interest is converged to disease-related proteins, it is replaced by protein production. The present need is a sample which yields crystals with high resolution diffraction quality. I believe that homogeneity in protein 3D structure is most important on which production conditions like host cell selection and vector design affect most.

```
Protein Production is a crucial step,
    formerly occupied by Protein Crystallization.

* What is a GOOD sample ? *
[ 100 % activity ]
[ single band with SDS-PAGE(electrophoresis)  ]
[mono-dispersion with Dyn, Light Scattering   ]
[ good folding with NMR-HSQC ]
[ good looking crystals  ]

Those do not guarantee
 " high resolution diffraction ".
Most important factor is
 " Homogeneity in protein 3D structure " ,
 on which production conditions affect most.
```

FIGURE 3. Protein Production is a crucial step.

A NEW DRUG DESIGN METHOD

The need of the present pharmaceutical industry is focused on "Lead Compounds" and "Lead Generation Method". The present drug design methods are based on "Docking" and they have not been able to satisfy the industrial needs. We have developed a new denovo drug design method and obtained good results.

Ideas of denovo Design Software

X-ray analysis of drug-protein complexes showed 2 major factors of the intermolecular interaction. As shown in Fig 4., they are ; 1) Complementarity of surface 3D structures, and 2) Complementarity of electrostatic potentials.

We have developed a drug design program as shown in Fig 5. and succeeded to show that the program could output known inhibitors for bovine trypsin as shown in Fig.6.

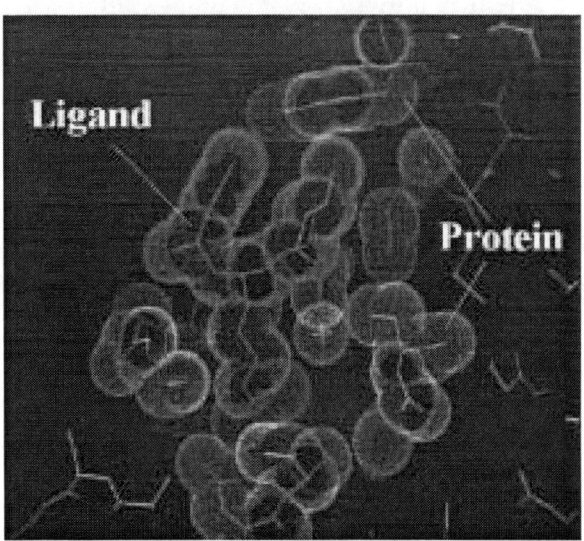

FIGURE 4. Idears of denovo design software.

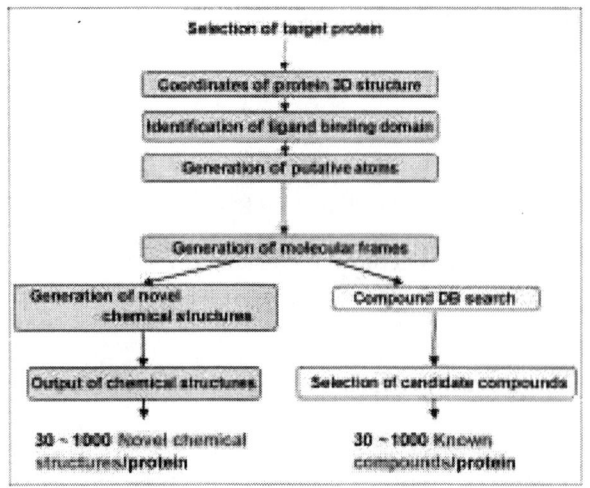

FIGURE 5. Flow of denovo design program.

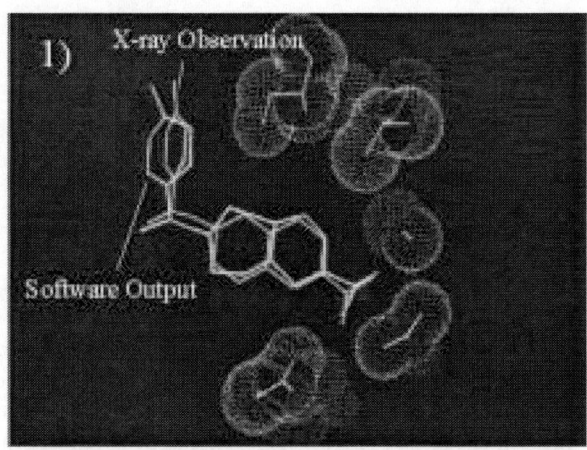

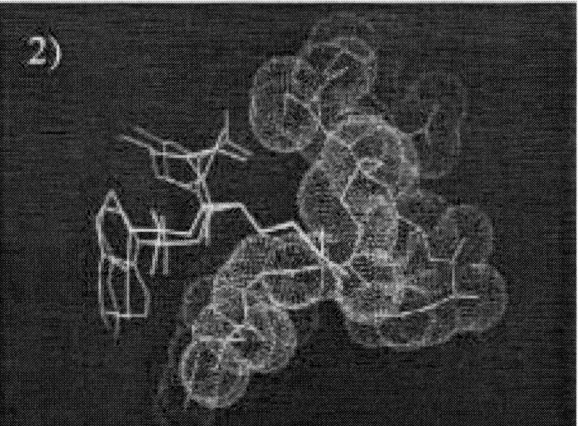

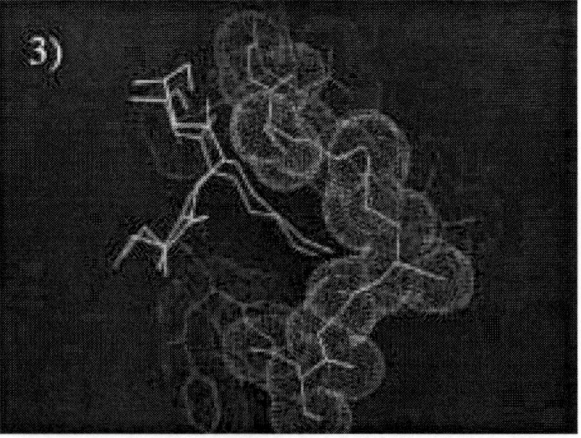

FIGURE 6. Output examples of the design program. 1) Ring molecule (ex. Trypsin/Futhan). 2) Ring and Linear (ex. Trypsin/Argatrovan). A molecule with 5 atoms was well produced. 3) Linear molecule (ex. Trypsin/BPTI(14-16))

Then, we applied the program to new target proteins. Fig. 7. shows a result of inhibitory activity measurement of PTP1B (Protein Tyrosine Phosphatase 1B).

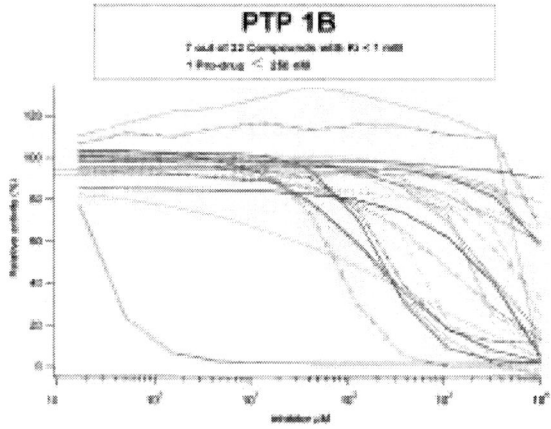

FIGURE 7. Result of in silico screening.

In this case, 2 compounds including a pro-drug have an inhibitory activity less than 100µM and 16 compounds have activity less than 10 mM. In other 2 cases, 13 to 60 percents have less than 100 µM activity and 90 percents have less than 10 mM activity. These rate are remarkably superior compared to previous "Docking methods".

ACKNOWLEDGMENTS

The autor thanks the collaborators, Drs. Masaaki Aoki, Toshiyuki Kohno, Shigetoshi Sugio, Ikuko Sugiura and Chieko Okumura.

REFERENCES

1. Aoki, M., Yokota, T., Sugiura, I., Sasaki, C., Hasegawa, T., Okumura, C., Ishiguro, K., Kohno, T., Sugio, S., and Matsuzaki, T., *Acta Cryst.,* D60, 439-446 (2004).

2. Sugiura, I., Sasaki, C., Hasegawa, T., Kohno, T., Sugio, S., Moriyama, H., Kasai, M. and Matsuzaki, T., *Acta Cryst.,* D60, in press (2004).

Dynamical Observations of Individual Protein Molecules using X-rays

Yuji C. Sasaki

Bio-medical Group, SPring-8/ Japan Synchrotron Radiation Research Institute, Mikazuki, Hyogo 679-5198, Japan
CREST Sasaki -team, Japan Science and Technology Corporation (JST), Tachikawa 190-0012, Japan

Abstract. We have successfully observed dynamical Brownian motions in an individual DNA molecule and other biological ones in real time with one-100th the atomic-scale precision (picometer-scale precision) using X-rays of the super photon ring-8 (SPring-8) for the first time.

INTRODUCTION

Most x-ray techniques are based on the averaged observations of many molecules, so the behavior of individual molecule cannot be determined. Therefore, we demonstrated the direct observations of the rotating motions of the individual single nanocrystal, which is linked to the individual specific sites in single protein molecules, using time-resolved Laue diffraction. In visible light wavelength λ, in vivo observations have greatly progressed due to the remarkable development of fluorescence single-molecule detection techniques [1, 2]. These single-molecular techniques have provided positional information at an accuracy of about $\lambda/100$, far below the optical diffraction limit ($\sim\lambda/2$). Now, we achieved time-resolved x-ray ($\lambda_{x-ray} \sim 0.1$ nm) observations of picometer-scale ($\lambda_{x-ray}/100$) slow Brownian motions of individual protein molecules in aqueous solutions. This single molecular detection system, which we call Diffracted X-ray Tracking (DXT), can monitor the local Brownian motions of individual molecules as shown in Fig. 1 [3, 4, 5].

In this work, we investigated the dynamical relationship between individual protein molecules and the labeled nanocrystals. For example, we controlled the number of the bonding sites of the labeled nanocrystals to individual protein molecules, chemical conditions, and the length between the individual protein molecules and the labeled nanocrystal. In order to control the number of the bonding sites on the labeled gold nanocrystals, we utilized both the periodical structure in Actin filaments and using the mercury compound. There are cysteine sites at 5.5 nm interval of individual Actin filaments. The diameter of gold nanocrystal is about 15-20 nm. Thus, the number of the bonding sites can be controlled to single bond or double ones using the mercury compound.

Actin is included the molecular building blocks of the eukaryotic cytoskeleton-like intermediate filament proteins and tubulin together with their respective associated proteins. Globular actin (G-actin) aggregates into two-stranded helical polymer filaments (F-actin) that consist of up to several-thousand monomers [6]. The mechanical properties of F-actin are very significant, given the central structural role played by actin filaments within muscles and the cytoskeleton.

EXPERIMENT

Using DXT method, it is possible to monitor the Brownian motions of adsorbed molecules or particles. Since the structure of F-actin is helical, we think the flexibility of F-actin is independent of the adsorbed orientation, which is parallel to the central axis of F-actin. Figure 2 shows a cross-sectional view of the adsorbed F-actin. To adsorb F-actin onto an amorphous gold substrate, we utilized the interaction between the reactive amino residue (Lys, Ser and Tyr)

CP716, *Portable Synchrotron Light Sources and Advanced Applications,*
edited by H. Yamada, N. Mochizuki-Oda, and M. Sasaki

in F-actin and the amorphous gold surface with a cross-linking reagent (Succinimidayl 6-[3-(2-pryidyldithio) -propionamido] hexanoate: LC-SPDP, Pierce Co., Ltd.) in 50 mM HEPES (pH 8.0) for six hours at 25° C.

Rabbit skeletal-muscle actin was prepared according to J. D. Pardee and J. A. Spudich [7] with a slight modification. An acetone powder (2.5 g) of rabbit back and leg muscle was extracted at 4ºC for 30 min by stirring by inversion with 50 ml 0.1mM CaCl$_2$/buffer A. Buffer A consists of 2.3 mM Tris-Cl/0.2 mM Na$_2$ATP(pH 8.0 at 25ºC). The extract was filtered and cleared by centrifugation at 18,000 g for 1 hour. KCl (to 50 mM) and ATP (to 1 mM) were added and the crude actin was allowed to polymerize without MgCl$_2$ for 2 hours. KCl was then added to 0.6 M, and the solution was gently stirred by inversion for 30 min. F-actin was collected as a pellet by centrifugation at 100,000 g for 2 hours, homogenized with 0.6 M KCl/1 mM ATP/0.1mM CaCl$_2$/buffer A, and centrifuged at 400,000 g for 20 min. The F-actin pellet was resuspended with 1 mM ATP/0.1mM CaCl$_2$/buffer A, and allowed to depolymerize by dialysis with 0.1mM CaCl$_2$/buffer A for 2 days. The dialysate was cleared by centrifugation at 400,000 g for 20 min, to obtain the purified G-actin containing bound Ca^{2+} (Ca-G-actin).

F-actin containing bound Mg $^{2+}$ (Mg-F-actin) was prepared essentially according to H. Strzelecka-Golaszewska, J. Moraczewska, S.Y. Khaitlina and M. Mossakowska [8]: Ca-G-actin was incubated with 0.2 mM EGTA/0.1 mM MgCl$_2$/buffer A at 25ºC for 10 min, then quickly allowed to polymerize with KCl(to 0.1 M) and ATP(to 0.5 mM).

Gold nanocrystals were fabricated by the sequential process shown in Fig.3. First, the gold elements were evaporated at thin film thickness (=10 nm) on the NaCl (100) substrate in vacuum conditions. We confirmed that the diameter of the gold nanocrystal is 20-30nm by scanning electron Microscopy (S-5000, Hitachi Ltd.). In order to disperse the evaporated gold in the aqueous solutions without aggregations, the surface of the substrate was dissolved by the detergent solution containing (3-[(3-Cholamidopropyl) dimethyl-ammonio] propanesulfonicacid: CHAPS, 50mM, pH=7.0).

We used the white X-ray mode (Laue mode) of beamline BL44B2 (RIKEN Structural Biology II, SPring-8, Japan) to record Laue diffraction spots from nanocrystals. Photon flux at the sample position is estimated to be about 10^{15} photon/s/mm^2 in the energy range from 7 kV to 30 kV. The x-ray focal beam size is 0.1 mm (horizontal) x 0.1 mm (vertical). A diffraction spot was monitored with an X-ray image intensifier (Hamamatsu Photonics, V5445P) and a CCD camera (Hamamatsu Photonics, C4880-82) with 656 x 494 pixels as shown in Fig. 3. The thickness of the solution on the substrate (~7μm) was controlled with the thickness of the spacer between the substrate and the polyimide film [4].

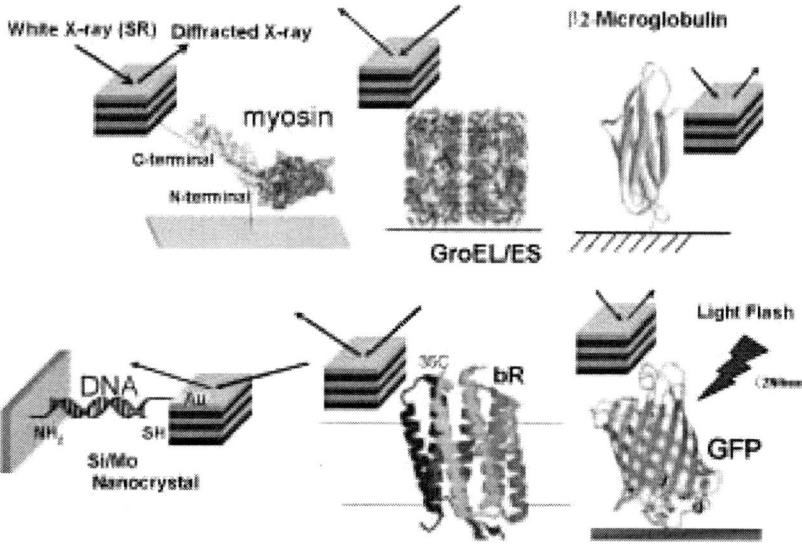

FIGURE 1. Schematic drawing of arrangements for single molecular tracking with x-rays (not to scale). **Diffracted X-ray Tracking (DXT)** monitors motions of a single nanocrystal with the guidance of a diffraction spot from the individual nanocrystal itself, which is labeled with the individual single molecular unit. For example, we observed dynamical single-molecule observations of myosin, GroEL/ES, β2-microglobulin, DNA, GFP, and membrane protein (in this case, bacteriorhodopsin in purple membrane).

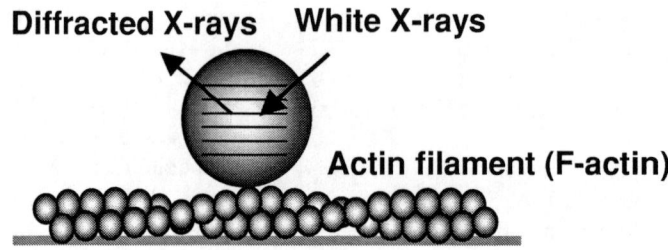

Diffracted X-rays **White X-rays**

Actin filament (F-actin)

FIGURE 2. Schematic drawing of the detection system for actin filaments (F-actin) with x-rays (not to scale). The N-terminal of G-actin is bound to the surface of the substrate. The C-terminal of G-actin is reacted with the single gold nanocrystal.

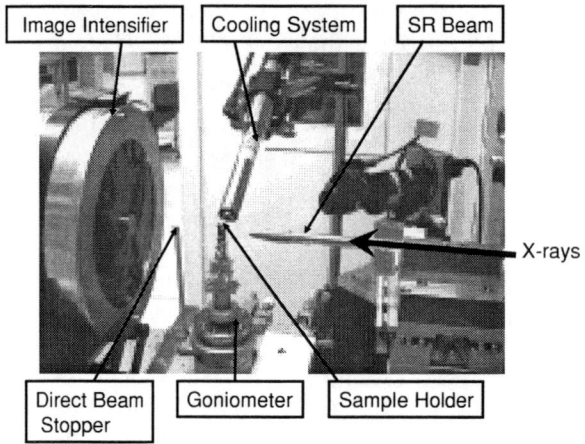

Image Intensifier Cooling System SR Beam

X-rays

Direct Beam Stopper Goniometer Sample Holder

FIGURE 3. The photograph of the instrumental arrangements for Diffracted x-ray tracking method.

RESULTS AND DISCUSSION

We could not detect the displacement ($\Delta 2\theta \sim 400$ mrad/36 ms) of the observed diffraction spots because of the size and characteristics of the x-ray detection camera. Thus, the gold nanocrystals that were not labeled with the actin filament were not detected by this DXT system. Additionally, we confirmed that the measurements of the diffraction angles (2θ) from the physical adsorbed nanocrystal on the substrate were stabilized within ($\Delta 2\theta_{limit}=$) 1.5 mrad during 1s. Thus, the presence of the detectable moving spots from the individual nanocrystals represents the interaction between the actin filament and the nanocrystals.

In order to characterize the motions of the individual labeled nanocrystals accurately, we developed a method based on plots of the mean-square displacement (MSD) against the time interval to analyze the trajectories of the individual diffraction spots obtained from DXT. In two-dimensional motions,

$D=<\Delta z^2>/(4\Delta t)$, where D is the two-dimensional diffusion coefficient, $<\Delta z^2>$ is the two-dimensional mean-square displacement for time-lags $\Delta t=(t_i-t_j)$, $<\Delta z^2>=(1/\Sigma)\ \Sigma z(t_i)-z(t_j))^2$, and $z(t_i)$ represents the position of the molecule at time t_i. In DXT, the relationship between the displacement of the observed diffraction angles $\Delta\theta$ and $<\Delta z^2>$ is $<\Delta\theta^2>d^2=<\Delta z^2>$, where d represents the distance between the C-terminal of F-actin and the rotating center (the surface of the gold substrate). Therefore, the values of $\Delta\theta$ from our DXT can be converted into values of Δz. Thus, the values of D can be determined from DXT when the rotational motion is regarded as Brownian motion on a two-dimensional flat surface. We now assume the value of d is 6 nm.

Figure 4 shows $<\Delta\theta^2>-\Delta t$ curves of a labeled gold nanocrystal using a mercury compound (p-chloromercuribenzoic acid: PCMB) in the Mg^{2+} aqueous solution with and without the presence of phalloidin. Surprisingly, the $<\Delta\theta^2>-\Delta t$ plots became linear curves under the presence of a mercury compound. In other words, the observed motions are assigned to free Brownian motions in each condition. We confirmed that the displacement of the $\langle\Delta\theta^2\rangle$-$\Delta t$ plots under no PCMB condition is corralled motions, when the number of reacted sites with the labeled nanocrystal is expected to total two to three in F-actin. We observe that the displacements ($\langle\Delta\theta^2\rangle$-$\Delta t$, $D_{non-phalloidin}$ = 0.18±0.08 mrad^2/s) of the observed spots without the presence of phalloidin are greater than those ($D_{phalloidin}$ = 0.058±0.01 mrad^2/s) where phalloidin is present. This result means the rigidity of F-actin is increased in the presence of phalloidin, and this is in fair agreement with many other experimental results [9].

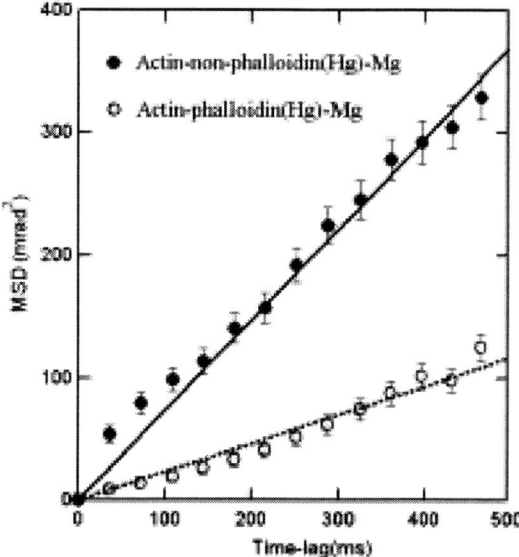

FIGURE 4. MSD $\langle\Delta\theta^2\rangle$ as function of time interval Δt. These data were obtained from about 60 diffracted spots. The exposure time was 1 s.

ACKNOWLEDGMENTS

The synchrotron radiation experiments were performed at the SPring-8 with the approval of the Japan Synchrotron Radiation Research Institute (Proposal no. 2002B0019-NL2-np)

REFERENCES

1. Basche, Th., Moerner, W. E., Orrit, M. and Wild, U. P., Single Molecule Optical Detection, Imaging, and Spectroscopy (Wiley-VCH, Munich,1997).

2. Weiss, S., Science, 283, (1999) 1676.

3. Sasaki, Y. C., Suzuki, Y., Yagi, N., Adachi, S., Ishibashi, M., Suda, H., Toyota, K. & Yanagihara, M. Phys. Rev. E. 62, 3843-3847 (2000).

4. Sasaki, Y. C., Okumura, Y., Adachi, S., Suzuki, Y. & Yagi, N. Nucl. Instr. Meth. Phys. Resea.A467-468, 1049-1052 (2001).

5. Sasaki, Y. C., Okumura, Y., Adachi, S., Suda, H., Taniguchi, Y. & Yagi, N. Phys. Rev. Lett. 87, 248102(1)-248102(4) (2001).

6. Steinmetz, M. O., Stoffler, D., Hoenger, A., Bremer, A.& Aebi U. J. Struct. Biol. 119, 295-320(1997).

7. Pardee, J. D., and Spudich, J. A, Methods Cell Biol. 24, 271-289, (1982)

8. Strzelecka-Golaszewska, H., Moraczewska, J., Khaitlina, S.Y., and Mossakowska, M. Eur. J. Biochem. 211,731-742(1993)

9. Page, R., Lindberg, U. & Schutt, C. E. J. Mol. Biol. 280, 463-474(1998).

Imaging and Medical Applications

Micro XAFS Study on Vanadium in Ascidians Alive Blood Cells Investigated by Fluorescence Scanning X-Ray Microscopy at ID21 at ESRF

K. Takemoto[1], T. Ueki[2], B. Fayard[3], A. Yamamoto[4], H. Sasaki[5], M. Salomé[3], J. Susini[3], H. Michibata[2], and H. Kihara[1]

[1]*Department of Physics, Kansai Medical University,18-89 Uyamahigashi, Hirakata, Osaka,573-1136, Japan*
[2]*Marine Biological Laboratory, Graduate School of Science, Hiroshima University, Mukaishima-cho 2445, Hiroshima, 722-0073, Japan*
[3]*European Synchrotron Radiation Facility (ESRF), BP220, 38043 Grenoble cedex, France*
[4]*Nagahama Institute of Bio-Science and Technology, 1266, Tamura-cho, Nagahama, Shiga, 526-0829, Japan*
[5]*Institute of DNA Medicine, Jikei University School of Medicine, 3-19-18, Nishi-Shimbashi, Minato-ku, Tokyo, 105-8471, Japan*

Abstract. X-ray microscopy microscope installed at the beam line ID21 at European Synchrotron Radiation Facility (ESRF) was used for the elucidation of micro-distribution of vanadium in ascidians. In order to examine chemical states of vanadium in the alive blood cell, a high pressure cryo-fixation technique was applied to fix its chemical states of vanadium. We have succeeded in observing uniform distribution of vanadium in the vacuole, in identifying the valency of vanadium in the cell, and more importantly in finding granules rich in vanadium of the 4th valency

INTRODUCTION

It has been widely known that vanadium is highly accumulated in blood cells of some kinds of ascidians (tunicates) [1]. It has also been elucidated the vanadium is not uniformly distributed but is selectively distributed in particular types of cells. These vanadium-containing cells are called as vanadocytes. Much attention has been paid to bioinorganic chemistry of vanadium in living ascidians, particularly to its intracellular distribution and chemical states in the cells.

Scippa *et al.* investigated the intracellular sites of accumulated vanadium by using a scanning transmission electron microscope with an energy dispersive X-ray spectrometer, and reported that vacuolated amoebocytes, signet ring cells, and some types of compartment cells were identified as vanadocytes [2, 3, 4, 5]. Frank *et al.* performed K-edge X-ray absorption spectroscopy (XAS) experiments of vanadium in the intact whole blood

cells from ascidians, and reported that endogenous vanadium was divided between V(III) ion and V(IV) ion [6, 7, 8].

We have applied X-ray microscopy to the elucidation of micro-distribution of vanadium in ascidians. X-ray microscope is a powerful tool to investigate micro-distribution of chemical species in living cells. In the present studies, we have used transmission and fluorescence scanning X-ray microscope installed at the beam line ID21 at the European Synchrotron Radiation Facility (ESRF), which covers energy range from 2 to 8 keV, and is thus capable of investigating local distribution and chemical states of vanadium, sulfur, and calcium. We have succeeded in visualizing the intracellular distribution of vanadium both in freeze-dried and wet specimens. In *Phallusia mammillata*, signet ring cells and vacuolated amoebocytes were identified as vanadocytes [9]. In *Ascidia sydneiensis samea*, signet ring cells were identified as vanadocytes [9].

CP716, *Portable Synchrotron Light Sources and Advanced Applications*,
edited by H. Yamada, N. Mochizuki-Oda, and M. Sasaki
© 2004 American Institute of Physics 0-7354-0195-0/04/$22.00

In order to examine chemical states of vanadium in the alive blood cell, we needed to fix its chemical states of vanadium by rapid-freezing, as vanadium is easily oxidized in the atmosphere and in addition cells are easily damaged by X-ray radiation at room temperature. To avoid the puncture of cells and to keep the valency of vanadium as is in the alive cell, we applied a high pressure freezing technique in the present study. By preparing the sample with this technique, we can prevent the puncture of the cell, and can preserve the native valency of vanadium as in the alive state.

EXPERIMENTAL PROCEDURES

Sample Preparation

Specimens of *A. sydneiensis samea* were collected in the Gulf of Yamada, Iwate, Japan. Ceolomic fluid was extracted and mixed with F12 medium (ICN Biomedicals, Inc., USA) containing 0.5 M NaCl [9]. Blood cells were isolated by centrifugation at 300 x g for 10 min at 4°C, re-suspended in the medium and stored at 4°C.

Samples for the measurement of XAS spectroscopy were prepared both with an immersion cryo-fixation by coolant, isopentan and propane, and with a high pressure cryo-fixation technique to prevent from the lysation of the cell due to the formation of ice crystals [10, 11]. In operation, specimens are subjected to a pressure of 2.1 GPa for a few milliseconds followed by jet freezing into liquid nitrogen under the same pressure. The water of the sample is vitrified under this regime and ice-crystal formation is retarded, thereby avoiding structural damage to the specimen. Samples up to 600 μm in thickness can be frozen without damage with the use of this machine [12].

One drop of blood cell suspension was placed on a copper mesh with a supporting film of collodion (Plastic substrate 200-A, Ohken Shoji, Japan). In case of the immersion cryo-fixation, the mesh was then dipped in coolant. Details of sample preparation is described somewhere else [9]. In case of the high pressure cryo-fixation technique, the mesh was jet-frozen under high pressure between two metal sample planchettes in the sample chamber of a high pressure cryo-fixation apparatus (HPM 010, BAL-TEC). The prepared sample meshes were stored in liquid nitrogen until observation.

X-ray microscopy

The scanning transmission x-ray microscope used in this study is operated at the ID21 beamline at ESRF [13, 14]. The ESRF storage ring is operated at 6 GeV with a maximum current of 200 mA. ID21 is a dedicated X-ray imaging and spectro-microscopy beamline in the energy range of 0.2-7 keV [13]. In the high-energy region (E > 2 keV), a fixed-exit Si(111) double crystal monochromator was used. Monochromatized beam is focused onto the sample position with the use of a zone plate as a focal element [14]. Expected energy resolution of the double crystal monochromator is about 10^{-4}.

The sample is raster scanned perpendicularly against the X-ray beam direction and synchronous signal detection allows image formation on a pixel-to-pixel basis. The transmission x-ray image was detected by a PIN photodiode detector. A germanium solid state detector of high energy resolution was placed at 90 degree versus the beam axis for fluorescence measurements. X-ray and fluorescence X-ray were obtained simultaneously. However, in this experiment, a transmission image could not observed because of a cryogenic equipment arrangement. A spatial resolution was as high as 0.5 μm. Samples were kept at -120 °C using a nitrogen flow cryo-stream during the measurement. Beam energy was fixed to 5.500 keV to ensure good fluorescence yield for vanadium. The "zero" of energy was calibrated with an absorption edge of vanadium metal foil [15].

RESULTS AND DISCUSSION

Figure 1 shows an electron microscope cross-section image of a signet ring cell. The nucleus and cytoplasm of a signet ring cell is seen in the cell periphery, due to the occupancy of the large fluid-filled vacuole in the central part of the cell. In previous study, vanadium was only detected in signet ring cells [9]. Therefore, the cell, in which vanadium was detected, was identified as signet ring cells and the object which should observe in the present study.

Figures 2 show fluorescence images of vanadium distribution in the ascidian cells. Figure 2a and 2b are images from the cell prepared by the immersion freezing method, and figure 2c by the high pressure rapid freezing method. They show cells with a diameter of about 10 μm. The size is the same as that of the intact signet ring cell, evaluated under an optical and electron microscope. When we see vanadium images in Figs. 2 carefully, we can easily find micro-granules in vacuoles. Averaged diameter of the micro-granules is about 3 μm.

Figures 3 show XAS spectra of vanadium K-edge in whole signet ring cells. The beam size is about 100 μm. Sharp pre-edges were observed near 3.5eV. Compare with them, remarkable change is not seen. Figure 4 shows micro-XAS spectra of vanadium K-edge in signet ring cells. The micro-XAS spectra were measured at various parts in the vacuole (solid line) and a granule (broken line). The beam size is 1 μm. In both figures, ordinates are normalized to 1 at the energy above 50 eV [15]. The figures demonstrate that the spectrum from the site of granule exhibits a sharp pre-edge at 3.5eV, whereas spectra from vacuole part do not show similar significant peaks around 3.5 eV, although some small peaks were observed. Pre-edge peak is sensitive to the oxidation state of vanadium [15, 16]. Frank *et al.* reported the normalized intensity of pre-edge peak at 4.8 eV due to vanadyl ion $[V^{IV}O]^{2+}$ is 0.4, whereas the peak at 3.0 eV of V(III) state shows 0.04 [6, 8]. Based on this experimental results, we can conclude vanadium in a granule and whole cells are mainly in vanadyl ion, whereas vanadium in vacuole is mainly in V(III).

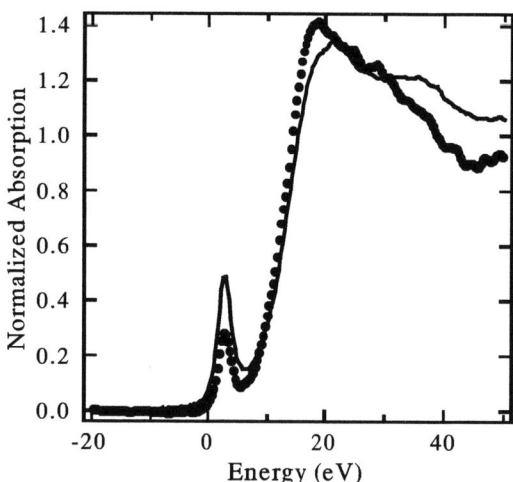

FIGURE 3. XAS spectra of vanadium K-edge of whole cells cryo-fixed by by the immersion freezing method with isopentane (solid line) and the high pressure freezing method. (broken line). Beam size is 100 μm.

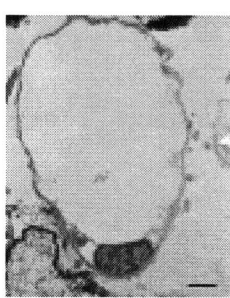

FIGURE 1. Electron microscope image of a signet ring cell. Scale bar is 1 μm.

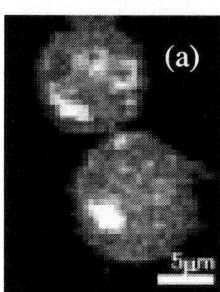
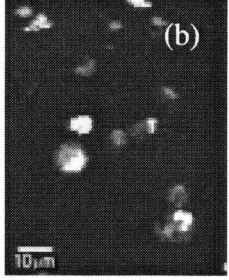
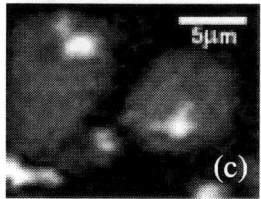

FIGURE 2. X-ray microscope images of signet ring cells. Cells cryo-fixed by immersion freezing method with propane (a) and isopentane (b), and by high pressure freezing method (c).

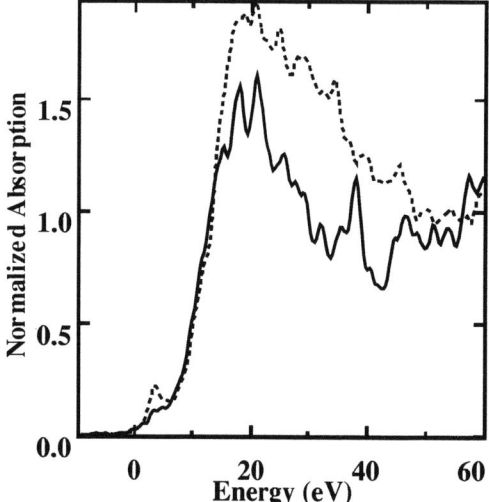

FIGURE 4. Micro-XAS spectra of vanadium K-edge in vacuole (solid line) and from the micro-granule (broken line). Beam size is 1 μm.

CONCLUSION

Using a combination of scanning X-ray microscopy and a high pressure freezing technique, direct intracellular accumulation and micro-XANES spectra of vanadium in a blood cell of *Ascidia sydneiensis samea* were taken successfully. This fluorescence image clearly demonstrates that vanadium was distributed uniformly in the vacuole of signet ring cell.

Furthermore, vanadium was selectively accumulated in the intravacuolar granule with a diameter of about 2.75 µm. Assuming that vanadium in vacuolar is identified as V(III) and vanadium in granule is identified as V(IV).

ACKNOWLEDGMENTS

Experiments were performed under approval of ESRF, proposal number: SC117. The present work was supported by the Grant-in-Aid for Encouragement of Young Scientists from the ministry of science, education and sports. We also thank Mr. T. Morita and other staff at the International Coastal Research Center, Ocean Research Institute, The University of Tokyo, at Otsuchi, Iwate, Japan, for their help in collecting adult ascidians.The authors are grateful to Dr. P. Frank at Stanford University for fruitful discussions.

REFERENCES

1. M. Henze, *Hoppe-Seyler's Z. Physiol. Chem.*, **72**, 494-501 (1911).

2. L. Botte, S. Scippa and M. de Vincentiis, *Experientia*, **35**, 1228-1230 (1979).

3. A.F. Rowley, *J. Mar. Biol. Assoc. U.K.*, **56**, 607-620 (1982).

4. S. Scippa, L. Botte, K. Zierold, and M. de Vincentiis, *Cell Tissue Res.*, **239**, 459-461 (1985).

5. S. Scippa, K. Zierold, and M. de Vincentiis, *J. Submicrosc. Cytol. Pathol.*, **20**, 719-730 (1988).

6. T. D. Tullius, W. O. Gillum, R. M. K. Carlson, and K. O. Hodgson, *J. Am. Chem.*, **102**, 5670-5676 (1980).

7. P. Frank, K.O. Hodgson, K. Kustin and W. E. Robinson, *J. Biol. Chem.*, **273**, 24498-24503 (1998).

8. P. Frank and K. O. Jpdgson, *Inorg. Chem.*, **39**, 6018-6027 (2000).

9. T. Ueki, K. Takemoto, B. Fayard, M. Salomé, A. Yamamoto, H. Kihara, J. Susini, S. Scippa, T. Uyama, and H. Michibata, *Zoological Science*, **19**, 27-35 (2002).

10. U. Riehle, Chemie-Ing. Techn., 40, 213-218 (1968).

11. H. Kanno, R. J. Speedy, and C.A. Angell, *Science*, **189**, 880-881 (1975) .

12. R. Dahl. and L.A. Staehelin, *J. Electron Microsc. Technique* , **13**, 165-174 (1989).

13. J. Susini, and R. Barrett, in X-ray Microscopy and Spectromicroscopy edited by. J. Thieme, G. Schmahl, D. Rudolph, and E. Umbach, Proc. of XRM96, Springer-Verlag, Heidelberg, 1998, pp I45-I54.

14. J. Susini, R. Barrett, B. Kaulich, S. Oestreich, and M. Salomé, in X-ray Microscopy, edited by W. Meyer-Ilse, T. Warwick and D. Attwood, AIP Proc. of XRM99, 2000, pp 19-26.

15. J. Wong, F.W. Lytle, R.P. Messmer, and D.H. Maylotte, *Phys. Rev. B,* **30**, 5596-5610 (1984).

16. C. Weidemann, D. Rehder, U. Kuetgens, and J. Hormes, *Chem. Phys.*, **136**, 405-412 (1989).

THz-wave parametric source and its imaging applications

Kodo Kawase

RIKEN, 2-1 Hirosawa, Wako, 351-0198, Japan

Abstract. Widely tunable coherent terahertz (THz) wave generation has been demonstrated based on the parametric oscillation using MgO doped LiNbO₃ crystal pumped by a Q-switched Nd:YAG laser. This method exhibits multiple advantages like wide tunability, coherency and compactness of its system. We have developed a novel basic technology for terahertz (THz) imaging, which allows detection and identification of chemicals by introducing the component spatial pattern analysis. The spatial distributions of the chemicals were obtained from terahertz multispectral transillumination images, using absorption spectra previously measured with a widely tunable THz-wave parametric oscillator. Further we have applied this technique to the detection and identification of illicit drugs concealed in envelopes. The samples we used were methamphetamine and MDMA, two of the most widely consumed illegal drugs in Japan, and aspirin as a reference.

INTRODUCTION

Recently, there has been a growing interest in developing terahertz (THz) techniques to a novel imaging modality [1]-[3]. The THz-waves, categorized between millimeter radio waves and far infrared light waves, exhibit properties of both sides of the electromagnetic spectrum. Like radio waves, they can be transmitted through a wide variety of substances such as paper, cloth, ceramics, plastics, wood, bone, fat, ice, various powders, dried food, and so on. In addition, like light waves, they can easily be propagated through space, reflected, focused and refracted using THz-optics. The short wavelength (several hundred μm), much shorter than that of micro waves, allows for a spatial resolution which is sufficient in many imaging applications. Furthermore, the THz-wave is nonionizing and nondestructive and hence is safe for human body. The range of potential applications is likely to expand even further with the increased availability of many absorption spectra (i.e. fingerprint spectra) peculiar to specific chemicals, including vitamins, sugars, pharmaceuticals, agricultural chemicals, discovered since last year in the THz-wave region.

The THz imaging techniques have been demonstrated for various practical applications, such as biomedical diagnostics, semiconductor wafer diagnostics, inspection of personal effects, quality control of packaged goods, inspection of artwork, and moisture analysis for agriculture. Most THz imaging methods have been used to identify the existence of a target. A transillumination THz image shows the sum of the absorption in each component when a sample includes various components. Therefore, the differences between components cannot be discriminated from THz images clearly.

In order to estimate the differences in the component spatial patterns, we applied principle component analysis to THz spectroscopic imaging [4]. We have separated the component spatial patterns of chemical samples in multispectral THz images, using the spectral fingerprints. Further, we have improved our method to separate clearly between component patterns and artifacts [5].

EXPERIMENTAL SETUP

The THz spectroscopic imaging system [4] consists of a Q-switched Nd:YAG laser, a THz-wave parametric oscillator (TPO) [6], imaging optics, an xy scanning stage, a detector, a lock-in amplifier, and a personal computer, as shown in Fig. 1. The Nd:YAG laser (wavelength 1.064 μm) pumps a nonlinear optical crystal MgO:LiNbO₃ (length 65mm), simultaneously generating a THz-wave (wavelength 120 - 300 μm,

CP716, *Portable Synchrotron Light Sources and Advanced Applications,*
edited by H. Yamada, N. Mochizuki-Oda, and M. Sasaki
© 2004 American Institute of Physics 0-7354-0195-0/04/$22.00

frequency 1 - 2.5 THz) and a idler wave (wavelength 1.068 - 1.074 μm) by non-collinear phase matched parametric oscillation. The idler beam was amplified in an oscillator consisting of flat mirrors with a half-area HR coating. The beam diameter, pulsewidth, and repetition of the laser were 1.5 mm, 25 ns, and 50 Hz, respectively. When the excitation intensity was 30 mJ/pulse, the intensity of the THz-wave was about 30 pJ/pulse (3 mW at the peak). The frequency tuning was achieved by rotating the oscillator slightly so that the phase-matching angle was changed. An array of six Si-prism couplers was placed on the y-surface of the crystal to couple out the THz-wave efficiently. The generated THz-wave was focused on the target by a polyolefin plastic lens (focal length 50 mm) producing a focal spot of about 0.5 mm in diameter. The target was continuously raster scanned by a x-z stage. The transmitted THz-wave was projected onto a pyroelectric or a Si-bolometer detector. The signal is separated from noise with a lock-in amplifier synchronized on the laser pulse frequency. The lock-in time constant was set to 100 ms. The output signal of the lock-in amplifier was stored in a personal computer after passing through an analog-to-digital conversion board during continuous motion of the sample. The stability of this imaging system was RMS = 2.3%. In this condition, the required minimum absorption per path was approximately 6.9% (at 3σ level).

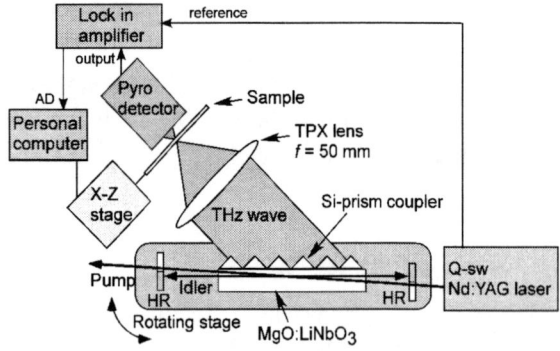

FIGURE 1. Schematic of THz spectroscopic imaging system using THz-wave parametric oscillator (TPO).

COMPONENT PATTERN ANALYSIS

The calculation procedure, component spatial pattern analysis [7,8], is based on the following principle: Consider that a target, which is composed of M substances having different spectral characteristics,

is imaged at N frequencies (or wavelengths). Each image is composed of L pixels, which are thought of as being rearranged one-dimensionally, for ease of calculation. When the imaging system is assumed to be linear, the transmitted intensity can be described by the following linear matrix equation:

$$[I] = [S][P] \tag{1}$$

where [I] is a N × L matrix of the N recorded images whose row vectors $i_1, \cdots, i_N$ represent each an L-pixel image taken at an individual frequency, [S] is a N × M matrix of the measured spectra of M drugs whose column vectors $s_1, \cdots, s_M$ represent the spectrum data set of each substance at N frequencies, and [P] is a M × L matrix of the spatial distribution of the M drugs whose row vectors $p_1, \cdots, p_N$ represent the spatial pattern of each drug with L pixels. The dimensions and elements of these matrices are shown in Fig. 2. When the matrix [I] and [S] are known, this matrix equation is solved easily. For the case when N = M, [P] is simply given by $[P] = [S]^{-1}[I]$. For the case when N > M, [P] can be solved using a least-squares method as

$$[P] = ([S]^t[S])^{-1}[S]^t[I] \tag{2}$$

where t denotes transpose. By this means, the spatial distribution of a specific component in a sample made up of several chemicals can be imaged.

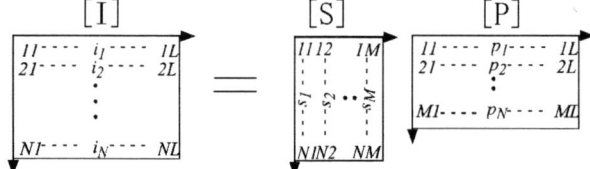

FIGURE 2. Schematic of Eq. (1).

We used pellets made from polyethylene powder mixed with 5-aspirin or palatinose. All samples were made with a pellet-maker at room temperature. The diameter and weight of a pellet were 13 mm and 0.2 g, respectively. These chemical samples have different spectral curves, as shown in Fig 3, measured using a TPO spectroscopic system. Here, these spectral data were observed using 50% concentrations of the pellets and then were subtracted the attenuation of polyethylene. I_0 and It are incident and transmitted intensities, respectively. To analyze the component spatial patterns, palatinose and 5-aspirin pellets were prepared in three (50, 40, and 20%) and two (50 and 20%) concentrations, respectively, and then fixed on a thin plastic plate using a double-face paper tape (Fig. 4). This plate was scanned at a translation speed of 2.0 mm/s in the horizontal direction with a 0.5-mm step in the vertical direction. The scanning area was 50 × 35 mm (horizontal × vertical), which corresponded to 100 × 70 pixels. The measurement time was approximately ten minutes. Eight transillumination THz images were measured by tuning from 1.3 to 2.0 THz in 0.1 THz steps, as shown in Fig. 5. These images are the log scale of the transmitted intensity divided by the incident intensity of the THz-wave. We can see the frequency-dependent attenuation at areas of the chemical samples.

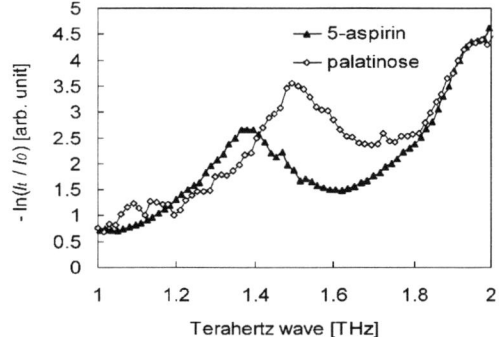

FIGURE 3. Spectral curves of 5-aspirin and palatinose.

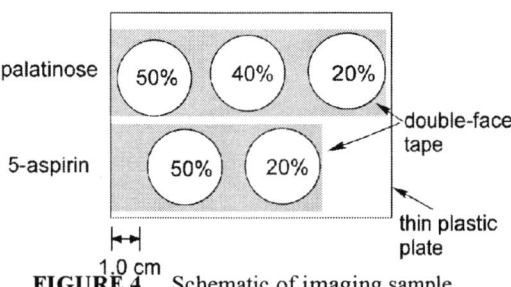

FIGURE 4. Schematic of imaging sample.

Using Eq. (2), we obtained the two component spatial patterns, as shown in Fig. 6. Here, the matrix [S] was observed at each THz frequency used from the spectral curves in Fig. 3. As both spectral curves of the palatinose and the 5-aspirin were observed using 50% concentrations of the pellets, the component patterns were multiplied by 50%. The component spatial patterns are clearly separated and the difference of concentrations is described in each image. To estimate the concentrations of the samples in the separated THz images, we set a ROI (region of interest) in each of the component patterns. The ROI was a circle with a diameter of 22 pixels, which is similar to the size of a

pellet. As the spectral curves of both palatinose and 5-aspirin were observed using 50% concentrations, the component patterns were multiplied by 50%. The estimated concentrations of palatinose were 44±5, 39±3, and 22±3%, respectively. The estimated concentrations of 5-aspirin were 46±4 and 17±2%, respectively. These values agree with the respective known concentrations.

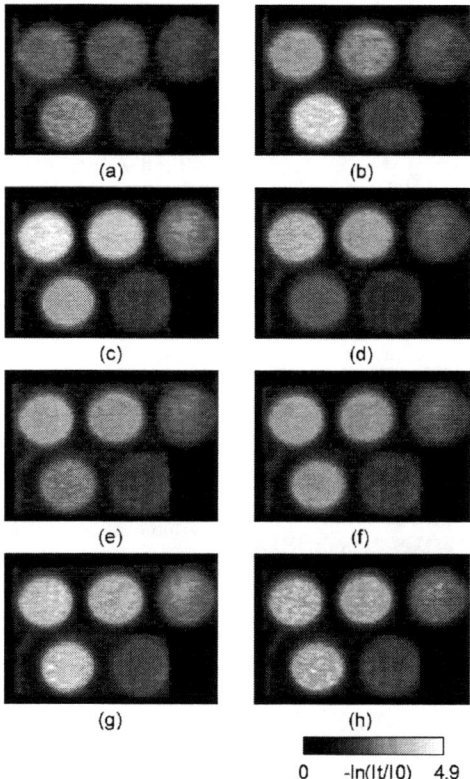

FIGURE 5. Transillumination THz images at (a) 1.3 THz, (b) 1.4 THz, (c) 1.5 THz, (d) 1.6 THz, (e) 1.7 THz, (f) 1.85 THz, (g) 1.9 THz and (h) 2.0 THz.

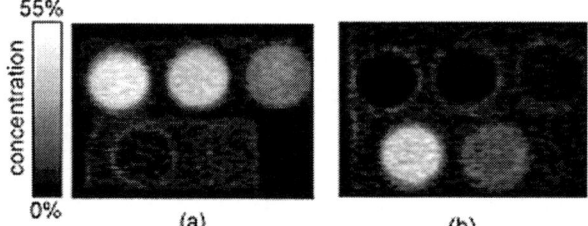

FIGURE 6. Component spatial patterns (a) palatinose (b) 5-aspirin.

NON-DESTRUCTIVE IMAGING OF ENVELOPE

The absence of non-destructive inspection technique for illicit drugs hidden in mail envelopes has

71

resulted in such drugs being not only smuggled across international borders but also transported from one jurisdiction to another within a country with surprising ease [9]. The situation must also be attributed to the inconvenience of having to obtain a search warrant to examine the contents every time the need arises. A majority of the legal systems in the world prohibit private letters, whether they be suspected or otherwise, from being examined without a search warrant. THz-wave is suitable for drug detection purposes, being able to screen the contents of envelopes and our measurement results having proven the existence of fingerprint spectra peculiar to illicit drugs in the THz region. Using the THz spectroscopic imaging system described above, the spatial distribution of the drugs inside the envelope was extracted from the multispectral images using the absorption spectra [10].

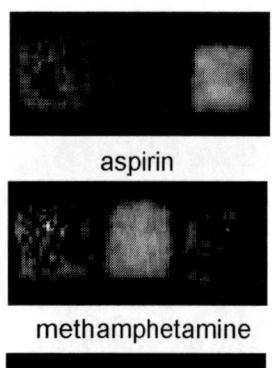

aspirin

methamphetamine

MDMA

FIGURE 7.
Extracted spatial patterns of aspirin, methamphetamine, and MDMA, using Eq. (2). The three drugs are clearly distinguished and corresponding spatial patterns obtained. The small polyethylene bags contain from left to right: MDMA, aspirin, and methamphetamine. The bags were hidden inside the envelope during imaging.

As samples we chose for this experiment three drugs that were: methamphetamine, currently the most widely consumed drug of abuse in Japan, MDMA, another drug of abuse becoming widespread on a global scale, and aspirin as a reference. Approximately ~20 mg of each substance were placed in a small 10×10 mm polyethylene bag. The three bags were then placed inside a usual airmail-type envelope. The target was raster scanned over a 20×38 mm^2 area. By changing the frequency emitted by the TPO within the 1.3 to 2.0 THz range, we obtained $N = 7$ multispectral images generating a matrix [I]. The absorption spectra of the three drugs were measured with the same TPO system as shown in Fig. 1. The corresponding absorption intensity values at the seven frequencies were extracted to obtain the matrix [S]. Although the spectra of methamphetamine and MDMA are similar, the difference between them enabled us to distinguish between the two using the component pattern analysis method. By substituting [I] and [S] thus obtained into Eq. (2), the spatial pattern [P] was calculated. Figure 7 shows the result of extracting the three components from this matrix, with each image corresponding to each of the sample drugs. As it is evident from these images, the three drugs have been clearly distinguished and the corresponding spatial patterns obtained.

ACKNOWLEDGMENTS
The authors wish to thank H. Inoue of National Research Institute of Police Science, Y. Ogawa, Y. Watanabe, M. Yamashita of our research unit, J. Nishizawa, H. Ito, T. Ikari, H. Minamide, Y. Ishikawa of Photo-dynamics Research Center, RIKEN, C. Takyu, T. Shoji of Research Institute of Electrical Communication, Tohoku University, and M. Hangyo of Research Center for Superconductor Photonics, Osaka University, for their valuable support. This work was partially supported by a Grant-in-Aid for Developmental Scientific Research (No. 15360197) from the Ministry of Education, Science, and Culture of Japan.

REFERENCES
1. D. M. Mittleman, G. Gupta, B. Neelamani, R. G. Baraniuk, J. V. Rudd, and M. Koch, *Appl. Phys. B* 68, 1085-1094 (1999).
2. B. Ferguson and X. C. Zhang, *Nature Materials* 1, 26-33 (2002).
3. P. H. Siegel, *IEEE T. Microw. Theory Tech* .50, 910-928 (2002).
4. Y. Watanabe, K. Kawase, T. Ikari, H. Ito, Y. Ishikawa and H.Minamide, *Appl. Phys. Lett.* 83, 800-802 (2003).
5. Y. Watanabe, K. Kawase, T. Ikari, H. Ito, Y. Ishikawa and H. Minamide, *Appl. Opt.* 42, 5744-5748 (2003).
6. K. Kawase, J. Shikata, and H. Ito, *J. Phys. D: Appl. Phys.* 35, R1-R14 (2002).
7. S. Kawata, K. Sasaki, and S. Minami, *J. Opt. Soc. Am. A* 4, 2101-2106 (1987).
8. K. Sasaki, S. Kawata, and S. Minami, *J. Opt. Soc. Am. A* 6, 73-79 (1989).
9. J. E. Parmeter, D. W. Murray, *Guide for the selection of drug detectors for law enforcement applications*, NIJ Guide 601-00, (National Institute of Justice, Washington, 2000).
10. K. Kawase, Y. Ogawa, Y. Watanabe and H. Inoue, *Opt. Exp.* 11, 2549-2554 (2003).

The Biological Effects on Cancer Cells by Synchrotron Radiation Generated from MIRRORCLE-6X

T. Ogata[*], T. Teshima[*,1], M. Matsumoto[*], A. Kawaguchi[*], Y. Suzumoto[*], D. Hasegawa[†], N. Mochizuki-Oda[†], H. Yamada[†], and N. Matsuura[¶]

[*]*Departments of Medical Physics & Engineering, and* [¶]*Clinical Laboratory Sciences, Osaka University Graduate School of Medicine, 1-7 Yamada-oka, Suita, Osaka 565-0871, Japan*
[†]*Synchrotron Light Life Science Center, Ritsumeikan University, 1-1-1 Nojihigashi, Kusatsu City, Shiga 525-8577, Japan*

Abstract. MIRRORCLE-6X, the unique portable synchrotrons, generates brilliant hard X-rays. The purpose of this study is to investigate the biological effects of synchrotron radiation on cancer cells to identify its effectiveness. A549 human lung adenocarcinoma and DU145 human prostate adenocarcinoma cells were used. We examined biological effects on cancer cells by colony formation assay, micronucleus assay, WST-1 method, and apoptosis detection of flow cytometry. Results of these assay revealed that the biological effects of X-ray generated from MIRRORCLE-6X on cancer cells are similar to those of 4 MV X-ray from linear accelerator (linac).

INTRODUCTION

In the clinic, ionizing radiation has been established as a highly effective modality used in the local control of tumor growth. The goal of radiation therapy is to maximize dose to the tumor while minimizing dose to healthy tissue. Synchrotron radiation is attractive in medical technology. Several experimental approaches with synchrotron radiation to aim at its goal enhance the effectiveness of radiation. For example, they include microbeam radiation therapy, photon activation therapy, and radiation dose-enhancement using a contrast agent.

Microbeam radiation therapy is a unique experimental technique first developed at the National Synchrotron Light Source (NSLS), Brookhaven National Laboratory [1]. Microbeam radiation therapy uses parallel arrays of thin (<100 μm) slices of synchrotron-generated X rays (microplanar beam or microbeams). Typically the width of a microbeam is 25 μm and distance between the beams 200 μm.

Experiment of rats bearing gliosarcoma with microbeam radiation therapy showed encouraging results without normal tissue damage such as white matter necrosis [2]. For animals inoculated subcutaneously with tumors, microbeam radiation therapy is superior to broad beam in preventive effects on complication such as desquamation [3-4].

Photon activation therapy uses synchrotron radiation to trigger a cascade Auger and photoelectrons in cancer cells. Cascades such as this can be quite damaging if they occur in very close proximity to DNA in the cell nucleus. Photon activation therapy is a two-step therapy. First, high-Z compound is introduced in tumor cell. Second, the irradiation with synchrotron radiation is targeted on K absorption to induce the photoelectric effect and its concomitant Auger cascades. Photon activation therapy using iodine enhanced therapeutic efficacy [5]. Irradiation with platinum compounds used widely for anticancer chemotherapy was performed to targeted on K shell edge of 78.39 keV [6].

The approach of radiation dose-enhancement using a contrast agent is carried out. Patients with metastatic brain tumor received radiation therapy delivered by

[1]Author to whom correspondence should be addressed at:
teshima@sahs.med.osaka-u.ac.jp

conventional computed tomography (CT) scanner [7]. Synchrotron radiation CT was performed on rats with glioma [8]. When contrast media is concentrated in tumors, the absorbed dose due to the photoelectric effect increased compared to normal tissue, increasing therapeutic ratio.

MIRRORCLE-6X in the Synchrotron Light Life Science Center of Ritsumeikan University is the world smallest synchrotron dedicated for high brilliant X-ray production that is comparable to synchrotron radiation. MIRRORCLE-6X generates high energy x-rays at high efficiency. Biological effects of X-ray generated from MIRRORCLE-6X might be somewhat different from those of X-ray from linear accelerator. We show biological effects with a focus on cell-killing of X-ray from MIRROCLE-6X compared to 4 MV X-ray from linac.

MATERIAL AND METHODS

Cell culture: A549 (American Type Culture Collection) were maintained in DMEM medium (Nihonseiyaku) with 10% fetal bovine serum (Gibco) and penicillin/streptomycin (Gibco) at 37°C in a humidified atmosphere of 10% CO_2 and 90% air. DU145 (American Type Culture Collection) were maintained in RPMI-1640 medium (Nihon-seiyaku) containing 10% fetal bovine serum and penicillin/streptomycin at 37°C and 5% CO_2.

Irradiation: MIRRORCLE-6X provides high brilliant X-ray on condition of 100 mA injector peak current and 400 Hz repetitions. Max dose-rate was approximately 6 Gy/min. Irradiation using 4 MV X-ray from linac was performed at Osaka University Graduate School of Medicine with a delivering rate of approximately 1.8 Gy/min.

Colony formation assay: Survival curves were obtained by standard colony formation assay. Irradiated cells were plated onto triplicate 60-mm-diameter plastic dishes aiming for 80 to 100 colonies per dish. After 10-12 days' incubation, colonies were fixed by 10% formalin and stained by crystal violet. Colonies with more than 50 cells were scored as a surviving colony. Surviving fractions against physical doses were plotted and fitted to surviving curves using the following linear-quadratic (LQ) model: SF = exp($-\alpha$D$-\beta$D^2), where SF is the surviving fraction and D is the physical dose.

Micronucleus assay: After irradiation, the culture medium was exchanged in all dishes, and 0.5 μg/ml of Cytochalasin B (Sigma) dissolved in

dimethylsulfoxide was added to block cytokinesis [9-10]. After several days' incubation, cells were rinsed with phosphate-buffered saline (PBS; Gibco) and fixed for 20 min with 99.8% methanol (Kanto Kagaku) at room temperature. After the dish was air-dried, cells were stained with 4, 6-diamidino-2-phenylindole (DAPI; 100 ng/ml in Tris buffer, pH 7.0; Sigma). Scoring was carried out using a fluorescence microscope (excitation maximum 344 nm, emission maximum 449 nm). At least 300 cells were assessed, and at least 50 binucleated (BN) cells were evaluated. The numbers of BN cells with micronuclei (MN) were counted, and MN numbers per single BN cell were also calculated.

WST-1 assay: The cells were collected just after irradiation with Trypsin-Ethylenediamine-tetracfic acid (EDTA) (Gibco), and 100 μl of the cells ($1\square10^4$/ml) in 96-well plate was added in each well. After culture for 24, 48, and 72 h in 10% CO_2 at 37°C, 10 μl of the WST [2-(2-methoxy-4-nitrophenyl)-3-(4-nitrophenyl)-5-(2, 4 disulfophenyl) -2H-tetrazolium, monosodium salt] reagent (Dojindo) was added, and incubated in the same condition. After 2 h, the number of cells was assessed with a microplate reader (measurement wavelength 415 nm and reference wavelength 630 nm).

Apoptosis detection of flow cytometry: Cells were assayed for apoptosis using propidium iodide staining and flow cytometry according to a modified technique from I. Nicoletti et al [11]. After 24 h irradiation, cells were washed in PBS, fixed with ice-cold 70% ethanol, and stored at -20°C. Pelleted cells were then stained with 1 ml of 25 μg/ml propidium iodide, 400 mg/ml RNase A, and 1% Triton X-100 at room temperature for 15 min. Fluorometric analysis was performed using a FACScan flow cytometer (Becton Dickinson). The percentage of sub-G1 cells was calculated by CellFit software (Becton Dickinson) and defined as apoptotic.

RESULTS

Clonogenic survival: We first examined clonogenic survival to detect cell-killing effect as reproductive cell death using the colony formation assay. Survival fractions at 2 Gy irradiation for MIRRORCLE-6X or 4 MV X-ray were 0.53 or 0.56 for A549 cells (Figure 1a), 0.68 or 0.71 for DU145 cells (Fig. 1b), respectively.

Micronucleus assay: The chromosomal damage effect of radiation was assessed by cytokinesis-block with cytochalasin B micronucleus assay. Ratios of binucleated (BN) cells with micronuclei (MN)

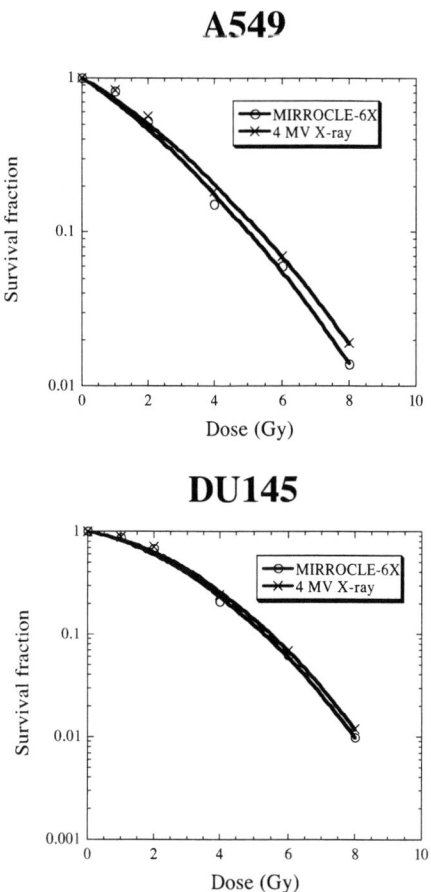

Figure 1. Clonogenic survival curves after X-ray from MIRRCLE-6X (○) or 4 MV linac (□) for A549 and DU145 cells. Data have been fitted according to the linear-quadratic model.

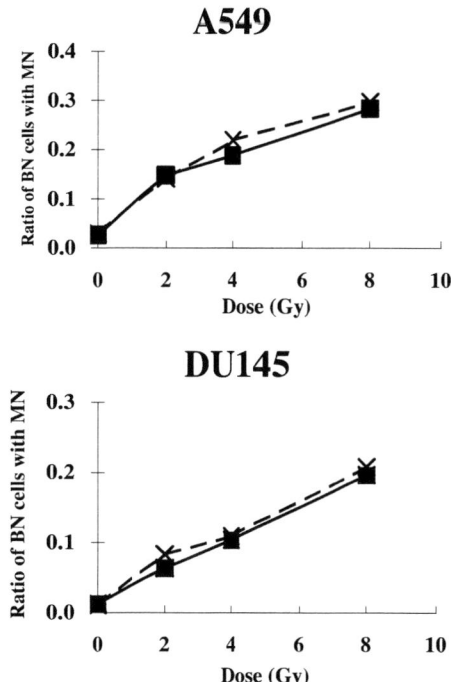

Figure 2. Results of micronucleus assay for A549 and DU145 cells irradiated with MIRRCLE-6X (□) or 4 MV linac (□). MN, micronuclei; BN, binucleated.

increased depending on the irradiated dose (Fig. 2). Ratios of BN cells with MN irradiated with 2 Gy of MIRRORCLE-6X or 4 MV X-ray were 0.15 or 0.14 for A549 cells (Fig. 2a), 0.07 or 0.08 for DU145 cells (Fig. 2b).

WST-1 assay: This assay based on the cleavage of the tetrazolium salt by mitochondrial dehydrogenases in viable cells was used to observe the effects of radiation on cell proliferation. For A549 cells irradiated with 2 Gy for MIRRORCLE-6X or 4 MV X-ray, cell proliferation was inhibited by 11.1% or 11.0% as compared with untreated controls at 72 h after irradiation, respectively (Fig. 3a). In DU145 cells at 72 h after irradiation, irradiation of 2 Gy for MIRRORCLE-6X or 4 MV X-ray inhibited cell proliferation by 3.8% or 3.5%, respectively (Fig. 3b).

Apoptosis detection of flow cytometry: The level of apoptosis programmed cell death was detected by flow cytometry after staining with propidium iodide. The percentage of apoptotic cells at 2 Gy irradiation for MIRRORCLE-6X or 4 MV X-ray were 4.9% or 5.3% for A549 cells (Fig. 4a), 6.1% or 6.4% for DU145 cells (Fig. 4b), respectively.

DISCUSSION

We examined biological effects of X-ray generated from MIRRORCLE-6X on cancer cells by colony formation assay, micronucleus assay, WST-1 method, and flow cytometry of apoptotic cells. We first examined biological effects with a focus on cell-killing effect of synchrotron radiation with high-energy and high-brilliance in the world. These data suggest that the biological effects of MIRRORCLE-6X irradiation are similar to those of 4 MV X-ray. We identified that synchrotron radiation generated from MIRRORCLE-6X is effective.

MIRRORCLE-6X generates high energy x-rays at high efficiency. Therefore, cancer therapy with high-energy x-ray and diagnosis with high spatial coherence are provided at same time. Diagnosis and treatment can be done with the same instrument. To perform radiation therapy precisely, the daily positioning

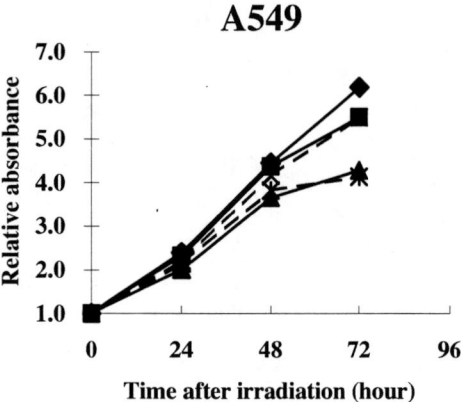

A549

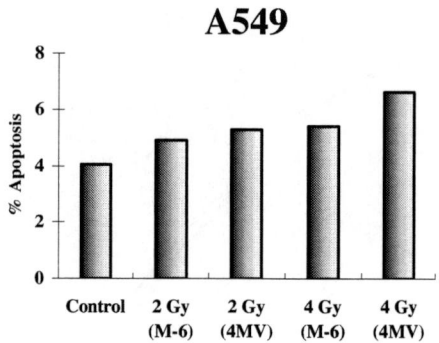

A549

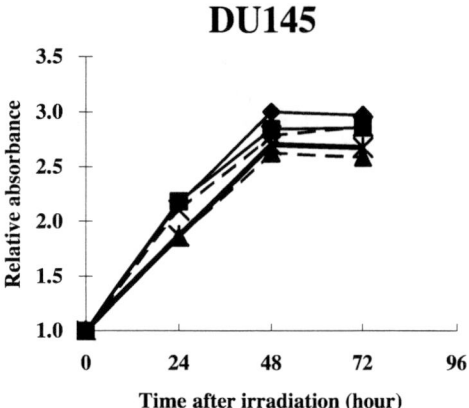

DU145

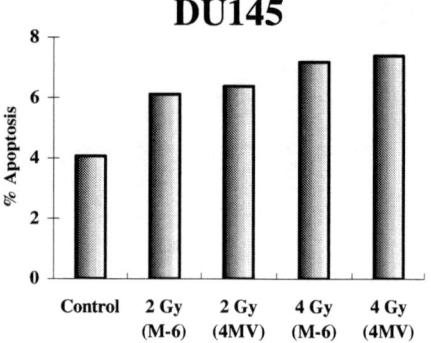

DU145

Figure 4. Percentage of apoptotic cells after treatment with radiation after 24 h irradiation. Cells were removed, stained with propidium iodide, and analyzed by flow cytometry for apoptotic cells. M-6, MIRRORCLE-6; 4MV, 4MV X-ray.

Figure 3. Effects of radiation on the proliferation of A549 and DU145 cells. Cancer cells were untreated (♦) or treated with 2 Gy for MIRRORCLE-6X (■), for 4 MV X-ray (×), 4 Gy for MIRRORCLE-6X (▲), or for 4 MV X-ray(✳). Each data point represents the relative absorbance in comparison with each sample at 0 h.

accuracy is one of the most important factors. Uematsu *et al* developed a fusion of computed tomography (CT) and linac (FOCAL) unit [12-14]. Radiation therapy was performed this new technique with the FOCAL unit has been gaining promising clinical results. We hope that these concepts will be applied to MIRROCLE-6X to perform radiation therapy precisely.

Patients bearing glioma with conventional radiation therapy occasionally experienced unacceptable damage to nearby vital central nervous system (CNS) tissues. Normal brain tissue of rats bearing gliosarcoma treated with microbeam radiation therapy is better tolerated than broad beams [2]. MIRRORCLE-6X can provide narrow beam size (10 μm-100 μm). It is considered that MIRRORCLE-6X is prospective facility for microbeam radiation therapy.

Experimental results of photon activation therapy using iodine showed that the effective radiation dose was enhanced by a factor of 2 [5]. MIRRORCLE-6X generates high brilliant X-rays of broad band from 1 keV to 6 MeV. If the irradiation with MIRRROCLE-6X is targeted on K absorption to induce the photoelectric effect and its concomitant Auger cascades, therapeutic efficacy might be enhanced.

MIRRORCLE-6X is foreseen as very promising for radiation therapy.

ACKNOWLEDGMENTS

We are indebted to Kouki Okita and all other staffers at Molecular Pathology Laboratory, Naomi Dejima and Kana Matsubara at Radiation Oncology Laboratory for excellent technical support.

REFERENCES

1. D. N. Slatkin, P. Spanne, F. A. Dilmanian, J.O. Gebbers and J. A. Laissue, *Proc Natl Acad Sci U S A* **92**, 1995, pp. 8783-8787.

2. J. A. Laissue, G. Geiser, P. O. Spanne, F. A. Dilmanian, J. O. Gebbers, M. Geiser, X. Y. Wu, M. S. Makar, P. L. Micca, M. M. Nawrocky, D. D. Joel and D. N. Slatkin, *Int J Cancer* **78**, 1998, pp. 654-660.

3. F.A. Dilmanian, G. M. Morris, N. Zhong, T. Bacarian, J. F. Hainfeld, J. Kalef-Ezra, L. J. Brewington, J. Tammam and E. M. Rosen EM, *Radiat Res* **159**, 2003, pp. 632-641.

4. N. Zhong, G. M. Morris, T. Bacarian, E. M. Rosen, and F. A. Dilmanian, *Radiat Res* **160**, 2003, pp. 133-142.

5. B. H. Laster, W.C. Thomlinson and R.G. Fairchild, *Radiat Res* **133**, 1993, pp. 219-224.

6. S. Corde, M. C. Biston, H. Elleaume, F. Esteve, A. M. Charvet, A. Joubert, V. Ducros, S. Bohic, A. Simionovici, T. Brochard, C. Nemoz, M. Renier, I. Tropres, S. Fiedler, A. Bravin, W. Thomlinson, J. F. Le Bas and J. Balosso, *Radiat Res* **158**, 2002, pp. 763-770.

7. J. H. Rose, A. Norman, M. Ingram, C. Aoki, T. Solberg and A. Mesa, *Int J Radiat Oncol Biol Phys* **45**, 1999, pp. 1127-1132.

8. J. F. Adam, H. Elleaume, A. Joubert, M. C. Biston, A. M. Charvet, J. Balosso, J. F. Le Bas and F. Esteve, *Int J Radiat Oncol Biol Phys* **57**, 2003, pp. 1413-1426.

9. M. Fenech and A. A. Morley, *Mutat Res* **147**, **1985**, pp. 29-36.

10. Y. Shibamoto, C. Streffer, C. Fuhrmann and Budach V, *Radiat Res* **128**, 1991, pp. 293-300.

11. I. Nicoletti, G. Migliorati, M. C. Pagliacci, F. Grignani and C. Riccardi, *J Immunol Methods* **139**, 1991, pp. 271-279.

12. M. Uematsu, A. Shioda, K. Tahara, T. Fukui, F. Yamamoto, G. Tsumatori, Y. Ozeki, T. Aoki, Watanabe M and Kusano S, *Cancer* **82**, 1998, pp. 1062-1070.

13. M. Uematsu, M. Sonderegger, A. Shioda, K. Tahara, T. Fukui, Y. Hama, T. Kojima, J. R. Wong and S. Kusano, *Radiother Oncol* **50**, 1999, pp. 337-339.

14. M. Uematsu, A. Shioda, A. Suda, K. Tahara, T. Kojima, Y. Hama, M. Kono, J. R. Wong, T. Fukui and S. Kusano, *Int J Radiat Oncol Biol Phys* **48**, 2000, pp. 443-448.

Homeostasis and Cancer Symptom in Elemental Concentration Profiles of Hair Observed by Fluorescent X-ray Analysis with Synchrotron Radiation

Jun-ichi Chikawa

Center for Advanced Science and Technology,
Kouto 3-1-1, Kamigori, Ako, Hyogo,678-1205

Hair samples of 37 donors including 12 patients of hepatocelluar carcinoma have been examined by fluorescent X-ray analysis using the SPring-8, which detected many kinds of trace elements in a single hair root. Homeostasis in concentrations of Ca, Fe, Cu, Zn and Sr was found to be in consistency with their concentrations measured for the serum in a healthy case. Unusual increases of [Cu] and/or [Fe] were observed for hair of the patients by disorder of the liver function to excrete these elements due to cancer. The unique behavior observed for [Ca] is discussed in relation to the "Calcium paradox", a phenomenon of increasing from the regulated Ca ion concentration in cytosol, which is caused by parathyroid hormone in the case of Ca deficiency due to many kinds of disease as well as insufficient intake and absorption of Ca. It is concluded that the analysis of hair is useful for screening serious diseases such as the cancer and osteoporosis.

INTRODUCTION

Detection of breast cancer from structures of hair was reported in 1999 by James, et al.[1] who observed two types of diffraction patterns with Debye-Scherrer rings and no rings for hair by the small angle X-ray scattering method (SAXS) using synchrotron radiation (SR). They concluded that breast cancer is responsible for the ring patterns and can be detected by examining whether hair shows the ring patterns. To confirm this conclusion, SAXS observations for hair were carried out in a number of SR facilities in the world; it was found that both the types of diffraction patterns were observed for hair samples from patients of breast cancer, and both also for healthy cases[2]. Vazina[3] made both SAXS and fluorescent X-ray analysis for hair samples and found that the hair samples showing the ring patterns of SAXS have higher concentrations of calcium.

Various elements can be detected for a piece of hair with a length of 0.2 mm by fluorescent X-ray analysis (FXA), as shown in Fig.1. Such a very high sensitive analysis with SR should be applied for health check. The purpose of the present paper is to give a scientific base for elemental concentration profiles of hair.

EXPERIMENT

Hair samples were provided from 37 donors. Eleven of them were patients of hepatocellular carcinoma (HCC), and their samples were encoded as H1 to H11. One was affected by both osteoporosis and HCC, and his samples encoded as OH-1, OH-2, OH3, and OH-4 taken with their respective intervals of 4 months, 1 month, and 5 months (OH-2u for the upper part at 1 mm from the root). The rest, 25 donors, were normal (healthy) and encoded as N1 to N25. For comparison, hair and blood samples encoded T1 to T8 were taken simultaneously from 8 of the normal donors.

Fluorescent X-ray analysis was carried out using synchrotron radiation at the Spring-8 BL-24 undulator beamline.[4] The beam was monochromatized at a photon energy of 20 keV by a silicon crystal. The beam size was 0.2X0.2 mm at the sample position. The beam axis was positioned at the center of the bulb of hair root. The fluorescent x-rays were measured by a silicon drift detector (SDD) and multi-channel pulse height analyzer.

When spectral peak heights, A and B, are observed for thin and thick hair samples, respectively, there is the relation of $A=\alpha B$, where α is the proportional constant. Since precise determination of α is difficult, we take a logarithmic scale; $\log A=\log \alpha B=\log B + \log \alpha$. If the concentration is the same, the two peaks, $\log A$ and $\log B$, can be superimposed with moving by $\log \alpha$. In this case, the concentration is $[\log P - \log S]=\log(P/S)$, where $\log P$ is the peak height and $\log S$ is the back-

CP716, *Portable Synchrotron Light Sources and Advanced Applications,*
edited by H. Yamada, N. Mochizuki-Oda, and M. Sasaki
© 2004 American Institute of Physics 0-7354-0195-0/04/$22.00

ground due to x-rays scattered by the sample. The concentration is expressed with the unit that gives the same intensity as the background height S. To show

deviations from the healthy standard $[\log P - \log S]_{st}$, the observed values $[\log P - \log S]$ were normalized by plotting M in a logarithmic scales according to the equation

$$[\log P - \log S] / [\log P - \log S]_{st} = \log M.$$

SERUM AND HAIR

Figure 1 shows FX spectra of hair and serum taken simultaneously from a healthy donor; the spectrum for serum is superimposed on that for hair by sliding vertically with multiplying the former by a factor of $\alpha = 0.5$ so as to have the same background level. Many peaks are seen by trace elements contained in both the samples. The peak for Zn in hair is much higher than that in serum, while Br is in the opposite relation. The peaks of Ca, Fe, Cu, and Sr for hair and serum are superimposed on each other. From the character of logarithmic scale, this means that the concentrations of these elements in hair are proportional to those in serum with the same proportional constant in a steady-state healthy case. It was found that the FX spectral

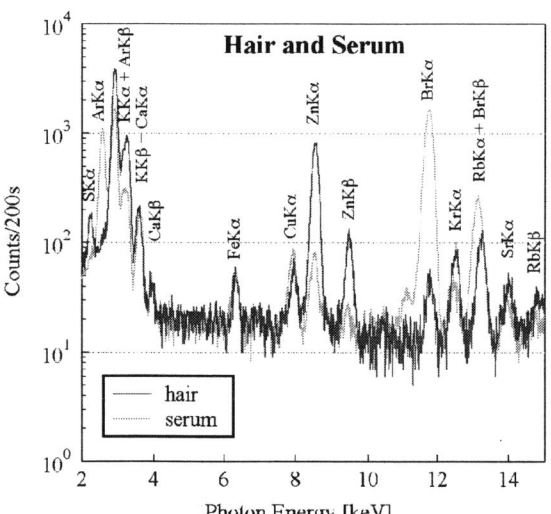

FIGURE 1. Comparison of fluorescent X-ray spectra for hair and serum taken simultaneously for a healthy donor. The spectrum for serum is superimposed on that for hair by sliding vertically; the former is multiplied by a factor of 0.5 so as to have the same background level.

profiles of serum were very similar for the 8 donors; almost all peaks superimpose on each other. This implies that homeostasis in concentrations of trace elements exists in serum. Therefore, hair has also constancy in concentrations of Ca, Fe, Cu, and Sr in the steady-state case, because of the relation between serum and hair seen in Fig.1. Their deviations will be symptoms of diseases.

CALCIUM IN HAIR

Calcium plays as a messenger in universal cellular signal transmission. Therefore, a strict balance is maintained between bone and circulating blood under a homeostatic control, and the Ca concentration in serum is always regulated to be at 10 mg per deciliter.[5] Ca concentration of bone is 4-orders of magnitude higher and that of cytosol is 4-oders of magnitude lower than that of serum; the concentration of Ca ion inside a cell must be so low that its flow into the cell may be detected as the signal.

When [Ca] in blood increases, the thyroid gland secretes calcitonin acting to deposit Ca in blood on bone. If [Ca] in blood decreases depending on intake, absorption and excretion of Ca, parathyroid gland secrets parathyroid hormone (PTH) which acts to draw Ca from bone into blood. The PTH also acts for kidney to suppress excretion of Ca into urine and to activate vitamin D for intestinal Ca absorption. In this way, the PTH contributes to recover the [Ca] deficiency in blood. Simultaneously, the PTH acts to increase [Ca] in cytosol and causes a toxic overflow of Ca into cells. This phenomenon is called "Ca paradox" because the overflow contradicts the Ca deficiency.[5]

The high [Ca] in cytosol inhibits the cellular signal transmission to cause various diseases. The Ca floods in vascular cells, brain cells, pancreatic islet cells, muscle cells, and epithelial cells are responsible for hypertension arteriosclerosis, Alzheimer's diseases, diabetes mellitus, muscular dystrophy, and malignancy, respectively, which are named generically "Calcium paradox diseases".[5]

It is also well known that malignant tumor such as breast cancer secretes parathyroid hormone-related protein (PTHrP) making actions similar to PTH and causing hypercalcemia.

Therefore detection of Ca deficiency is very important and can be done by measuring concentrations of PTH in blood (Note that [Ca] in blood is always constant). However, the PTH measurements have not

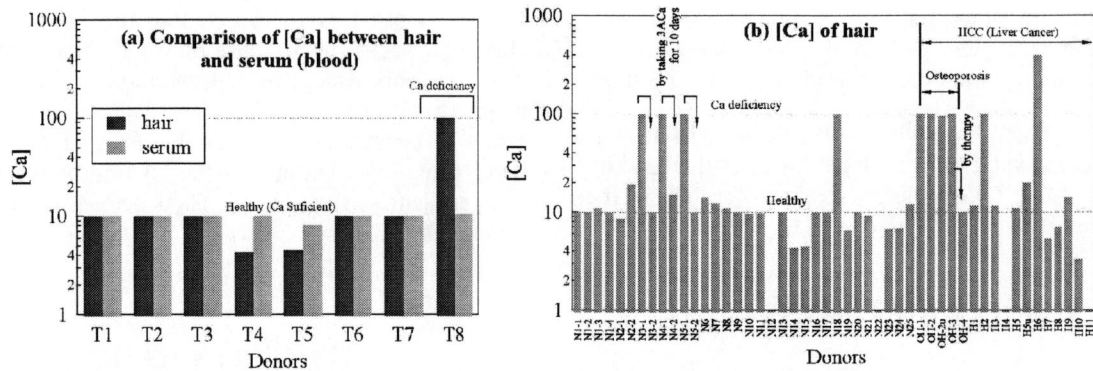

FIGURE 2. Calcium concentrations of hair roots normalized to be 10 for the healthy case. (a) Comparison of hair and serum. (b) [Ca] in hair from 37 donors. For N12, N22, H4, and H11, [Ca] ~ 0 due to fallen hair

been made in the conventional blood test. A simple method to detect Ca deficiency is required. The Ca flood is expected to take place in hair.

[Ca] observed in hair

Ca concentrations, $[Ca]_H$ in hair and $[Ca]_S$ in serum, are compared for the 8 donors from T1 to T8 in Fig.2. The peak heights of $CaK\beta$ were measured for spectra of hair and serum like the example of Fig.1 and were found to be superimposed on each other in normal cases. In fig.2 (a), this peak height of (log *P*–log *S*) is normalized to be 10 in a logarithmic scale. The constancy of $[Ca]_S$ is seen except T5. The peak height of $[Ca]_H$ agrees with that of $[Ca]_S$ except T4, T5, and T8. This means that $[Ca]_H$ is proportional to $[Ca]_S$ in the normal case. For T8, $[Ca]_H$ is much higher than $[Ca]_S$, although the $[Ca]_S$ is normal. This high value of $[Ca]_H$ was found to be due to Ca deficiency by experimental results for hair samples of the 37 donors shown in Fig.2 (b). The concentrations of hair for most of the donors distribute around the level of 10 in the normalized logarithmic scale. The hair samples of the donors denoted by N3, N4, N5, N18, OH and H2 show concentrations as high as 100. It should be noted that they have the same height; the peaks are superimposed precisely. By taking a supplement called "Active Absorbable Algal calcium" (3ACa)[6] with 900mg of Ca per day for 10 days, the high values of $[Ca]_H$ were reduced to the low level as seen from the examples donated by N3, N4 and N5. The donor denoted "OH" had been affected by osteoporosis, and his hair samples from OH-1 to OH-3 taken for a year had the high concentration, which was also reduced by cancer therapy and supplimentation of 3ACa. These results show that the high $[Ca]_H$ is for the case of Ca deficiency.

It is seen from Fig.2 that Ca deficiency is detected easily by FXA of hair; (1) In the case of Ca enough (healthy), $[Ca]_H$ in hair distributes around the low level which is proportional to the total concentration $[Ca]_S$ of serum. (2) In the case of Ca deficiency, $[Ca]_H$ increases to that proportional to the square of the $[Ca]_S$.

What does Ca in hair tell us?

Serum contains many kind of proteins such as albumin, globulin, fibrinogen, etc. The concentration $[Ca]_S$ in serum is regulated at 10 mg/dl. Its half is contained in the protein (mainly in albumin), which is also regulated to be 4 ~ 5 g/dl, having a concentration of (Ca/Protein) = 1mg/g. Since alubumin has a molecular weight of 66000, the number of Ca atoms per molecule is 1 to 2 (Ca atomic weight=40). The other half exists as Ca^{++} ions. The total $[Ca]_S$ is the sum of the ion concentration $[Ca]_I$ and $[Ca]_P$ carried by the protein, i.e., $[Ca]_S = [Ca]_I + [Ca]_P$. The Ca ions play the role of the signal messenger, and $[Ca^{++}]$ is kept constant, though the $[Ca]_P$ somewhat varies. On the other hand, a cell has Ca pumps and Ca^{++} ion channels in the cell membrane. The pumps always work to pump out Ca from the cell, and the cytosol is kept at a nearly zero concentration of Ca^{++} with closed Ca^{++} channels in the healthy case. Since the Ca floods take place in the case of Ca deficiency, the pumps are not so powerful. Hair is grown with a growth rate of about 0.3 mm/day by hair matrix cells in a follicle which is surrounded with capillary blood vessels. Generally there are Ca storage sources such as endoplasmic reticulum and mitochondria in a cell. In the steady–state hair growth, however, the concentration of an element in hair must be balanced with its supply from blood, without contributions of the inner sources in hair matrix cells. The hair growth is slow enough to make the cytosol in equilibrium with the serum. Here we consider Ca concentrations of hair phenomenalogically. We introduce the number of Ca atoms N_S per protein molecule and the concentration $[PM]_S$ of protein molecules, i.e. $[Ca]_P=N_S[PM]_S$. For convenience sake, we express the Ca ion concentration in serum by $N_S[PM]_S=[Ca]_I$ (since $[Ca]_I \sim [Ca]_P$), although they are free ions in serum.

In the healthy case of Ca enough, Ca^{++} channels are closed, and only protein carrying a half of $[Ca]_S$ in serum is incorporated into hair matrix cells; the concentration $[PM]$ of protein molecules in the cytosol is in equilibrium with that of the serum $[PM]_S$, i.e., $[PM]$ is kept constant in the cytosol. For simplicity we assume that $[PM] = [PM]_S$. In the cytosol, a part of Ca atoms carried by the protein molecules resolves into the cytosol, and the number of Ca atoms N_E per protein molecule is in equilibrium with the Ca ion concentration $N_L[PM]$ in the liquid (cytosol), where $N_L+N_E=N_S$ and $N_S[PM]_S=[Ca]_P$. The reaction rate to form protein molecules with N_E Ca atoms is proportional to the collision probability $N_L[PM][PM]$ which is the product of $[Ca^{++}]=N_L[PM]$ and $[PM]$ in the cytosol, and the dissociation rate of protein molecules having N_E Ca atoms is proportional to $N_E[PM]$. In chemical equilibrium, both the rates must be equal, i.e., $rN_E[PM]=qN_L[PM][PM]$ with the proportional constants q and r. Then, using $N_L=N_S-N_E$,

$$N_E = N_S[1-1/\{1+(q/r)[PM]\}]$$

The $[Ca]_H$ of hair is determined by the flow of protein having N_E atoms per molecule toward hair from the hair matrix cells, we have

$$[Ca]_{HE}=kN_S[1-1/\{1+(q/r)[PM]\}][PM], \quad (1)$$

where k is the proportional constant. Although Eq. (1) means that some part of Ca ions leave in the cytosol, the $[Ca^{++}]$ increases up to $N_S[PM]$, and $[Ca]_{HE}$ in steady-state growth of hair must agree with the supply from the serum, i.e.,

$$[Ca]_{HE}=kN_S[PM]_S=k[Ca]_P. \quad (2)$$

In the case of Ca deficiency, the PTH opens the Ca^{++} ion channels. Since the chemical potential of Ca ion is different between the cytosol and serum, we introduce the Ca ion concentration $N_L[PM]$ which is always in equilibrium with the concentration $N_S[PM]_S$ in the serum. Now we calculate the number of Ca atoms, N_D, accommodated per protein molecule in the sytosol in equilibrium with the ion concentration $N_L[PM]$. Similarly to the previous case, the chemical equilibrium gives $rN_D[PM] =qN_L[PM][PM]$. We have the following equations:

$$N_D=(q/r)N_L[PM], \quad (3)$$

$$[Ca]_{HD}=kN_D[PM]=k(q/r)N_L[PM]^2. \quad (4)$$

If we take $N_D=N_S^2$, eq. (4) turns out to be

$$[Ca]_{HD}=kN_S^2[PM]^2 \quad (5)$$

Equation (5) agrees with the experimental result in Fig. 2 that $[Ca]_{HD}=[Ca]_{HE}^2$. By taking into account the equilibrium between $N_D[PM]=N_S^2[PM]$ and

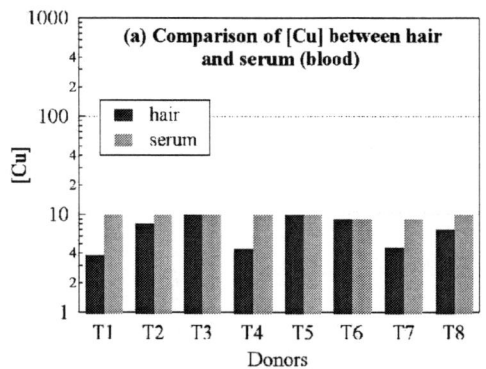
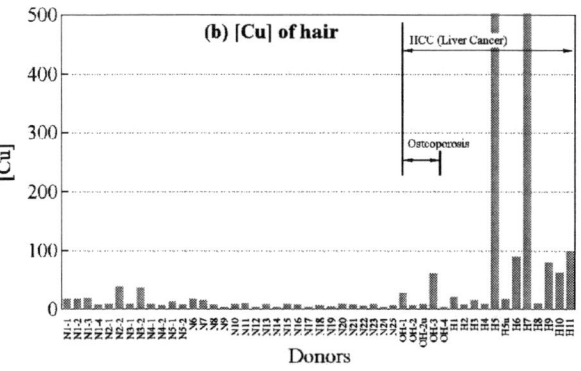

FIGURE 3. Copper concentrations of hair roots normalized to be 10 for the healthy case. (a) Comparison of hair and serum. (b) [Cu] in hair from 37 donors including 12 HCC patients. The size of cancer in mm: OH 40, H1 30, H2 30x20 & 25, H3 12, H4 40, H5 12 & 12, H6 35, H7 35, H8 30, H9 15, H10 36, H11 10 & 15.

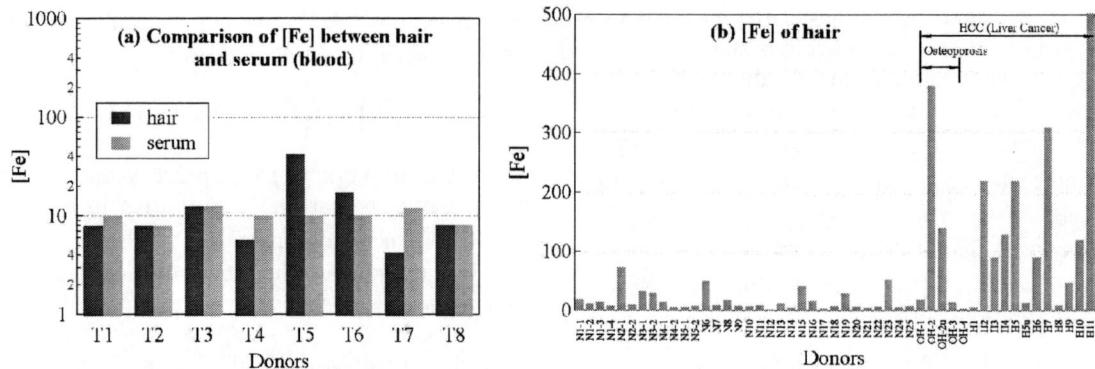

FIGURE 4. Iron concentrations of hair roots normalized to be 10 for the healthy case. (a) Comparison of hair and serum. (b) [Fe] in hair from 37 donors including 12 HCC patients.

$[Ca]_I = N_S[PM]$, we obtain $N_L = N_S$. This conflicts with the difference in chemical potential of Ca ions between cytosol and serum, and may be due to the fact that the Ca ion channels act in one way from the outside to the inside of a cell. One may conclude that the one-way Ca ion channels in the cell membrane keep the $[Ca^{++}]$ in cytosol at the same as that in serum, which determines the [Ca] in the protein phase responsible for the $[Ca]_{HD}$ in the unhealthy case of Ca deficiency.

It is clear from the Ca behavior stated above that Ca exists in both the phases of the liquid and protein in serum and cytosol. The $[Ca^{++}] = N_S[PM]$ in serum is controlled to be constant. Therefore the $[Ca]_{HD}$ determined by the $[Ca^{++}]$ in serum becomes constant as seen in Fig. 2. Whereas $[Ca]_P$ often fluctuates, so that $[Ca]_{HE}$ varies in some extent as seen in Fig. 2.

In the above statement, $[Ca]_{HE}$ and $[Ca]_{HD}$ were calculated with a fixed [PM]. However, the [PM] in serum somewhat varies; in the case of larger [PM], the number of Ca atoms per molecule decreases with unchanged [Ca] in the protein phase. Because the number of protein molecules necessary for making a unit mass of hair is fixed, a high [PM] in serum results in a lower $[Ca]_{HE}$. This is the case for T4 in Fig. 2(a). The same result is seen for Cu and Fe in T4 of Figs. 3(a) and 4(a). This implies that Cu and Fe are carried into hair by protein, similarly with the case of Ca. In T5 of Fig. 2(a), both the $[Ca]_S$ and $[Ca]_{HE}$ are lower than their normal values and can be explained by a reduction of Ca carried per protein molecule due to the PH different from that of the normal serum; since $[Ca]_I$ is kept at the normal value, a reduction of 20% in the total $[Ca]_S$ turns out to be 40-% reduction in $[Ca]_{HE}$. Also, hypercalcemia can be detected as seen for H6 in Fig.2 (b).

The unusually high [Ca] in hair is considered to be due to PTHrP secreted from tumor. This suggests a strong possibility to detect breast cancer accompanied by secretion of PTHrP.

COPPER AND IRON IN HAIR

Copper and iron are elements necessary for living, and are excreted after the use as bile by the liver. Hepatocellular carcinoma (HCC) degrades the excre-tion[7] and increases [Cu] and [Fe] in serum and hair. The FXA results are shown in Figs. 3 and 4. Similarly to Fig.2, the peak heights for CuKα and FeKα are compared for serum and hair in the healthy case in Figs. 3(a) and 4 (a). Constancy of [Cu] and [Fe] in serum is seen and defines the healthy level, which is normalized to be 10 in the log scale. The distributions of [Cu] and [Fe] observed for 37 donors are shown in linear scales in Fig.3 (b) and 4 (b), respectively. It is seen that the hair samples from the donors having HCC shows unusually high [Cu] and/or [Fe], except donors H1 and H8. For H8, an unusual value of $[Zn]_H$ was detected. Furthermore, all the hair samples from HCC donors showed very low [Rb] and [Sr] compared with healthy cases. It should be noted that the size of cancer for the donors were 10 ~ 40mm which is the general size when cancer first detected, because HCC gives no subjective symptoms. It is concluded that HCC can be detected by FXA of hair.

CONCLUSION

Concentrations of trace elements such as Ca, Fe, Cu and Sr in serum were found to be constant under a homeostatic control, and their concentrations in hair are proportional to those in serum in healthy cases. In unhealthy cases, however, their deviations from the normal values are amplified in hair, although there is a tendency to suppress such deviations in serum under the homeostatic control. It is remarkable that the observed concentration profiles for hair can be treated by the simple chemical equilibrium with serum. FXA of hair can detect the Ca deficiency responsible for the Ca paradox diseases as well as cancer such as HCC.

ACKNOWLEDGMENTS

The author would like to express his sincere thanks to Professor T. Fujita for his advice such that the hair analysis should be made before and after taking the AAACa. He is also indebted to Professor Y. Kagoshima, Himeji Institute of Technology, for his great help on the experiment in Spring-8.

REFERENCES

1. James V., Kearsley T., Amemiya Y., and Cookson D., Nature **398**, 33-34 (1999).

2. Amenische H. et al. *Synchrotron Radiation News*, **Vol.12**, No.5, 32-34 (1999)

3. Vazina A. A., Private communication. Also, see Aksirov, A.M. et al. Nucl. Instr. Meth. **A470**, 380-387 (2001).

4. Kagoshima Y., Koyama T., Wada I., Niimi T., Tsusaka Y., Matsui J., Kimura S., Kotera M. and Takai K., to be published in the proceedings of SRI2003, AIP conference proceedings.

5. Fujita, T., J. Bone Miner.Metab. **16** 195-205 (1998). Fujita, T. and G.M.A. Palumieri, J. Bone Miner. Metab. **18,** 109-125 (2000).

6. Fujita T. J. Bone Miner. Metab. **14**, 31-34 (1996).

7. Nakayam, A. Fukuda, H. Ebara M. Hamasaki, H. Nakajima, K. and Sakurai H., Biol. Pharm. Bull. **25**(4) 426-431 (2002).

Infrared Free Electron Laser Induced Angioplasty for Arteriosclerotic Region of Blood Vessels

Kunio AWAZU and Yuko FUKAMI*

Institute of Free Electron laser, Osaka University, 2-9-5 Tsuda-yamate, Hirakata, Osaka 573-0128, Japan
*Dept. of Electrical and Electronic Engineering, National Defense Academy,
1-10-20 Hashirimizu, Yokosuka 239-8686, Japan

Abstract. In order to estimate the optimum laser conditions for efficient dissociation of cholesterol ester in an arteriosclerotic region of blood vessels, we have investigated the relationship between laser wavelength and power density on cholesterol ester dissociation using a mid infrared free electron laser (MIR-FEL). In this study, cholesteryl oleate, which is a typical cholesterol ester found in arteriosclerotic regions, was irradiated with 5.75-μm-FELs, which cause vibration of ester bonds. Two results were obtained. (1) Ester dissociated depending on the absorption coefficient, and the macropulse duration was shorter than the thermal relaxation time, showing that ester bonds dissociated into carboxylic acid and cholesterol by macropulse-induced thermal effects without accompanying thermal diffusion, (2) Using a wavelength of 5.75 μm, the maximum ester dissociation ratio was achieved under the optimum laser conditions of a macropulse energy density of 0.4-1.0 J/cm^2. We conclude that MIR pulsed-lasers with a wavelength of 5.75 μm can be useful for removal of cholesteryl ester in an arteriosclerotic region of blood vessels.

INTRODUCTION

Cholesterol esters of fatty acids, such as oleic acid, are known to accumulate in arteriosclerotic regions [1,2]. It is important in the medical treatment of arteriosclerosis to remove cholesterol esters selectively without causing side effects. Atherosclerotic plaques are currently treated by either surgical removal, or injection of a chemical solvent to dissolve the plaque. These procedures, however, are both stressful and relatively inefficient. Therefore a less invasive treatment would be desirable. Laser angioplasty employing conventional lasers (e.g. UV eximer, IR CO$_2$ lasers) and an optical fiber catheter have been used to break atherosclerotic plaques [3-7]. Unfortunately this treatment tends to damage the surrounding tissue resulting in potentially serious side effects.

Generally it is known that, free cholesterol, which is not an ester compound, can be removed from the cells to the liver, owing to the metabolism such as the process of reverse cholesterol transport (RCT) [8]. However, most of the cholesterol in the arteriosclerotic region consists of cholesterol ester bound with fatty acids, which can exist stably and accumulate [1,2]. The atherosclerotic plaque is covered with endothelial cell layer with the thickness of 10-20 μm. Considering these facts, here we aim at the less invasive therapy for arteriosclerosis, which consists of dissociating the ester bond of cholesterol ester without causing the thermal damage to the peripheral normal tissue, endothelial cell layer.

In order to minimize the side effects, we have employed mid infrared free electron lasers [9] (MIR-FELs) [10-13]. The sharp absorption peaks of many materials are distributed in the MIR region, allowing selective targeting by altering the wavelength. This can be utilized in the treatment of arteriosclerosis where effective removal of cholesterol esters can be achieved by selecting the wavelength which satisfies the absorption conditions for the cholesterol ester but will not adversely affect the normal tissue. Cholesterol esters have a specific absorption at ~5.5 μm which corresponds to the ester bond stretching vibration. However, the main contents of normal tissue are water and protein, which has a large absorbance at ~3 μm, 6-7 μm, and longer than 7 μm, assigned to C-H/O-H stretching, group vibrations of amide, and single bonds

CP716, *Portable Synchrotron Light Sources and Advanced Applications,*
edited by H. Yamada, N. Mochizuki-Oda, and M. Sasaki
© 2004 American Institute of Physics 0-7354-0195-0/04/$22.00

between C/O/P, respectively. Using a wavelength of ~5.5 μm, the requirements described above can be satisfied. Furthermore this wavelength interacts far less efficiently with normal tissue, thereby limiting the damage to the surrounding area.

Based on such considerations, we have investigated the irradiation effects on cholesterol oleate, a typical cholesterol ester, using MIR-FEL having wavelength tunability in the MIR region. In previous studies, we have obtained the following results [12, 13]. (1) By irradiation of 5.75-μm-FEL, this is highly absorbed by ester bonds, the ester bond of cholesteryl oleate non-invasively dissociated into cholesterol and carboxylic acid, which are familiar to the normal tissues. (2) Cholesteryl oleate dissociation could be discussed quantitatively using the concentration of dissociated ester, obtained from mass spectra of the sample. (3) Ester dissociation was a laser-induced thermal effect and was dependent on the amount of laser energy absorbed (i.e. mainly on the absorption coefficient) rather than on the excited mode.

In this study, we investigated the FEL parameters suitable for cholesterol ester dissociation whilst at the same time considering practical clinical applications.

MATERIALS AND METHODS

Cholesteryl Oleate

The target material was cholesteryl oleate (Sigma, #C-9253, FW = 651.1), a typical cholesteryl ester associated with arteriosclerotic regions. The molecular structure is shown in Fig. 1. In the MIR region at 5.75 μm, cholesteryl oleate has a sharp absorption peak which can be assigned to the stretching vibration mode of the ester bond. Though the largest absorption exists at around 3 μm, considering the constraint of not affecting the normal tissue, the wavelength of 5.75 μm, which is hardly absorbed by the normal tissue, is suitable.

The thermal relaxation time, which indicates the time required for the heat to diffuse out of the laser absorbed volume (= $s\delta$, product of laser spot size and optical penetration depth), is estimated to be several tens of μs [14], which is longer than a macropulse duration, τ. Comparing the value of τ and thermal relaxation time, the heat caused by absorption of laser energy can be confined in the laser absorbed volume and the thermal effect can be confined only within the laser absorbed volume.

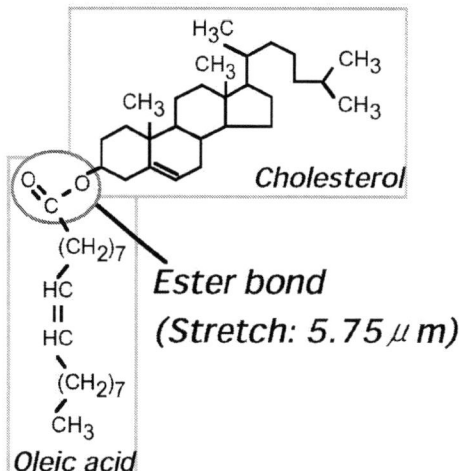

FIGURE 1. Molecular structure of cholesteryl oleate. Cholesteryl oleate molecule consists of cholesterol and oleic acid, which are connected by an ester bond.

For FEL irradiation experiments, a ZnSe sample cell was used. 1 mg of cholesteryl oleate was spread on a 2-mm-thick ZnSe disk of 32-mm-radius in an area of 0.2 cm^2.

MIR-FEL Irradiations

The MIR-FEL at the Institute of FEL, Osaka University has wavelength tunability in the MIR region. FEL is a pulse laser, having a double pulse structure, a macropulse and a micropulse as shown in Fig. 2(a). Here, we mainly focus on the macropulse energy density, E_M [J/cm^2], because the heating effects are mostly initiated by the macropulse, which is delivered from one tenth of the average power density. The peak power density of the macropulse, P_M, is of 10^3-10^4 W/cm^2 order and is expressed as $P_M = E_M / (\tau \, 10^{-6}) = 10^6 \, E_M / \tau$ [W/cm^2], using the macropulse duration, τ [μs].

In this study, we employed the wavelength of 5.75 μm, which is highly absorbed by cholesteryl oleate, while it is less absorbed by the normal tissues such as protein and water, as described in the introduction paragraph. And in order to investigate optimum laser conditions for cholesterol ester dissociation, we observed the dependency of the following parameters, macropulse energy, E_M, and incident macropulse number, n.

The optical path used in FEL irradiation experiments is shown in Fig. 2. An FEL raw beam was focused via a parabolic mirror (60 mm radius, f~300 mm) onto the sample in the cell. A polarizer was used to control the incident average power. Two

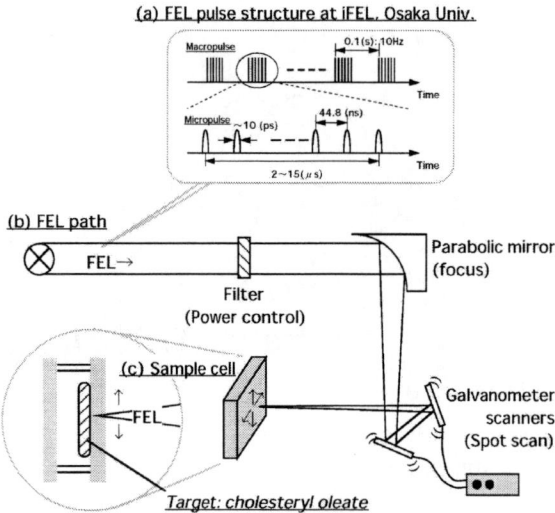

FIGURE 2. Pulse structure of FEL (a), optical path used in FEL irradiation experiments (b), and ZnSe sample cell (c).

galvanometer optical scanners were inserted between the focusing mirror and the sample to move the focal spot two-dimensionally on the area where the sample was spread. After FEL irradiation, the cell was dismantled and the sample was dissolved in chloroform.

Analysis Using Mass Spectroscopy

After FEL irradiation the sample was dissolved in chloroform and the solution was analyzed using a mass spectrometer (JEOL, JMS-AX500). The ionization method consisted of fast atom bombardment using 3-nitrobenzyl alcohol as a matrix. The mass analyzer was a sector MS. From the data of mass spectroscopy, the dissociation ratios of the FEL irradiated samples were estimated, as follows. The mass spectrum of cholesteryl oleate has a characteristic mass signal at $m/z = 365-375$, while cholesterol has two signals, at 365-375 and 378-390. Using the ratio of corresponding signal intensities at two regions, cholesterol content (i.e. concentration of dissociated ester in the irradiated sample) R_d was estimated [12].

RESULTS

Macropulse energy density was varied within the range of $E_M = 0.1-1.6$ J/cm^2 by two means, macropulse width, τ, and macropulse peak power, P_M. In Fig. 3, the ester dissociation ratio, R_d, was plotted as a function of macropulse energy density, E_M. The

wavelength was tuned to 5.75 μm, and the incident pulse number, n, was set to 200. By the left vertical axis, two kinds of data are shown: The circles represent data where the macropulse peak power, P_M, was varied and the crosses where the macropulse width, τ, was varied. As shown in these dependencies on the macropulse energy density, a similar trend was observed in the two cases. From these results, it was found that ester dissociation ratio was a function of the macropulse energy, regardless of the macropulse peak power or macropulse width, in the parameter ranges used in these experiments.

From the study of R_d as a function of E_M, the R_d reached a maximum value of 80% at $E_M = 0.4$ J/cm^2, but does not increase any further despite the power increase. This relationship can be divided into three distinct regions, (1) $R_d < 10\%$ at $E_M < 0.2$ J/cm^2, (2) $R_d = 80\%$ at $E_M = 0.4-1.0$ W/cm^2, and (3) $R_d = 80\%$ and ablated at $E_M > 1.2$ W/cm^2. Thus it was found that with 5.75-μm-FEL, macropulse energy density of $E_M = 0.4-1.0$ J/cm^2 could give $R_d = 80\%$ at the laser absorbed volume, without accompanying ablation.

Furthermore, this ester dissociation can be considered to be the heat confinement effect. As shown by the right vertical axis in Fig. 3, each FEL macropulse was estimated to cause a temperature rise of several hundred deg. C, which causes carbonization when the sample is heated statically. FEL macropulse shorter than the thermal relaxation time raised the temperature of the laser absorbed volume in the sample to dissociate ester bonds, and diffused before sequential reactions to the carbonization during the thermal relaxation time. Thus, in order to increase the ester dissociation ratio without carbonization, it is useful to increase the pulse number and reproduce the

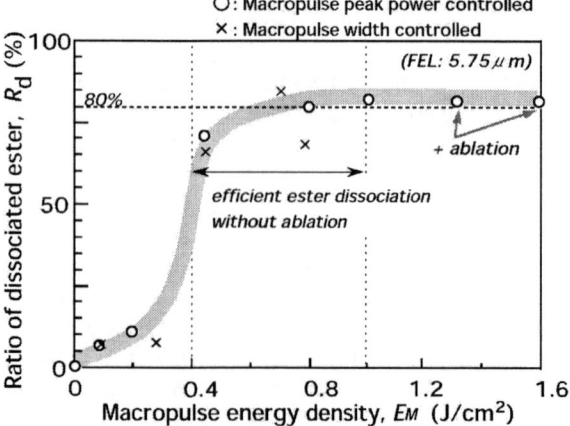

FIGURE 3. Ester dissociation ratio R_d, as a function of macropulse energy density, E_M.

heat confinement effects.

Next, the ester dissociation ratio, R_d, was observed as a function of the incident macropulse number, n. The wavelength was tuned to 5.75 μm, and the macropulse energy was set to E_M = 1.0 J/cm^2. In the region of less than 20 pulses, as n increased, R_d increased up to 80%, and R_d remained unchanged. The dependency on n can be fitted by an exponential function: $R_d = 0.8(1-0.78^n)$. This relationship indicates that within the interacted volume (= laser absorbed volume) of the sample, the remaining cholesteryl oleate was dissociated at the rate of 80(1-0.78) = 18 % by each macropulse. Thus for the laser parameters used here, it was found that the total energy required for the maximum dissociation ratio was of the order of 10^1 J/cm^2, several tens of macropulses.

DISCUSSIONS

The results obtained in these experiments helped us to identify suitable parameters for cholesterol ester dissociation in arteriosclerotic regions. (1) The wavelength was 5.75 μm, which interacts strongly with cholesterol esters, but not with normal tissues. (2) The required energy per macropulse was within the range of E_M = 0.4-1.0 J/cm^2, which caused ester dissociation without ablation. (3) Regarding pulse structure, pulse duration was less than the thermal relaxation time realizing the heat confinement (τ = ~10 μs), and the pulse interval was longer than the thermal relaxation time, allowing heat to diffuse out of the laser absorbed volume (> 1 ms). (4) The pulse number was of the order of several tens, ex. ~20 pulses for 5.75-μm- and 1.0-J/cm^2-FEL, showing that the total energy required was of the order of 10^1 J/cm^2.

In the actual arteriosclerotic regions, cholesterol ester does not exist in the pure condition; it is surrounded by other components of tissue, mainly water. Here, the FEL macropulse induced laser energy confinement within the laser absorbed volume and the heat effect was limited in the area. From these discussions, the surrounding water can be considered not to affect the reaction of ester dissociation but attenuates the laser energy incident on the accumulated cholesterol ester. The effect of this attenuation, we estimated in the following paragraphs.

To discuss the practical treatment of cholesterol oleate removal from the atherosclerotic region, we considered the case in which FEL irradiated the 2-layered-model where the cholesterol ester is covered with normal tissue of thickness D. In the case where the laser irradiation was from outside the normal tissue, i.e. cholesteryl oleate was irradiated via the normal tissue; the following points have to be considered. (1) Whether enough power can be delivered to the lower cholesteryl oleate layer. (2) Whether the temperature rise in the upper normal tissue layer induces any thermal damage. Answers to these questions may vary depending on the incident macropulse power, E_M, and thickness of the normal tissue, D. We briefly estimated the conditions to satisfy these points, using the properties of water for the upper soft normal tissue, of which as much as 80 wt% is water. Here we ignored the absorption of protein, one more main element of normal tissue, since there is little absorption of protein at the wavelength of 5.75 μm [10].

From these estimates, we propose the model of 5.75-μm-FEL-irradiation shown in Fig. 4. The results obtained in this study indicate that effective cholesteryl oleate dissociation could be achieved using 5.75-μm-FEL-irradiation with E_M = 0.5 J/cm^2 via a water layer with D = 13 μm onto cholesteryl oleate. Since the thickness of the endothelial cells is typically 10-20 μm, the light source must be introduced through the blood vessel using a fiber catheter with the edge being held close to the endothelial cells to prevent interference from the blood. This provides a non-invasive method of efficiently initiating cholesterol ester dissociation and removal from arteriosclerotic regions. Further it was also revealed that the ~80% of cholesterol ester in the irradiated region could be dissociated by several tens of macropulses, equal to several seconds in irradiation time.

Using this method, the risk of wrong-irradiation also can be reduced. In the case that there no cholesterol ester is present at the irradiated area, no thermal damage will be caused, since the temperature rise is less than caused in the endothelial cells. The effect of temperature rise of the laser absorbed

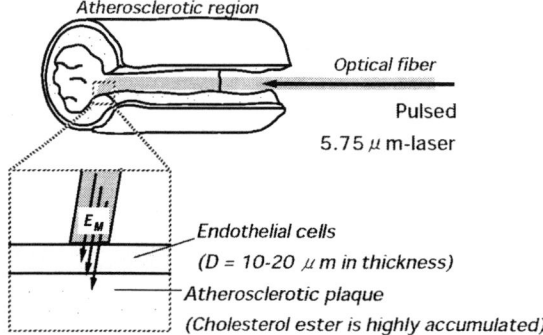

FIGURE 4. Proposed irradiation setup using 5.75-μm-FEL for clinical use.

cholesterol ester can also be ignored, since the heated volume is confined to the marked local area and the amount of heat energy is small compared to the surrounding volume.

When some thermal effect under 100 deg. C is allowed, the geometric limits can be reduced and the requirement of the distance between the accumulated region and the fiber edge can be up to 80 μm. Incidentally, if the request for removal efficiency is given the priority over the noninvasiveness, cholesterol ester can be removed mechanically using both, the ablation of cholesterol ester itself, or water explosive vaporization, which is carried out by the same method using conventional laser, with high energy laser pulse of, $E_M = 1.5$ J/cm^2 and 9.0 J/cm^2, respectively.

CONCLUSIONS

In order to estimate the optimum laser conditions for efficient dissociation of cholesterol ester in an arteriosclerotic region of blood vessels, in this study, cholesteryl oleate was irradiated using 5.75-μm-FELs, which cause vibration of ester bonds. The following factors are important when considering the medical applications of this technique. (1) The irradiation products are non-toxic for the surrounding tissue (2) The temperature at the laser absorbed volume can rise up to hundreds of deg. C without thermal damage to the peripheral tissue. (3) Cholesteryl oleate in the arteriosclerotic region below endothelial cells of ~10 μm thickness can be dissociated up to 80 % by the transmitted 5.75-μm-FEL. Thus we conclude that MIR pulsed-lasers with a wavelength of 5.75 μm can be useful for the removal of cholesterol esters from an arteriosclerotic region of a blood vessel. For future work, irradiation experiments using the samples including the water and protein to confirm the estimation given here, and *in vivo* observation to determine how long it takes for the removal of dissociated ester from the accumulated area using living tissues are required.

REFERENCES

1. Takano, T., Ananuma, K., Ohkuma, J., and Kanasaki, K., *Acta. Histochem. Cytochem.* **19,** 135 (1986).

2. Voet, D., and Voet, J. G., *Biochemistry,* New York, John Wiley & Sons, 1995.

3. Arai, T., Mizuno, K., Kikuchi, M., Sakurada, M., Miyamoto, A., and Kurita, A., *Proc. Annu. Int. Conf. IEEE Eng. Med. Biol. Soc.* **3,** 1598 (1993).

4. Takano, T., and Mineo, C., *J. Phermabio-Dyn.* **13,** 385 (1990).

5. Hayashi, J., Saito, T., and Aizawa, K., *Therapeutic Res.* **16,** 3064 (1995).

6. Saito, T., Hayashi, J., Sato, H., Kawabe, H., and Aizawa, K., *Med. Electron. Microsc.* **29,** 137 (1997).

7. Hayashi, J., Saito, T., and Aizawa, K., *Lasers Surg. Med.* **21,** 287 (1997).

8. Nakano, S., *Cholesterol,* Tokyo, Shinsei-shuppansha, 1997.

9. Brau, C., A., *Free-Electron Lasers,* New York, Academic Press, 1990, p. 51.

10. Awazu, K., Nagai, A., and Aizawa, K., *Lasers Surg. Med.,* **23,** 233 (1998).

11. Fukami, Y., Maeda, Y., and Awazu, K., *Nucl. Instrm. Methods,* **B144,** 229 (1998).

12. Awazu, K., and Fukami, Y., *Nucl. Instrm. Methods,* **A475,** 650 (2001).

13. Fukami, Y., Heya, M., Matsubara, H., and Awazu, K., *Jpn. J. Appl. Phys. Suppl.,* **41,** 132 (2002).

14. Niemz, M., H., *Laser-Tissue interactions,* Berlin, Springer-Verlag, 1996.

Pressure-tuning FTIR Spectroscopy: Applications to Biomedical Research and Diagnosis

Patrick T. T. Wong

Department of Biochemistry, Microbiology & Immunology and Department of Pathology, Faculty of Medicine, University of Ottawa, Ottawa, Ontario Canada K1H 8M5

Abstract. A pressure-tuning FTIR spectroscopic technology for the investigation of the structural and dynamic properties at the molecular level in biological cells and tissues has been developed in our laboratory. This allowed us to study the molecular basis of various biomedical events including structural and dynamic changes of bio-molecules in diseased tissues and cells. After a brief introduction of this technology and a summary of various biomedical applications of this technology, details of the biomedical applications to the study of structural changes in bio-molecules of human tissues during the neoplastic transition and to the screening of human cervical cancer and precancerous lesions including clinical statistics are given.

INTRODUCTION

While molecular biology has made revolutionary contributions towards our understanding of the life and regulation of biological cells, additional approaches will be needed to study biological phenomena with enhanced resolution. Pressure-tuning infrared (FTIR) spectroscopy has been shown to be a powerful method to study not only the structure of molecules but also their relationship to surrounding molecules [1]. We have resolved the technical and methodological problems [2, 3] in sample preparation of biological tissues and cells for optimal spectra acquisition and developed this technology for the study of structural properties at the molecular level in biological tissues and cells. This advance allowed us to investigate the molecular basis of a wide range of biological and biomedical problems.

THE TECHNOLOGY

Pressure-tuning FTIR spectroscopy refers to as FTIR spectroscopy combined with pressure tuning on the infrared spectral features such as band frequencies, absorbance, splitting and resonance. There are many advantages to apply pressure as a variable in this technology. The most important one is that a change in pressure changes only the space available to the molecules without changing their kinetic energy. Consequently, the degree of intermolecular interactions can be directly varied with pressure and thus new knowledge on intermolecular interactions and structural properties can be obtained.

To record pressure-tuning FTIR spectra, samples of tissues or cells are placed into the sample compartment of a metal gasket in a high pressure diamond anvil sample cell [1]. Since the diameter of the sample compartment on the gasket is about 0.5-0.6 mm, the amount of sample for each pressure-tuning FTIR spectroscopic study is extremely small. Relatively weak thermal infrared light source is used in the conventional FTIR spectroscopy. Therefore, a very long data acquisition time and an infrared beam focus mechanism are required to obtain good quality spectra. By the advance of the Synchrotron infrared radiation, the problems arising from all the disadvantages of the conventional infrared light source in the pressure-tuning FTIR spectroscopy technology can be resolved.

Figure 1 shows the FTIR spectrum of a human colon tissue in the frequency region of 950-1800 cm^{-1}. It consists of many bands with different frequencies and intensities, which are the results of transitions from the ground state to the first excited state of vibrations of various functional groups in tissue molecules. The assignment of most of these infrared bands has been well established in the literature by the study of the infrared spectra of isolated DNA, RNA, cell nuclei,

CP716, *Portable Synchrotron Light Sources and Advanced Applications,*
edited by H. Yamada, N. Mochizuki-Oda, and M. Sasaki

skeletal proteins, membrane lipids, and other bio-molecules from tissues and cells [4-15].

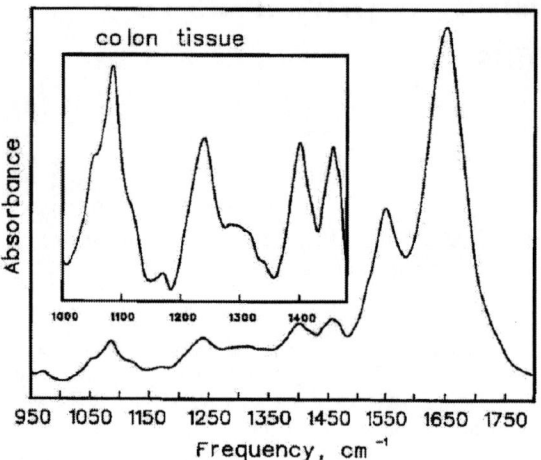

FIGURE 1. Infrared spectra of a colon tissue

TABLE 1. Structural Properties in Bio-molecular Assemblies

- Conformational Structure: Membrane Lipids, Two dimensional solutions of Lipid Bilayer, Proteins, Polypeptide, DNA and RNA
- Orientational Fluctuations
- Interchain Packing and Orientation
- Degree of Interchain interaction
- Location and Strength of Hydrogen bonding
- Mechanism of Structural Phase Transitions
- Nature and Strength of Binding: Protein/Protein, Protein/Lipid, Protein/Metal, Protein/Nucleic Acid, and Ligand/Receptor
- Location and Binding Strength of Exogenous Molecules (including Drugs) in Membrane Lipids, DNA, and Proteins
- Degree of Unsaturation in Membrane Lipids
- Fluidity and Trans/Gauche Population in Membrane Lipids
- Interdigitation in Membrane Lipids
- Micellar to Lamellar conversion
- Reversible and Irreversible Protein Denaturation
- Correlation between Conformational States and Kinetic Process: Hydrogen Isotope Exchange Kinetics, Enzyme Activity Reaction Kinetics
- Salt Bridge Formation and dissociation in Proteins
- Rigidity and Disorder of Conformational Structure in Proteins
- Effects of Ligands on the Molecular structure in Proteins
- Base Interaction in Nucleic Acids

In our laboratory, we have done a significant amount of basic spectroscopic work over the years and identified various spectral parameters such as frequencies, intensities, band shapes and band splitting,

which are related to the structural properties in biological systems [4, 6, 7, 16-21]. A list of structural and dynamic properties in bio-molecular assemblies that we have studied by pressure-tuning FTIR spectroscopy is given in Table 1.

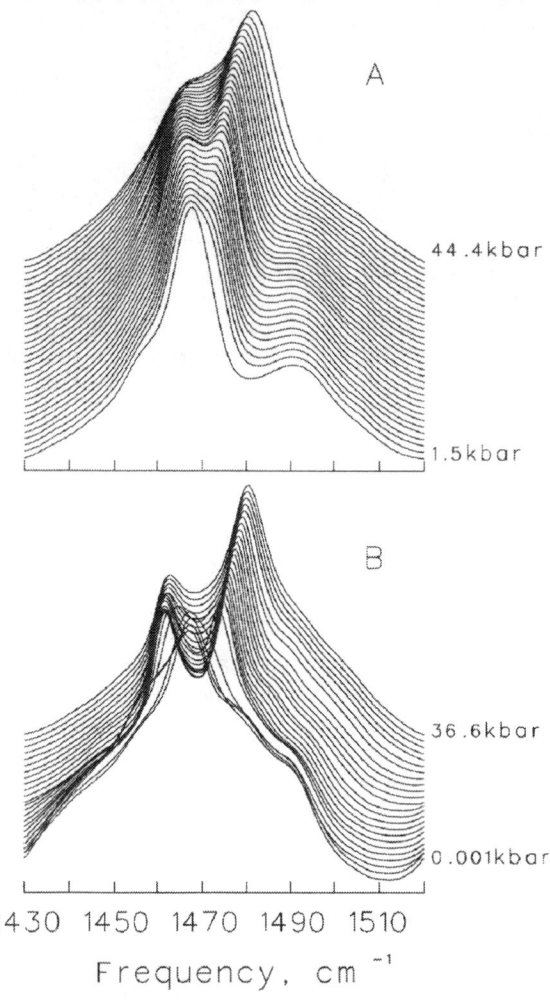

FIGURE 2. Stacked contour plots of the infrared spectra of the δCH2 mode of lipid bilayers.

Structural and dynamic properties of lipid bilayers in biomembranes derived from the pressure-induce correlation field splitting of their vibrational modes are given here to illustrate the correlation between the pressure-tuning spectral parameters and the structural properties in bio-molecular assemblies. Fig. 2 shows the pressure-tuning FTIR spectra of the bending mode of methylene chains in lipid bilayers. The mode undergoes a correlation field splitting into two bands at high pressure. The spectra in Fig. 2A is from a conventional lipid bilayers whereas those in Fig 2B from an interdigitated lipid bilayers.

According to the first order perturbation theory [18], the exciton level of a lipid bilayer with two orientationally nonequivalent methylene chains per unit cell will split into two branches and thus each vibrational mode will split into two with the frequencies given in equations (1) and (2) [18].

The first term in these equations is the potential energy changes of vibrational modes of individual chains arising from the perturbation of neighboring stationed chains. The second and the third terms are the potential energy changes due to the perturbation of intramolecular displacements of orientaionally equivalent and nonequivalent neighboring chains, respectively. The difference of these two frequencies is the correlation field splitting as shown in equation (3), which only contains the third terms in equations (1) and (2). If the orientation of all methylene chains in a lipid bilayer is parallel to each other, the third terms in equation (1) and (2) are zero. Consequently, there is no correlation field splitting in the vibrational modes of methylene chains in this lipid bilayer. Also, there is no correlation field splitting in the spectra in other two circumstances. They are: i) The conformation of the methylene chains is highly disordered due to the presence of a large number of gauche bonds, and thus the coupling of the vibrational modes between neighboring chains is random and weak, and ii) The methylene chains are conformationally ordered and fully extended, but the orientation of these fully extended chains is disordered due to the reorientational fluctuations and the torsion/twisting motions of the chains. In these two circumstances, only shift in frequency and broadening in the spectral bands of methylene chains will be observed.

The perturbation function U is dependent on the interchain distance and relative interchain orientation. Therefore, by following the modification of the correlation field splitting, the structural changes in lipid bilayers of biomembranes can be monitored. It has been demonstrated [18] that the conformational disorder, the reorientational fluctuations and the torsion/twisting motions of methylene chains in lipid bilayers can be ordered and dampened by an external pressure. Consequently, at high enough pressure, these disorder structures can be removed and thus the correlation field splitting in the vibrational modes of the methylene chains would appear provided that the equilibrium orientations of neighboring chains are nonequivalent. For orientationally more disordered chains, a higher pressure is required to stop these disorder fluctuations and motions and thus the correlation field splitting pressure at which the splitting starts to appear is higher. Therefore, the order/disorder dynamics of the methylene chains in lipid bilayers can be determined by the magnitude of the correlation field splitting pressure. Other structural properties in biomolecules can be derived from other spectral parameters in a similar manner.

After we have further developed the pressure-tuning FTIR spectroscopic technology to monitor the structural and dynamic properties of bio-molecules in intact biological tissues and cells [20], we were able to understand the molecular basis of various biomedical problems. Table 2 lists the molecular basis of various biomedical events that were investigated by pressure-tuning FTIR spectroscopy in our laboratory [22 and references therein]. In the present paper, only some results of biomedical research on human cancers are given.

$$\nu_{\mathrm{I}} = \nu_0 + \frac{1}{8\pi^2 \nu_0 c} \left(\sum_{j=1}^{2N} \left(\frac{\partial^2 U_0}{\partial q_j^2} \right)_0 + \sum_{j=1}^{N} \left(\frac{\partial^2 U_1}{\partial q_a \partial q_{a_j}} \right)_0 + \sum_{j=1}^{N} \left(\frac{\partial^2 U_2}{\partial q_a \partial q_{b_j}} \right)_0 \right) \quad (1)$$

$$\nu_{\mathrm{II}} = \nu_0 + \frac{1}{8\pi^2 \nu_0 c} \left(\sum_{j=1}^{2N} \left(\frac{\partial^2 U_0}{\partial q_j^2} \right)_0 + \sum_{j=1}^{N} \left(\frac{\partial^2 U_1}{\partial q_a \partial q_{a_j}} \right)_0 - \sum_{j=1}^{N} \left(\frac{\partial^2 U_2}{\partial q_a \partial q_{b_j}} \right)_0 \right) \quad (2)$$

$$\Delta \nu = \frac{1}{4\pi^2 \nu_0 c} \sum_{j=1}^{N} \left(\frac{\partial^2 U_2}{\partial q_a \partial q_{b_j}} \right)_0 \quad (3)$$

TABLE 2. Biomedical Applications

- Liver Cirrhosis
- Infected Thymus Gland, Salivary Gland, and Hela Cells
- Aging Tissues
- Griseofulvin Induced Liver Tumor in Mice
- Effect of Carcinogen and Hepatectomy on Liver
- Anesthetic Reactions with Nerve Membranes
- Function of Toxins on Biomembranes
- Protein Aggregation in Rosenthal Fiber of Brain Diseases and in Mallory Body of Liver Diseases
- Structural Changes at the Molecular Level in Human Cancers: Colon, Liver, Cervix, Vagina, Endometrium, Ovary, Stomach, Skin, Esophagus, Breast, Brain, Lung and Testis
- Mechanism of Anti Cancer Drug Hexadecyl-Phosphocholine, Mitoxantrone and Porphyrin Derivatives
- Resistant and Sensitive Cancer Cells and Tumors to Cisplatin and to Radio-Therapy
- Premalignant Lesions: Adenoma of Colon, dysplasia of Cervix and Barrett's Esophagus
- Cervical Tissues and Cells with various diseases and Conditions
- Mechanism of Photodynamic Therapy for Cancers
- Neuroblatoma
- Uptake of Anti Cancer Drugs in Kidney
- Calcium Binding on the Human reproductive Function of Sulfogalactosylceramid in Sperm
- Pressure Effects on Hemostasis and Blood Coagulation
- Mechanism of Side effects by Anti Osteoporosis Drugs.

HUMAN CANCER RESEARCH

Fig 3 shows the infrared spectra of normal and malignant tissues of human cervix in the frequency region 950-1500 cm^{-1}. It is evident that the spectrum of cervical tissue changes dramatically from normal tissue to cancer tissue. Many other spectral changes are also observed in other regions of the spectra [6, 23]. The infrared spectra of normal and caner tissues of other human organs are similar to those of the cervix, except for the strong band at 1025 cm^{-1} in the spectrum of normal cervical tissue. This band is due to the vibrational mode of glycogen. Therefore, this band is only observed in the spectra of glycogen rich tissues.

The dramatic changes observed in the cancer spectra, reflects significant changes in the structural properties at the molecular level in cancer, which can be derived by pressure-tuning FTIR spectroscopic study. Table 3 lists the structural changes which are common to all cancers that we have studied to date.

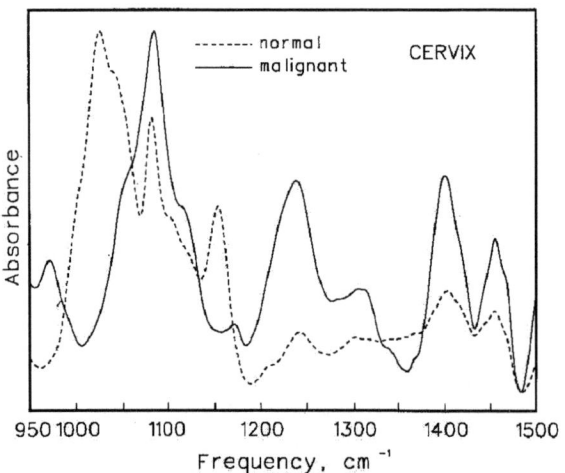

FIGURE 3. Infrared spectra of normal and malignant cervical tissues.

For illustration, the structural changes in the nucleic acids in cancer cells derived from the changes in the asymmetric P-O stretching band are discussed here. The spectra of the infrared band at 1240 cm^{-1} of normal and malignant samples are enlarged and superimposed in Fig. 4. Fig. 4A shows the original spectra and Fig. 4B shows the corresponding third power derivative spectra [24]. It is evident in Fig. 4 that this band consists of two overlapping bands at 1240 cm^{-1} and 1224 cm^{-1} and the band intensity at 1224 cm^{-1} is extremely week in the spectrum of a normal sample whereas it becomes much stronger in the spectrum of a malignant sample. This spectral feature is common to all types of cancers that we have studied.

It is well known that the frequency of this phosphate band of nucleic acid is very sensitive to the interaction between the phosphate groups and water molecules [8, 9, 11]. The frequency of this band is at about 1220 cm^{-1} when the phosphate group is fully hydrogen-bonded to water and is at about 1240 cm^{-1} when it is anhydrous. Therefore, in the malignant cells the number of hydrogen-bonded phosphate groups of nucleic acids is dramatically increased as indicated by the increase in the intensity of the 1224 cm^{-1} band.

TABLE 3. Structural Changes in Cancers

- Increase in the Nuclear Size
- Increase in the Number of Hydrogen-Bonded Phosphodiester Groups in DNA
- Increase in the Intermolecular Interaction and Packing in Nucleic Acids
- Increase in the Conformational and Reorientational disorder in the Methylene Chains of Membrane Lipids
- Increase in the Membrane Fluidity
- Decrease in the Methyl-to-Methylene Ratio
- Decrease in the Number of Hydrogen-Bonded C-O Groups in Carbohydrates and Protein Residuals
- Changes in the Conformational Structure in Cytoskeleton Proteins
- Increase in the Ordering and rigidity in the Structure of Cytoskeleton Proteins

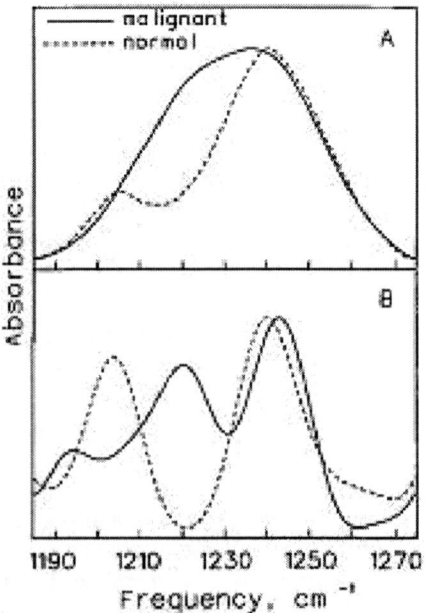

FIGURE 4. Infrared spectra of cervical tissues in the asymmetric phosphate stretching region. (A) Original spectra from a normal and a malignant sample. (B) Corresponding third-power derivative spectra with a break point of 0.35.

The pressure dependences of frequencies of these two component bands are shown in Fig 5. The frequency of the 1224 cm^{-1} band decreases with increasing pressure, which provides further evidence that this band is due to the P-O mode of hydrogen-bonded phosphate backbone of nucleic acids.

This pressure behavior of hydrogen-bonded molecules could be illustrated in Fig. 6. When an oxygen atom is hydrogen-bonded to a hydrogen atom of water, the electron density of the lone pair electrons on the oxygen atom migrates towards the hydrogen atom. As a result, the electron density between the oxygen atom and the phosphorous atom would migrate towards the oxygen atom. Consequently, the electron density between the oxygen and the phosphorous atoms and thus the P-O bond order is decreased. This results in a weakening of the bonding and the force constant between the oxygen atom and the phosphorous atom. Therefore, the stretching frequency of the P-O groups is decreased.

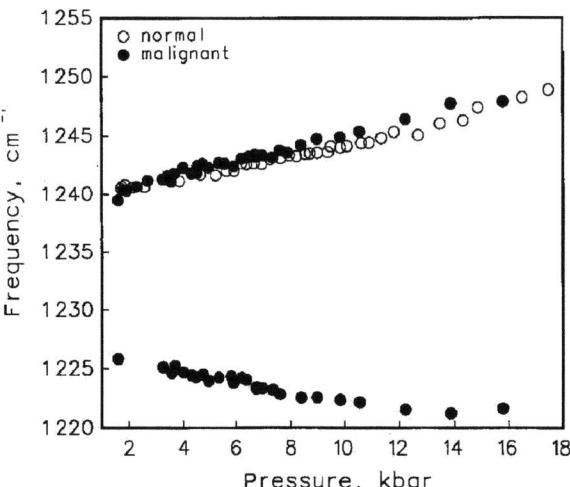

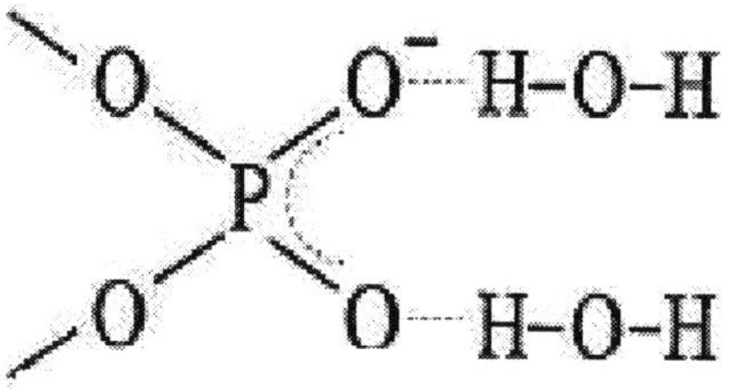

FIGURE 6. Hydrogen–bonds between PO_4^- and H_2O.

SCREENING OF CERVICAL CANCER AND PRECANCEROUS LESIONS

An important and practical biomedical application of the FTIR spectroscopy is the screening of cervical cancer and precancerous lesions, In this application, only IR spectra at atmospheric pressure is sufficient and pressure-tuning is not required. A special sample holder has been designed for this study [25], in which the sample compartment is about 5 mm in diameter. This development made it possible to obtain extremely high quality spectra within very short period of spectral acquisition time.

The present screening tool for cervical abnormalities is the Papanicolaou smear test (Pap test). Due to the wide-spread cervical screening program in developed countries by the Pap test, the mortality from cervical carcinoma has been greatly reduced in recent year in these countries. The Pap test is a 60 years old method that relies on the human eye to detect the morphological changes of cells under a microscope. It is the predominant approach in both developed and developing countries, and relies heavily on trained cytopathologists and cytotechnicians to manually inspect the cells. The process is tedious and labor intensive as only a handful of abnormal cells are present in each positive Pap Smear sample from among hundred of thousands of cervical cells. Lastly, the test is ultimately limited by the subjective nature of visual examination of morphological changes. As a result, it has a mean sensitivity of 58% and a mean specificity of 69%, according to the World Health Organization [26].

Computer-assisted Pap smear tests, such as the Papnet test, have been developed with the aim of reducing the false results of the Pap smear test. The Papnet test screens most of the morphologically normal cervical cells in a Pap smear slide by a computerized image analysis and then identifies the 128 most suspicious cells with neural network software for subsequent rescreening by cytotechnologists and cytopathologists. There are serious problems in the Papnet test and other computer-assisted Pap smear tests.

FIGURE 5. Pressure dependence of the frequencies of the asymmetric phosphate stretching mode. The band from malignant tissue has two component bands.

Pressure is known to increase the strength of hydrogen-bonding. Therefore, the frequency of this mode is further decrease with increasing pressure. The increase in the frequency of the 1240 cm^{-1} band with increasing pressure in Fig. 5 is the result of pressure enhanced intermolecular interactions and is determined by the first and second terms in equation (1) and (2).

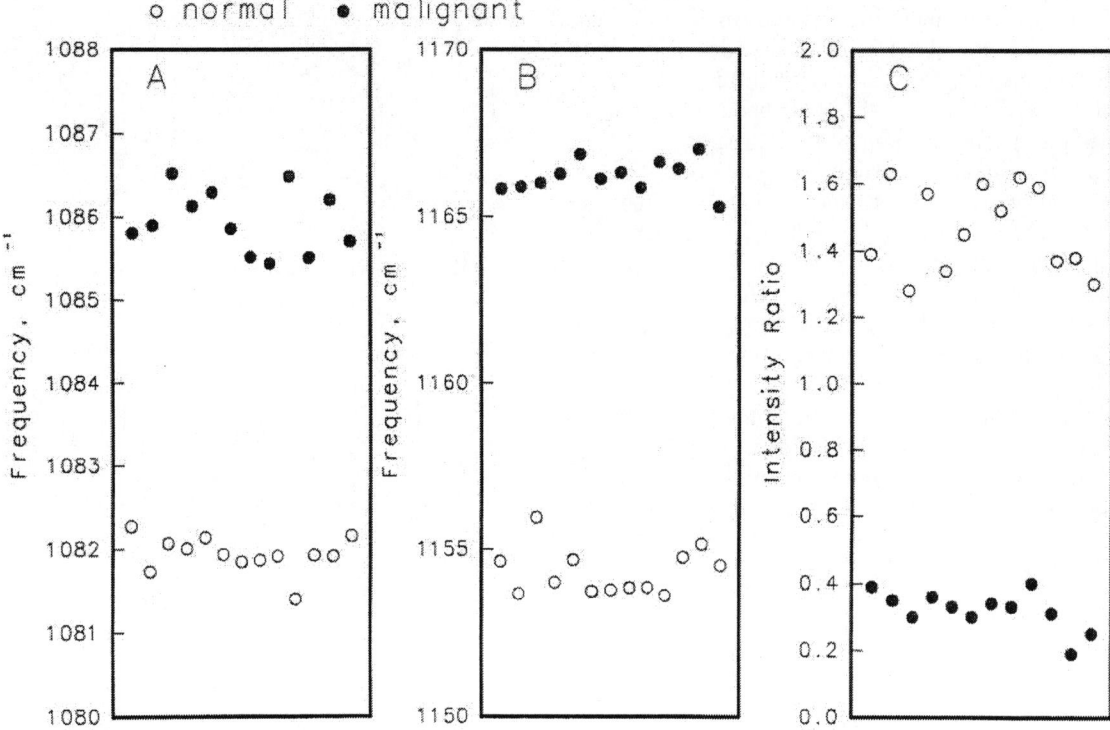

FIGURE 7. Three spectral parameters in normal and malignant cervical samples. Intensity ratio refers to the ratio of peak intensity of 1025 cm^{-1} and 1082 cm^{-1} bands. Frequency in A is the peak position of the symmetric P-O stretching band and that in B is the peak position of the C-O stretching band. Each sample was obtained from a single patient.

These methods fail to identify and isolate the images of diagnostic cells, which are clustered and superimposed Pap smear slides. This occurrence is very common in Pap smear slides with abnormal cells. It is difficult to establish a prior set of algorithmic rules to deal with an infinite variety of images for each category of cells usually seen in cervical smears. Under or over-staining and improper dehydration in the preparation of Pap smear slides will give rise to false errors. Another major drawback of the computer-assisted Pap smear tests is that they are too expensive to be used as a primary cervical screening program for the general public. Currently, it is only used in few countries for secondary screening to check Pap smears that have already been examined by an experienced cytotechnologist working under the supervision of a cytopathologist.

In many developing countries, there is no cervical cancer screening program due to the shortage of well trained cytolopathologists and cytotechnologists. Therefore, it is now realized that additional approaches for the screening of cervical cancer and precancerous lesions are needed

The infrared spectra of cervical cells in Fig. 3 show clear cut differences between normal and cancerous cervical specimens [6]. Several selective spectral parameters of a series of normal and malignant cervical cells specimens (the symmetric P-O stretching frequency, the C-O stretching frequency and the intensity ratio between the 1024 cm^{-1} and 1082 cm^{-1} bands) are shown in Fig. 7. It is clear that cervical cancer can be easily detected by these infrared spectral parameters. There are many other spectral parameters in the spectra also show clear cut differences between normal and invasive carcinoma of the cervix. Therefore, infrared spectroscopy is undoubtedly a powerful method to detect cervical cancer. However, in addition to cancer, there are many other lesions and diseases, such as the precancerous lesion, dysplasia, various infections and inflammations. In our studies, we found that the infrared spectra of cervical specimens with precancerous lesion and other diseases and conditions also show different spectral parameters from those of normal and cancer cervical specimens [6, 27].

TABLE 4. Statistics based on biopsy as gold standard

	Cancer	HSIL	LSIL	Negative (Inflammation)	Negative`
Histology Positive (217)	12	98	107	0	0
Histology Negative (84)	0	0	0	67	17
Pap Test Positive (196)	9	87	92	1(HSIL), 3(LSIL), 3(ASCUS)	1
Pap Test Negative (105)	3	11	15	60	16
Infrared Positive (215)	12	98	104	0	1
Infrared Negative (86)	0	0	3	67	16

Pap Test / Infrared	False Negative	Sensitivity	False Positive	Specificity
Pap Test	13.4 %	86.6 %	9.5 %	90.5
Infrared	1.4 %	98.6 %	1.2 %	98.8 %

TABLE 5. Infrared spectroscopic statistics based on Pap smear test as gold standard

Pap Test	WNL 530				WNL (S. Infl.) 126			
	NL	WNL	S. Infl.	Positive	NL	WNL	S. Infl.	Positive
FTIR Test	170	261	89	10	22	51	36	17

FTIR Test	No. False Positive IR	False Positive Rate	Specificity
	27	4.1%	95.9%

NL: Normal. WNL: Within Normal Limit. S. Infl.: Severe Inflammation

Therefore, our studies have demonstrated that FTIR spectroscopy is a promising tool for the screening of anomalies in the cervix. After we resolved the problems arising from several confounding factors, such as the effects of cellular degradation, polymorphs, columnar cells, metaplastic cells, mucus, red blood cells and others [28], we have obtained excellent clinical statistics of the FTIR technology. Both the sensitivity and specificity of the FTIR spectroscopy test are much better than the Pap test [28, 29]

Analysis of the cervical specimens of 301 patients for Pap test and FTIR spectroscopy relative to the gold standard of histology of biopsy is shown in Table 4. The sensitivity and specificity were 86.6 and 90.5% for Pap test and 98.6 and 98.8 for FTIR spectroscopy. The false-negative rate and negative predictive values between the two were significantly different in favor of FTIR spectroscopy. The Pap test showed a false negative of 13.4%, a false positive of 9.5%, a negative predictive value of 72.3%, and a positive predictive value of 95.9%. The FTIR spectroscopy showed a false negative of 1.4%, a false positive of 1.2%, a negative predictive value of 96.5%, and a positive predictive value of 99.5%.

Although the false positive rate for FTIR spectroscopy was much better than that for Pap test in this study, the values of false positive rate from this study was not accurate due to the small sample size of the positive specimens. In another study, 656 cases of negative Pap test were compared with their FTIR spectroscopic results in Table 5. The false positive rate and the specificity of the FTIR spectroscopy were 4.1% and 95.9%, respectively, assuming that there was no false negative in the Pap test of these 656 cases.

The FTIR spectroscopy technology detects the structural changes at the molecular level of cells and tissues, and thus provides greater sensitivity and specificity, resulting in a significantly improved accuracy. Moreover, many studies in our laboratory and other laboratories [30] have demonstrated that structural changes at the molecular level occur earlier than morphological changes in abnormal cervical cells. Therefore, the FTIR technology can detect anomalies in cervical cells before changes in cell morphology detected by the traditional methods. There were no affordable methods to conduct cervical cancer screening of the general population in developing countries, given the lack of trained cytopathologists and cytotechnologists, and complex supporting infrastructure required. Lower cost and no specialized skills required to apply the FTIR technology enable some developing countries to start a cervical cancer screening program by means of the FTIR technology.

REFERENCES

1. P. T. T. Wong, "Vibrational Spectroscopy Under High Pressure," in Vibrational Spectra and Structure, edited by J. R. Durig, Elsevier, New York, 1987, pp 357-445.
2. P. T. T. Wong, US Patent No. 4,970,396, (1990).
3. P. T. T. Wong, US Patent No. 4,980,551, (1990).
4. B. Rigas, S. Morgello, I. S. Goldman, and P. T. T. Wong. Roc. Ntl. Acad. Sci. USA 87, 8140-8144 (1990).
5. P. H. Takahashi, S. W. French, and P. T. T. Wong. Alcoh. Clin. Exp. Res. 15, 219-223 (1991)
6. P. T. T. Wong, R. K. Wong, T. A. Caputo, T. A. Godwin, and B. Rigas. Ntl. Acad. Sci. USA 88, 10988-10992 (1991).
7. M. Fung Kee Fung, M. K. Senterman, N. Z. Michael, S. Lacelle and P. T. T. Wong. Biospectroscopy, 2, 155-165 (1996).
8. H. Susi, "Infrared Spectra of Biological Macromolecules and Related Systems" in Structure and Stability of Biological Macromolecules, edited by S. N. Timashell, and C. D. Fasman, Marcel Dekker, New York, 1969, pp. 575-663.
9. Parker, FS, Application of Infrared Spectroscopy in Biochemistry, Biology and Medicine, 1971 Plenum, New York.
10. U. P. Fringeli and HsH, Gunthare, "Infrared Membrane Spectroscopy" in Membrane Spectroscopy, edited by E. Grell, Springer-Verlag, New York, 1981, pp 270-330.
11. F. S. Parker, Application of Infrared, Raman and Resonance Raman spectroscopy in Biochemistry, 1983, Plenum, New York.
12. J. M. Sanchez-Ruiz,, and M. A. Martinez-Carrion, Biochem. 27, 3338-3342 (1988).
13. J. H. Crowe, F. A. Hoekstra, L. M. Crowe, T. J. Anchordoguy and E. Drobins, Cryobiology 26, 76-84 (1989).
14. M. S. Cadrin, W. French, and P. T. T. Wong,, Exp. Mol. Pathol. 55, 170-179 (1991).
15. P. T. T. Wong, E. D. Papavassiliou, and B. Rigas, Appl. Spectrosc. 45, 1563-1567 (1991).
16. P. T. T. Wong, M. Cadrin and S. W. French. Experimental and Molecular Pathology 55, 269-284 (1991).
17. P. T. T. Wong, S. M. Goldstein, R. C. Grekin, T. A. Godwin, C.Pivik and B. Rigas. Cancer Research 53, 762-765 (1993).
18. P. T. T. Wong. Biophysics J., 66, 1505-1514 (1994).
19. P. T. T. Wong, "High-pressure Studies of Biomembranes by Vibrational Spectroscopy", in High Pressure Chemistry and Biochemistry, edited by R. Van Eldik and J. Jonas, D.Reidel Publishing Co., Boston, 1986, pp 381-400.
20. P. T. T. Wong, "High Pressure Vibrational Spectroscopic Studies of Aqueous Biological systems: From Model Systems to Intact Tissues" in High Pressure Chemistry, Biochemistry and Materials Science, edited by R. Winter and J. Jonas, Kluwer Academic Publishers, Boston, 1992, pp 511-543.
21. P. T. T. Wong, "Correlation Field Splitting of Chain Vibrations: Structure and Dynamics in Lipid Bilayers and Biomembranes: in High-Pressure Effect in Molecular Biophysics and Ensymology, edited by J. L. Markley, D. B. Northrop and C. A. Royer, Oxford University Press, New York, Oxford, 1996, pp 256-273.
22. P. T. T. Wong, Can J. Appl. Spectrosc. 40, 143-151 (1995).
23. P. T. T. Wong, R. Wong and M. Fung Kee Fung, Appl. Spectrosc. 47, 1058-1063 (1993).
24. D. G. Cameron, D. J. Moffatt, J. Testing Eval., 12, 78-85 (1984).
25. M. K. Wong and P. T. T. Wong, US Parent No. 5,463,223 (1995).
26. R. Sankaranarayanan, A. Budukh, and R. Rajkumar, Bulletin of the World Health Organization. 79, 954-962 (2002).
27. H. M. Yazdi, M. A. Bertrand, and P. T. T. Wong, Acta Cytologica, 40, 664-668 (1996).
28. P. T. T. Wong, M. K. Senterman, P. Jackli, R. K. Wong, S. Salib, C. E. Cambell, R. Feigel,

W.Faught, M. Fung Kee Fung, Biopolymers (Biospectroscopy) **67**, 376-386 (2002).

29. M. Fung Kee Fung, M. K. Senterman, P. Eid, W. Faught, N. Z. Mikhael, P. T. T. Wong, Gynecol. Oncol. **66**, 10-15 (1997).

30. S. R. Lowry, Cellular and Molecular Biology, **44**, 169-177 (1998).

Imaging of the atheroma in a frozen-section of human atheroscleotic tissue by FT-IR microscopy

Norio Miyoshi, Tetsushi Yamada*, Toru Ogawa*, Ken-ichi Akao**, Takao Nanba***

Department of Tumor Pathology and

*Dentisty and Oral Surgery, University of Fukui, Matsuoka, Yoshida-gun, Fukui, 910-1193,

** Department of Development, JASCO, Ishikawa-cho, Hachio-ji city, Tokyo, 192-8537,

***Department of Physics, Faculty of Science, Kobe University, Kobe, Japan.

Abstract. There is a long history of standard H. & E. staining for pathological diagnostic samples, where the organism is fixed with formaldehyde. However, one should not be neglect that the pathological diagnostic sample has an artificial, secondary protein structure, differing from the structure of just frozen-sectioned organism. We observed quantitatively the cholesterol distribution in the atheroma of a frozen-section of autopsied aorta by imaging, for the first time, using the Fourier transform infrared microscopy (FT-IRM) technique. We detected the cholesterol distribution and found that the protein structure changed to a β–sheet conformation, from an α–helix conformation on the inside edge of the calcified area.

INTRODUCTION

Recently, FT-IRM has been applied[1-4] to observe calcified tissues. The amide-I peak (band at 1655 cm^{-1}) of the spectrum of proteins, lies in the N-H stretching and vibrational mode region of the spectrum (i.e., the molecular finger prints region) and is a reflection of the protein's secondary structure which is altered by calcification in the aorta.

EXPERIMENTS

FTIRM has been used to biologically image a frozen section (10 μm thikness) of the autopsied aorta containing both healthy and atheroscleotic tissue. This was accomplished by using FT-IRM to detect; the protein structure throughout the section, cholesterol (C-H vibrational modes of CH_2 and CH_3 at 2728-2991 cm^{-1}) accumulated in the atheromatous tissue, and phosphodiester (stretch mode of phospholipids at 1086 cm^{-1}) typically accumulated in the calcified region of the atheromatous tissue.

RESULTS AND DISCUSSION

We have imaged a series of frozen sections of the aorta using FT-IRM mapping of an untreated section by integrating the area under the peak 2728-2991 cm^{-1} in Fig.1C, Curve B. Data was collected over an area equivalent to 21x21=441 points, with an aperture size of 40x40 mm^2 per point (Fig. 2A). FT-IRM images were compared to those obtained from microscopy of sections stained by either H. & E. (Fig. 1A) or Sudan-IV (Fig. 1B).

The cholesterol derivative calcified at the aorta was determined to be a cholesterol oleiniate (standard sample; Fig.1C; Curve A) by comparing the FT-IR spectrum of this aorta with the inside of a normal one (Fig.1C; Curve E). The FT-IRM image of the aorta (Fig.2A) clearly shows the concentration dependent distribution of cholesterol in the aorta. This image correlates well with that obtained using Sudan-IV staining (Fig.1B) without the paraffin block, however, unlike the FT-IRM image, the staining is not quantitative.

Furthermore, we used an analytical program (IR-SSE; JASCO Co. Ltd.) for the determination of the protein secondary structure to obtain mapping spectra in order to detect the conformational difference of the

CP716, *Portable Synchrotron Light Sources and Advanced Applications*,
edited by H. Yamada, N. Mochizuki-Oda, and M. Sasaki
© 2004 American Institute of Physics 0-7354-0195-0/04/$22.00

secondary structure of the protein in the aorta tissue. The results of the conformational information, shown as a relative percentage, were plotted as a function of spatial coordinates to create images that depend on the kind of secondary structure, as shown in Fig.2B. It was shown that the a-helix percentage increased in the area of high cholesterol and that there is an extremely high distribution of the b-sheet conformation on the inside edge of the calcified tissue (Fig.2C at 1086 cm^{-1}) as shown in Fig.2D. This suggests that the protein conformation on the inside edge of the aorta was damaged by calcification of phospholipids and fibrosis.

In the future, it is necessary to build a bridge between pathological diagnosis and an endoscopic diagnosis. Observing viable organs by "molecular finger-print technologies" such as FT-IR and Raman spectroscopy, allows diagnosis of samples in the infrared region and is a first step in that direction.

ACKNOWLEDGEMENT

This work was supported by a Grant-in-Aid for Scientific Research (C) (11672293) and (B) (14370793) from Japan Society for the Promotion of Science (JSPS) Foundation. Furthermore, it was also supported by R & D Promotion of Research from Hokuriku-Sangyou-Kattuseika- Center in Kanazawa.

Fig. 1A.   **Fig. 1B.**

Fig. 1C

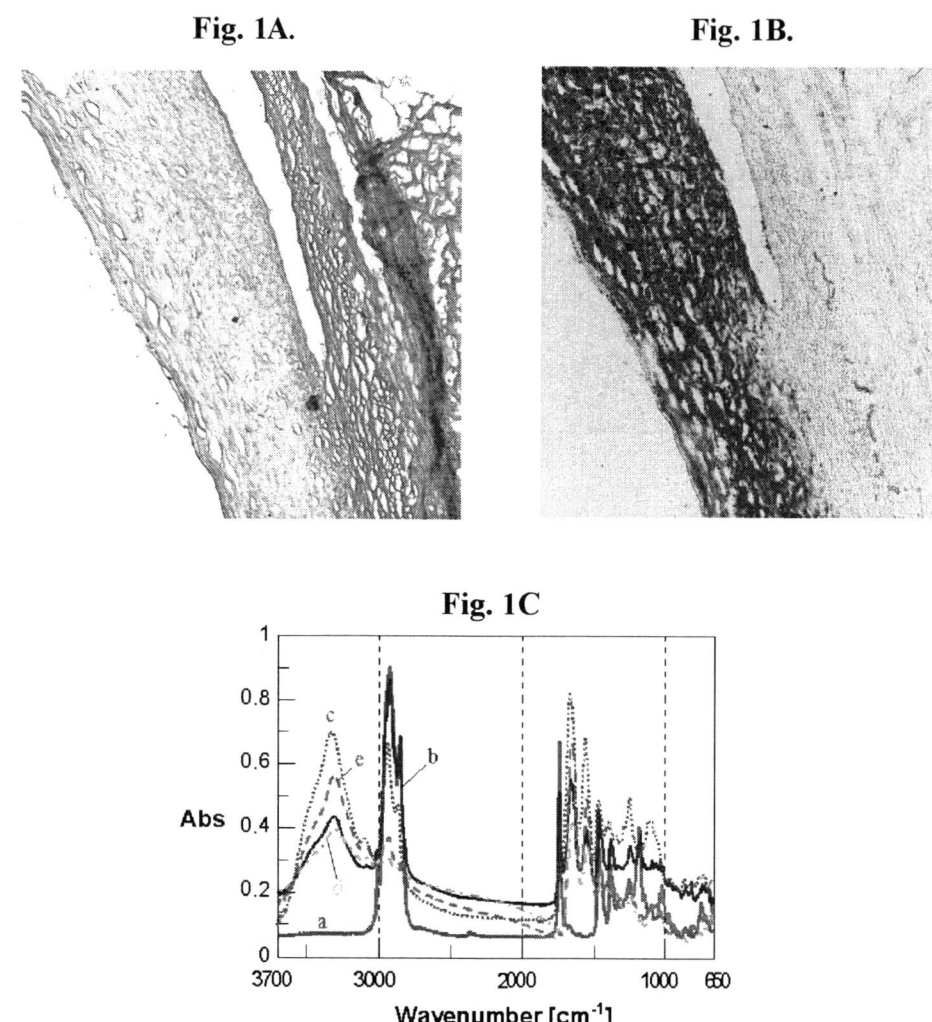

Fig.1: HE **(A)** and Sudan-IV **(B)** staining images and FTIR spectra **(C)** of the calcified and normal areas. Curves **a:** standard cholesterol; **b:** calcified area of the aorta; **c:** inside wall of the calcified aorta; **d:** periphery of the calcified aorta and **e:** inside of the normal aorta (not shown in the mapping), respectively.

Fig. 2A

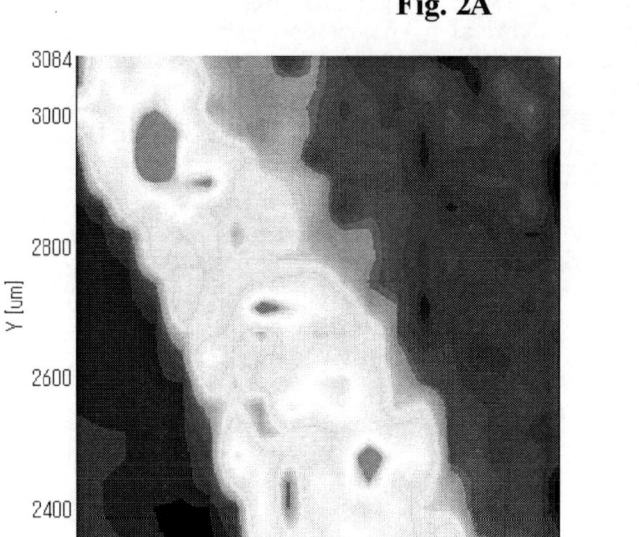

Fig. 2B

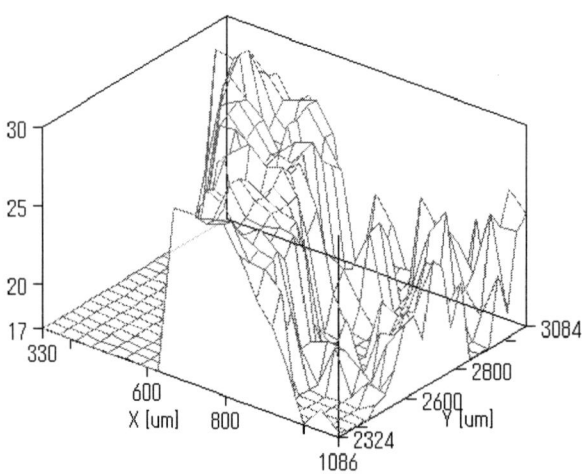

Fig.2: FT-IRM imaging of calcified aorta. **(A)** 2D image of C-H stretching mode of CH_2 (2955 cm^{-1}) of cholesterol; **(B)** Plotting 3D Image of α–helix component (%) of the protein secondary structure analyzed by our estimation program with curve-fitting for amide-I absorption band against the same X-Y image A. (C) 2D image of P-O stretching-vibrational mode (1086 cm^{-1}) for phospholipids; (D) β-sheet component (%) of the protein secondary structure analyzed by our estimated soft curve fitting for amide-I absorption band against the same X-Y image C, respectively.

Fig. 2C

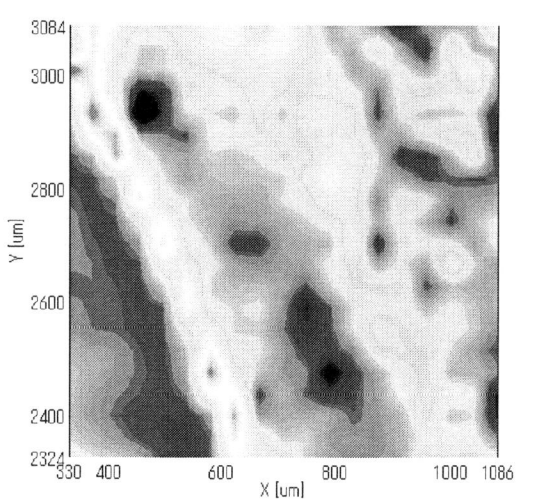

Fig. 2D

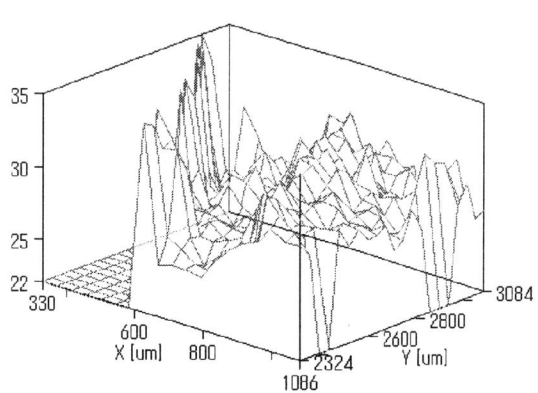

Fig.2: FT-IRM imaging of calcified aorta. (A) 2D image of C-H stretching mode of CH₂ (2955 cm⁻¹) of cholesterol; (B) Plotting 3D Image of α–helix component (%) of the protein secondary structure analyzed by our estimation program with curve-fitting for amide-I absorption band against the same X-Y image A. (**C**) 2D image of P-O stretching-vibrational mode (1086 cm⁻¹) for phospholipids; (**D**) β-sheet component (%) of the protein secondary structure analyzed by our estimated soft curve fitting for amide-I absorption band against the same X-Y image C, respectively.

REFERENCES

1 R. Manoharan J. J. Baraga, R. P. Rava, R. R. Dasari, M Fitzmaurice, M. S. Feld, *Atherosclerosis.* **103,** 181-193 (1993).

2 J. Hayashi, T. Saito, K. Aizawa, *Lasers Surg. Med.* 21, 287-293 (1997).

3 W. Volker, A. Schmidt, E. Buddecke, *Atherosclerosi*s. **77,** 117-130 (1989).

4 N. Miyoshi, T. Ogawa, T. Yamada, K. Akao, K. Abe, F. Kaneuchi, K. Sano. *ITE Letters Batteries New Tech & Med.* **1,** 154-156 (2000).

5 J. J. Wang, C. W. Chi, S. Y. Lin, Y. T. Chem, *Anticancer Res.* **17,** 3473-3478 (1997).

Current Status and Perspectives of Hyperthermia in Cancer Therapy

Masahiro Hiraoka, Yasushi Nagata, Michihide Mitsumori, Masashi Sakamoto, Shin-ichiro Masunaga

Department of Therapeutic Radiology and Oncology, Kyoto University Graduate School of Medicine, 54 Shogoin Kawaharacho, Sakyo-ku, Kyoto 606-8507, Japan.

Abstract. Clinical trials of hyperthermia in combination with radiation therapy or chemotherapy undertaken over the past decades in Japan have been reviewed. Originally developed heating devices were mostly used for these trials, which include RF (radiofrequency) capacitive heating devices, a microwave heating device with a lens applicator, an RF intracavitary heating device, an RF current interstitial heating device, and ferromagnetic implant heating device. Non-randomized trials for various cancers, demonstrated higher response rate in thermoradiotherapy than in radiotherapy alone. Randomized trials undertaken for esophageal cancers also demonstrated improved local response with the combined use of hyperthermia. Furthermore, the complications associated with treatment were not generally serious. These clinical results indicate the benefit of combined treatment of hyperthermia and radiotherapy for various malignancies. On the other hand, the presently available heating devices are not satisfactory from the clinical viewpoints. With the advancement of heating and thermometry technologies, hyperthermia will be more widely and safely used in the treatment of cancers.

INTRODUCTION

During the past two decades, hyperthermia in combination with radiotherapy or chemotherapy has been investigated basically and clinically as a new cancer treatment modality (1). Numerous biological experiments demonstrate strong biological rationale for the use of hyperthermia in cancer therapy (2). Contrary to those promising results for biology, physics and engineering of hyperthermia including the delivery of heat and the measurement of temperature is not sufficiently developed and still offers great limitations for clinical application of this new treatment modality. Nevertheless, clinical experiences with combined locoregional hyperthermia and radiation therapy and/or chemotherapy have been rapidly accumulating over the past decade. In Japan, clinical application of hyperthermia has been intensively carried out. There are special features in those Japanese trials. First, they employ heating devices which have been developed in Japan. Secondly, trials for deep-seated tumors are more common than those for superficial tumors. Thirdly,

most trials are phase I, II trials, and phase III randomized trials have rarely been undertaken. At Department of Therapeutic Radiology and Oncology of Kyoto University Graduate School of Medicine, the clinical application of locoregional hyperthermia started in 1979 and more than 800 patients with a malignant tumor have been treated. In this article, we review current status of clinical hyperthermic trials in Japan emphasizing our experiences.

HEATING METHOD

Hyperthermia is basically divided into whole-body hyperthermia and locoregional hyperthermia depending the volume heat energy delivered. Locoregional hyperthermia is furthermore classified into external, intraluminal, and interstitial heating. External heating methods of locoregional hyperthermia include external heating with ultrasound or electromagnetic waves using microwave and radiofrequency(RF) devices, and hot fluid irrigation in the cavity such as urinary bladder, thoracic cavity and

CP716, *Portable Synchrotron Light Sources and Advanced Applications,*
edited by H. Yamada, N. Mochizuki-Oda, and M. Sasaki

pelvic cavity. The problem is that each technique has its own inherent advantages and disadvantages, and none of them are not satisfactory from the clinical viewpoints. External heating administers heat to the tumor through various body structures and the ideal heating device should be capable of raising the whole tumor volume to a therapeutic temperature without overheating adjacent normal tissues. One possible method for the selective heating of deep-seated tumors is the use of focused ultrasound. Major problems of ultrasound are that it is reflected at the soft tissue/air and bone/soft tissue interfaces and is attenuated rapidly in bone; both these factors make ultrasound difficult to apply to tumors adjacent to air or bone. However, a preliminary experience with a newly-developed ultrasound has indicated some promise for the treatment of deep-seated tumors. The most commonly used deep heating method is so-called regional heating. Since regional heating techniques apply energy to the adjacent deep-seated tumors in an unfocused manner, energy is also delivered to the adjacent normal tissues. Under such conditions, selective heating of tumors over the normal tissue is only possible when heat dissipation by blood flow is more predominant in the normal tissue. An annular phased array system delivering 60-80 MHz electromagnetic waves and RF capacitive heating apparatus are examples of regional heating devices. The former system has the advantage in that subcutaneous fat is not excessively heated and it is thus suitable for obese patients. However, this method causes systemic symptoms such as tachycardia and malaise which result from the use of large-sized applicators. Systemic stress is reported to be more severe in patients with abdominal tumors than in those with pelvic tumors or tumors of the extremities (3), suggesting the limited usefulness of this heating modality for tumors in the upper abdomen. RF capacitive heating is the most widely used method for the external hyperthermia of deep-seated tumors. An capacitive heating device (Thermotron RF8, Yamamoto Vinita Co., Osaka, Japan) have been developed with a grant from Research Development Corporation of Japan (4). The size of the electrodes to be used is determined according to the size and location of the tumor, and a pair of equal-sized electrodes 20-28 cm in diameter is usually employed. When the tumor is eccentrically located, a pair of different-sized electrodes is employed with the smaller one being placed on the skin close to the tumor so that the high temperature area is shifted toward the lesion. Each electrode is covered with a water pad and a temperature-controlled salt solution is perfused into the water pad to avoid excessive heating of the skin and the subcutaneous fat. The advantages of RF capacitive heating are its wide applicability to various

anatomical sites and the relatively minor systemic stress it causes. Its main disadvantage is the excessive heating of subcutaneous fat and patients with a subcutaneous fat thickness of more than 1.5-2 cm are difficult to treat with this modality (5). A 430 MHz microwave heating device with a lens applicator (HTS-100, Tokimec Co. Ltd., Tokyo, Japan) has been employed for superficial or sub-surface tumors (less than 5 cm in diameter and less than 5 cm in depth) since January 1988. The details of the heating apparatus and the method for temperature measurement were described previously (6). A four-aperture lens applicator with total aperture size of 212x80 mm was used. The applicator was covered with a water bag, in which deionized water was circulated. The skin surface temperature was controlled by changing the temperature of circulating water from 10-40 C. Sugimachi et al. (7) have developed an RF intraluminal hyperthermia system (Endoradiotherm 100A, Olympus Optical Co., Tokyo). Very localized heating is possible with this device by inserting an endotract electrode into lumens in the human body, such as the esophagus,the rectum and the uterine cervix. A wide counter electrode is placed on the skin surface of the body so that the RF flux concentrates around the endotract electrode. Various types of electrodes are available according to the size of the lumen and the site of the lesion. An electrode is connected to the RF system which is operated at the frequency of 13.56 MHz with maximum power of 250 watts.

CLINICAL APPLICATIONS OF HYPERTHERMIA

Hyperthermia for superficial tumors

A large number of clinical experiences with combined hyperthermia and radiation therapy have been reported for superficial and subsurface tumors. These tumors have several advantages over deep-seated tumors as a target of the combined treatment: relatively well-developed heating devices, accessibility of thermometry, and easy assessment of response to heat in both tumors and normal tissues. Therefore, they are considered to be a good model to investigate the combined effects of hyperthermia and radiation. Additionally, not a few superficial tumors including malignant melanoma, soft tissue tumors, locally advanced tumors, and recurrent tumors following radiation therapy are still refractory to conventional treatments. The combined hyperthermia and radiation therapy has potential clinical significance for the treatment of these tumors. Usefulness of thermo-radiotherapy was initially demonstrated by several trials including us in patients with two or more comparable tumors (matched tumors) (8). Approximately a twofold increase in local response rates was shown in thermoradiotherapy than in radiotherapy alone. In addition to these studies, prospective randomized trials have been recently performed. A trial carried out by the Radiation Therapy Oncology Group (RTOG) in the United States failed to show a difference in the response rate between radiation alone and radiation plus heat when the tumors treated were analyzed all together (9). When the tumor response was assessed according to the tumor size, a significantly higher response rate was achieved with the combined hyperthermia and radiotherapy for tumors with maximum diameter less than 3 cm, but not for tumors more than 3 cm in diameter. The combined treatment also showed a substantially higher response rate in breast tumors but not in head and neck tumors or other tumors in comparison with radiotherapy alone. Since smaller tumors and breast lesions are easier to heat, it is suggested that heating limitation may be the reason for the lack of enhanced effects of combined treatment for large tumors and non-breast lesions. The other randomized trials including an international trial for breast cancer(10), an ESHO(European Society for Hyperthermia Oncology) trial for malignant melanoma(11), a JASTRO(Japanese Society for Therapeutic Radiology and Oncology) trial for Superficial tumors(12) have demonstrated the improvement of local control rate with the use of hyperthermia.

Hyperthermia for deep-seated tumors

Using a originally developed RF intracavitary heating device, Sugimachi and his collegues have applied hyperthermia in combination with radiotherapy and chemotherapy to patients with esophageal carcinoma. The long-term results were compared between two groups of patients treated with hyperthermo-chemo-radiotherapy (HCR) and those with chemo-radiotherapy (CR). The 5-year survival rates of patients with resectable carcinoma, given preoperative HCR or CR, were 43.2 % and 14.7 %, respectively (p<0.05). The 2-year survival rates of those with unresectable carcinoma and receiving HCR or CR were 15.5 % and 1.2 %, respectively (13). A prospective randomized trial was carried out to examine the effects of hyperthermia given preoperatively. Sixty-six patients with esophageal cancer underwent subtotal esophagectomy following either preoperative HCR therapy or CR therapy. The incidence of lack of viable cancer cells in the resected

specimens was 25 % in HCR group and 5.9 % in CR group. The cumulative 3- year survival rate was 50.4 % in the HCR group and 24.2 % in the CR group (14).

It has been shown that RF capacitive heating devices can effectively raise temperatures of lung tumors which invaded or were in contact with the chest walls, to which no other heating modality has been successfully applied. We reported clinical results of 20 patients with lung cancer treated by thermoradiotherapy (15). The mean of Tmax, Tave, and Tmin was 42.9 C, 41.6 C, and 39.7 C, respectively. Of 12 tumors treated by a curative intent, 2(17%) achieved CR, 7 (58%) PR and 3 (25%) NC. The side effects associated with hyperthermia were pain in 12 patients (60%) and dyspnea in 3 (15%), all of which resolved after termination of treatment. Improvement of local response rate (16) and survival rate (17) has been demonstrated with the combined treatment of regional hyperthermia and radiation therapy in non-randomized trials.

REFERENCES

1. Abe M and Hiraoka M: Review; Localized hyperthermia and radiation in cancer therapy. Int. J. Radiat. Biol. 47: 347-359, 1985.

2. Streffer C, Vaupel P and Hahn GM; Biological basis of oncologic thermotherapy, Springer-Verlag, Berlin, 1990

3. C.A. Perez et al., "Clinical results of irradiation combined with local hyperthermia," vol. 52, pp. 1597-1603, 1983.

4. Sapozink, M. D., Gibbs, F. A., Gates, K. S. and Stewart, J. R. (1984). Regional hyperthermia in the treatment of clinically advanced, deep seated malignancy: Results of a pilot study employing an annular array applicator. Int. J. Radiat. Oncol. Biol. Phys. 10 : 775-786

5. Abe M, Hiraoka M, Takahashi M et al.: Multi-institutional studies on hyperthermia using an 8 MHz radiofrequency capacitive heating device (Thermotron RF8) in combination with radiation for cancer therapy. Cancer 58, 1589-1595: 1986.

6. Hiraoka M, Jo S, Akuta K et al.: Radiofrequency capacitive hyperthermia for deep-seated tumors. I. Studies on thermometry. Cancer 60: 121-127, 1987.

7. Nikawa Y, Kikuchi M, Mori S et al.: Development and testing of a 2450 MHz lens applicator for localized microwave hyperthermia. IEEE Trnas. MTT-33, 1212-1216, 1990.

8. Sugimachi K, Inokuchi K, Kai H et al.: Endotract antenna for application of hyperthermia to malignant lesions. Gann (Jpn. J. Cancer Res.) 74: 622-624, 1983.

9. Overgaard J.: The current and potential role of hyperthermia in radiotherapy. Int. J. Radiat. Oncol. Biol. Phys.,16:535-549,1989.

10. Perez CA, Pajak T, Emami B. et al.: Randomized phase III study comparing irradiation and hyperthermia with irradiation alone in superficial measurable tumors: Final report by the Radiation Therapy Oncology Group. Am. J. Clin. Oncol.14:133-141,1991.

11. Vernon-CC; Hand-JW; Field-SB; Machin-D; Whaley-JB; van-der-Zee-J; van-Putten-WL; van-Rhoon-GC; van-Dijk-JD; Gonzalez-Gonzalez-D; Liu-FF; Goodman-P; Sherar-M; Radiotherapy with or without hyperthermia in the treatment of superficial localized breast cancer: results from five randomized controlled trials. International Collaborative Hyperthermia Group. Int. J. Radiat. Oncol. Biol. Phys. 1996 Jul 1; 35(4): 731-44

12. Overgaard-J; Gonzalez-Gonzalez-D; Hulshof-MC; Arcangeli-G; Dahl-O; Mella-O; Bentzen-SM; Hyperthermia as an adjuvant to radiation therapy of recurrent or metastatic malignant melanoma. A multicentre randomized trial by the European Society for Hyperthermic Oncology. Int-J-Hyperthermia. 1996 Jan-Feb; 12(1): 3-20, 1996

13. JASTRO Study Group: A randomized phase III trial of hyperthermia in combination with radiotherapy for superficial tumors. J. Jpn. Soc. Ther. Radiol. Oncol. 10, 161-164, 1998.

14. Kuwano H, Matsuura H, Mori M. et al.: Hyperthermia combined with chemotherapy and irradiation for the treatment of patients with carcinoma of the oesophagus and the rectum. In: Cancer Treatment by Hyperthermia, Radiation and Drugs, T. Matsuda (Ed.) Taylor & Francis, London, 353-364, 1993.

15. Kitamura K., Kuwano H., Watanabe M. et al.: Prospective randomized study of hyperthermia combined with chemoradiotherapy for esophageal carcinoma. J. Surg. Oncol. 60: 55-58, 1995.

16. Hiraoka M, Masunaga S, Nishimura Y et al.: Regional hyperthermia combined with radiotherapy in the treatment of lung cancers. Int. J. Radiat. Oncol. Biol. Phys. 22: 1009-1014, 1992.

17. Terashima H, Nakata H, Yamashita S et al.: Pancoast tumour treated with combined radiotherapy and hyperthermia- a preliminary study. Int. J. Hyperthermia 7: 417-424, 1991.

18. Karasawa K; Muta N; Nakagawa K et al: Thermoradiotherapy in the treatment of locally advanced nonsmall cell lung cancer. Int. J. Radiat. Oncol. Biol. Phys. 1994 Dec; 30(5): 1171-1177

Poster Presentations

Quantum Mechanical Approach to the Meaning of Existence, Will and Life*

Hironari Yamada

Ritsumeikan University, COE Research Organization, Synchrotron Light Life Science Center

Abstract. Author has given an interpretation of quantum mechanics in totally new approach named Dialogue Principle. Author point out how similar is the living phenomena and human behavior to the movement of elementary particles, if we agree with that the wave function implies the existing probability of elementary particles in the topological space. In the dialogue principle, the reasons for this similarity is explained by introducing common concept "character" and "will" among living specimens as well as elementary particles. Author describes that "will" is the origin of all strange behavior among elementary particles as well as human beings. "Will" is the subject of 21st Century to be studied scientifically. Author proposes that the concept of "will" should be added to the base of modern science to gives unified understanding of all existence including elementary particles, human beings, and universe. In this paper the digest of the dialogue principle is given, and approaches to the meaning of life by the philosophy of quantum mechanics.

PHENOMENA IN MICROSCOPIC WORLD IS SO SIMILAR TO THAT OF HUMANS BEHABIOR

We have long been aware of the phenomenon in microscopic is surprisingly similar to human's behavior. For instance electrons in the extremely narrow electric wire, where electrons are aligned in one line, demonstrate density wave. This phenomenon is quite similar to the drivers or cars on a highway. When traffic congestion appears you often experience such a wave appears. In the case of human this is due to the drivers will. We often quote human behavior to explain physics phenomena.

EXISTENCE IS DEFINED BY THE DIALOGUE, THUS CAN BE DEFINED IN THE WAVE FORM

A condition in which only one electron exists in the universe cannot be defined. If only one electron appears in the universe, in quantum mechanics a sole electron's wave function Ψ spreads infinitely in the universe. Since it is infinite if the space is not limited,

the existence probability $\psi\psi^*/\int\psi\psi^*dv$ becomes zero.

Meaning of the existence appears in quantum mechanics in self-consistent manner. Other electrons surrounding define the existence of some electron, which makes the boundary. In same manner, when a human exist in the universe solely, its existence cannot be defined. Human only understand himself and his or her existence by the mirror image created by the reflection from others. We are not talking about how to eat or how to live, but how to recognize. In reverse, when we think over the meaning of existence, an individual must always exist in plural as species and class.

Interference and resonance are the essence of wave

Interference is an essential nature between individuals in all classes, which has a will. It is evident among human being. An interference phenomenon is a result of dialogue, and is expressed in the waveform. Interference is the essential relation among individuals, but wave nature is only the appearance and mathematical formalism. Electron is not wave but a particle. Behavior of electron is expressed in the waveform in quantum mechanics.

CP716, *Portable Synchrotron Light Sources and Advanced Applications,*
edited by H. Yamada, N. Mochizuki-Oda, and M. Sasaki
© 2004 American Institute of Physics 0-7354-0195-0/04/$22.00

DIALOGUE PRINCIPLE: 13 ARTICLES

ARTICLE 1: The world is constituted with individuals and the associated fields.

- It is an individual (in Japanese KOTAI 個体), not a solid (in Japanese also KOTAI 固体) or a particle (in Japanese RYUSI 粒子). It is a generalization of particle's concept. It is all the existence, which has peculiar form distinguished as an individual such as an electron, a proton, a neutron, an atom, a molecule, ··· a protein, a DNA, ··· a cell, ··· an organ, ··· a spore, a seed, a plant, a human, ··· the earth, the sun, the solar system, ··· the universe, ···. Let's name electron in the following as a representative of elementary particles.
- We start with the evidentially existing world, which exists without human's observation regardless of way of thinking. We start with the objective existence philosophy (客観的実在 in Japanese), which must be the foundation of science.
- None of materials exist without associated fields. Meaning of the field appears in the later article after some more preparation.

ARTICLE 2: An individual always exist in plural form as specie and class.

- Meaning of existence of all individuals is given here for the first time in physics. The plural form is the key of every existence. Quantum mechanics is the beautiful system, which accommodate the meaning of the existence in the self-consistence manner.
- A condition in which only one electron exists in the universe cannot be defined. If only one electron appears in the universe, in quantum mechanics a sole electron's wave function Ψ spreads infinitely in the universe. Since it is infinite if the space is not limited, the existence probability $\Psi\Psi^*/\int\Psi\Psi^* dv$ becomes zero.

  Meaning of the existence appears in quantum mechanics in self-consistent manner. Other electrons surrounding define the existence of some electron, which makes the boundary. In the same manner, when a human exist in the universe solely, its existence cannot be defined. Human only understand himself and his or her existence by the mirror image created by the reflection from others. We are not talking about how to eat or how to live, but how to recognize. In reverse, when we think over the meaning of

existence, an individual must always exist in plural as species and class.
- The existence of class and species is evidence known for long time.
- The reason why class has appeared is an interesting research subject to be solved. It is thought to relate closely to the existence of quantum. Since quantum has minimum energy and discreet mass, there appears class and between classes it is stepwise.

ARTICLE 3: The field does not exist independently from the individual's body. Consequently field has a class.

- In the dialogue principle this statement is provided for distinguishing our principle from non-scientific argument. In the future, however, we need to improve the Dialogue Principle to accommodate the fact that particles are generated in the strong field. But this field must also associated with other individuals.
- Concept of the field (in Japanese BA 場) is generalized including the field associated with human society which appears in interaction between individuals as called "human beings", but not only the field attended with the four forces in physics. When there is no deliver or receive of energy as a result, but information's are delivered or received and generates internal changes occur to individual. Field appears when peoples are gathered. We call such field among human society for log time BA in Japanese. In human relations or society, peculiar field is generated based on exchange relations in economy. Also field appears when particles are exchanged in physics. An opinion stating that there is a class in the field forms the basis of the following logical development.
- Field is the concept based on dialogue (meaning is an interaction or a communication), which will be discussed later.

ARTICLE 4: An exact same individual never exist, even if it is of the same kind and same class. We call this each individual has a character.

- Here, the concept of character is generalized. None of electron stays in the same phase space. We call this is due to the character of electron.
- All the individuals at least exist in different phase space and have different momentum and different energy. We can name this the character. We cannot deny that electron has further character than the difference in the phase space. There is no reason proved that electron does not have a character. Even

simple electron is yet unknown and mysterious. Each electron behaves differently from others. However, because of uncertainty principle of quantum mechanics, that cannot be observed. Quantum mechanics deals with this difference as the "wave". Human beings recognize each other because they are in a same class, but they are different. We know this is the character. Electrons may recognize each other, since they are in the same class. But from another class it will be difficult to observe the difference. Difference in human beings is based not only on complexity of internal structure but also on the difference of topological space as human individuals. Also, complexity of human being is the result of the reflection of many electron's character, atom's, molecule's, protein's, or DNA's character.

- To be exact, Fermions have character, but Bosons do not. That is indicated in the difference between Fermi statistics and Bose statistics. Bose particles can share the same sate in plural individuals. Photon, which is Boson, can exist as a medium of dialogue as discussed later.

ARTICLE 5: Individual has a "will". Consequently, "will" is different in the different class.

- Here, concept of will is generalized. We state the generalized will a "will". Note that "will" is different from consciousness. Consciousness is a concept of memory storage referred to when decision is made. "Will" belongs to the body making a decision.
- We must understand that if human has will, DNA must have will. Electron is not to far from DNA. If so why electron does not have "will". "Will" is not the property of a higher animal. A higher animal makes up his mind according to his complicated decision process. However, the observed action from others is quite simple. Action is made up by "will".
- We use the word character in science, but we have never use the word "will" in science. The concept of "will" is, however, very important in philosophy, sociology, and economy. "Will" is the key issue in 21st century to unify science, philosophy, sociology, economy and all art. Or we can say "will" must be a subject of science in 21st century. Our theory now steps into new category. We use, however, "will" in rather limited definition in the following in the present stage. We can replace the word "will"

in every sentence to the word character without any distortion, but "will" gives a little further insight in to individual. Character is an appearance of "will".

- "Will" decides the place to be there. Without "will" none of individuals can not decide his position in the topological space, otherwise they will be disintegrated and separated. In this principle "will" is a kind of force, which unifies the individual. At this point, we do not refer to whether there exist "will" to select or we call the generation of selection a "will".
- Selected character is called "will". Selection occurs in the manner ruled by statistics in every class. We see that there is a strong correlation between the statistical selection in human being and probability phenomenon in the microscopic world. We can say that an essence of "will" is a probability phenomenon and probability phenomenon occurs by the "will". Will decision of human beings is done randomly when there is no potential or no incentives. Human often uses word "prefer" or "love", when he makes decision without reasons. We name this free will. In the same sense, wondering circular motion of birds after enough eating is known to be random as same as Brawn motion. Electron motion is the same.
- Individuals belong to class with specific field associated. Every field has potential or incentives. The field in each class rules "will" or the field is generated by the "wills". Note that specific motions are generated by the field, which is generated by "wills" belonging to some class. When we call reason in human society, this is generated by the society. In the same sense, electrons in the class generate electric potential.
- "Will" of plants is to decide by itself where to grow branches and when to germinate. Trees of completely the same shape never exist. This is not only due to the complexity but also due to the "will" of tree. Diode, which is fabricated in exactly same shape never exist. Electrons in the same phase space never exist. Atoms in exactly the same state never exist. They must be different in the phase of electrons involved and different in the "will" of nucleus. That is the generated character and here we call it has "will".
- Today, we cannot state scientifically whether "will" has a purpose or not. We think one of the purposes of individuals "will" is to

maintain individual and the seed. Individual try to maintain closer seed, because no individual can exist alone. To develop evolution and evolutionary selection cannot be a self-evident purpose.

ARTICLE 6: Each individual do "dialogue" through the field. Consequently, "dialogue" depends on class.

- Here, concept of dialogue is generalized. Not only higher animals do dialogue, it is a method to recognize others for all individuals. All individuals in every class do dialogue. None of materials ever exist without dialogue.

- The dialogue as already stated in ARTICLE 2 defines existence of all individuals. In the other word dialogue forms the part of existence. In physics dialogue is called an interaction, in engineering it is called a communication, human makes conversation. Here, abstract concept common in all the classes is called dialogue.

- Essence of dialogue is based on the fact that there do not exist the same individual (individual in the same state). If each individual is completely the same, dialogue is not necessary. Existence of individual is always regulated by dialogue. Own topological space is decided based on dialogue. That is a meaning of existence and a meaning of species.

- Dialogue becomes more difficult as classes apart further.

- Observation problem in quantum mechanics is a dialogue between different classes. Result of dialogue between different classes controls or destroys the behavior of more primitive class.

- Time is a concept associated with dialogue. Time is the duration necessary for dialogue. You will easily understand the special relativity theory of Einstein in this manner. In other word the special relativity agree well with this statement. Special relativity is the dialogue between the individuals moving at near the speed of light. There is no concept of universal or common time in the dialogue principle as stated in the special relativity theory. Time evolution depends on the place as well as class where individuals belong. The time necessary to recognize others is different for different person as is usual. Electrons are in a same situation. This is the origin of uncertainty related to time. Time is a concept associated with the field, consequently differs according

to class. Time is not independent from individuals. That means time depend on class.

- Individuals change their will through the dialogue.

ARTICLE 7: Interference phenomenon appears as a result of dialogue.

- Interference is an essential nature between individuals in all classes, which has a will. It is evident among human being. Interference between electrons is expressed as wave. The wave nature is a phenomena or an appearance, but not essential.

- Interference appears only between the individuals of the same class, which does not appear between the individuals of the different class. It does generate among human beings. It generates when particles of different class get together and form one class like an atom.

- An interference phenomenon is a result of dialogue, and is expressed in the waveform. Interference is the essential relation among individuals, but wave nature is only the appearance and mathematical formalism. Electron is not wave but a particle. Behavior of electron is expressed in the waveform in quantum mechanics.

- For example, the aligned electrons in the extremely narrow electric wire, and the drivers on a highway in the lane both demonstrate density modulation, and modulation propagates as a wave. We understand this is both due to the result of dialogue between electrons or between drivers. We know that driver's behavior is wave like, but driver is not wave. This is same in electron. Electron is not wave. Electron has a will.

ARTICLE 8: In the way of dialogue, at least photon is one of mediums.

- Because photon can be shared by any individuals in any class.

- Photon delivers information of individuals. It does not necessarily exchange energy. The fact that electrons deliver and receive photons under regular condition, which is a concept of the second quantization in quantum mechanics.

- We cannot deny the fact that there might be a dialogical method other than photons. We cannot understand the strange phenomena in microscopic world without assuming another way of dialog, in which information is delivered at the speed higher than that of the light speed. However since time evolution depends on class, we need deeper

112

consideration. There might be a tacit consent between individuals. In the case of life, there is a tacit consent that DNA owns. There might be one in electrons. Electrons might share the common memory of when they were generated at the beginning of the universe in the big ban.

ARTICLE 9: Uncertainty principle is indefiniteness of dialogue and it is based on the fact that individuals have character and "will".

- Uncertainty principle is a result of dialogue in any class. It includes the fact that individuals change the "will" by dialogue. Individuals change their topological place by the dialogue.
- Uncertainty principle includes both the indefiniteness of dialogue and uncertainty due to different classes.
- Observation problem in quantum mechanics is a latter problem. Observation problem is a dialogue between different classes. As a result of dialogue, hierarchical relation is generated between classes. Dialogue controls or destroys more primitive class.

ARTICLE 10: Conclusion lead by dialogue deeply reflects character of individuals, however, average behavior follows the rule in physics or society.

- The difference from the average behavior is called uncertainty. It is controlled by statistical rules.
- Which way an electron will pass cannot be predicted by uncertainty principle. That is because dialogue urges the change of individual's "will". However, the way of most electrons will pass in average can be predicted.
- Which way the particular human will pass cannot be predicted, but you can predict a behavior of a group or an average behavior. You cannot predict human behavior of the next moment, but you can predict where he/she is heading to in average.
- Classical mechanics and Maxwell's electromagnetic theory describes average nature.

ARTICLE 11: Quantum mechanics could be a general rule concerning dialogue and may describe the structure of dialogue between all the individuals.

- We have already stated in ARTILCE 2 that quantum mechanics includes the meaning of existence, which is defined by the dialog. Wave function shows an essence of dialogue. The existence is defined by the reflection from others. The complex conjugate for leading the existence probability is a message or reflection from others.

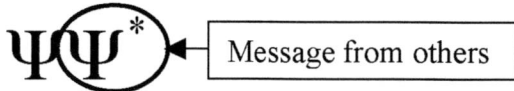

The wave function is not always formed from one dialogue. It may be a conclusion of indefinite number of dialogue or of some dialogues. Balanced state as a result of dialogue is indicated as a wave function.

- Wave function localizes when the boundary is limited by the potential or by the field. Human being never spread uniformly all over the earth. They usually localize and form colony or city. Colony could be an Eigen state. Electrons density modulation is the same. Those phenomena are all originated by the will. Eigen function indicates the structure of the field generated by dialogues between the same kind individuals.
- There are some trials introducing quantum mechanics to solve the distribution of colony of bacteria or animal.
- Hilbert space describes the structure of the field generated by the dialogue. It is a description of structure of dialogue and all the possible routes (defined as vector) caused by dialogue. Each individual select a route following the "will". Formalism of Hilbert space is common to all class, but the used parameter is different.
- The meaning of operator is indeed an operation by individual's "will". Essentially, the way of operation is different in each individual; however, average number is used in quantum mechanics.
- The fact that quantum mechanics is probabilistic and statistical, and that social phenomenon is statistical is not a coincidence. They are both the conclusion of Dialogue Principle. It means that the structure of the field is similar no matter which class. Mathematical formalism of Hilbert space can be general in every class, however, parameter used and operator is peculiar in the class. If we found the way of dialogue in the class of interest we can construct the wave function.
- Quantum phenomena should appear in the macroscopic world, because the macroscopic world is formed by the microscopic particles. Indeed we see many quantum phenomena such as super fluid, superconductivity, and so on. Living phenomena must be the quantum

phenomena.

ARTICLE 12: Resonant phenomenon is a unification of the "will" generated as a result of dialogue. Therefore, it is a phenomenon common in every class.

- When resonance appears in a group movement, state vector or wave function contracts and becomes one condition and character will be unified. It is called coherent.
- Resonant phenomenon of human society and of peculiar particles and atoms is the same kind in dialogue principle. In the resonant condition all humans are cooperating, and all particles and atoms are cooperating.
- The simplest resonant phenomenon is an electric current in which a group of electrons flows into the same direction. A stream of rivers, super conductivity, super fluid, electromagnetic wave, and a stream of automobiles, etc. In army the will is unified, and solder loss their character. This is a sort of resonance.
- Photons in resonance are called laser. Again wave is a structure of the field made by photon, but it is not an attribute of photon.
- Classical mechanics is a mechanics under resonant condition; it is a condition in which group of state vectors is shrunk into one. Movement of planet is the one in which all the structured individuals are cooperating. Electrons in the current shares the same information, thus such condition cannot be changed by the observation. We can describe the movement of planet as a superposition of all materials in the planet, which has the same wave function with regard to the motion regarding to the center of its circular motion. This is called the state of macroscopic or the classical world.

ARTICLE 13: Every thing migrates as a result of dialogue.

- The meaning of migration of nature is clarified here. The same state never appears. The "god" under the Newton mechanics does not decide where to migrate. It is certain that where all the individuals in the universe would migrate is decided as a result of repeated dialogue.
- We cannot expect the destination of the migration, because of not complexity but because change in the will is involved due to the dialogue.
- If the earth is supposed to be one of the individuals in this principle, "will" of the earth is decided by the "will" of all the individuals constituting the earth. Human beings are included as well. Will of humans is participating in deciding where the earth would migrate. Human's "will" might control the earth or it might perish. "Will" of the earth is not decided previously according to monotheism, it is decided by the entire structuring individuals consensus. This is quite a natural conclusion. And it is concluded that human beings have responsibilities to the future of the earth. Majority would think that the future of earth is decided a priory without human's will. Let say again that this is not true. This is an important result of the Dialogue Principle.
- It is also concluded that artificial we call, but this is also the part of the nature. Created materials or created machines are also individuals in this principle and involved in the discussions of the future.
- As a conclusion Dialogue Principle never contradict with the daily human common sense and agree very well with the observation of microscopic world.

CONCLUSION

In such way, Dialogue Principle is not inconsistent with daily human's sensitivity or feeling, and it agrees with the common knowledge of microscopic world. The above ARTICLE 13 is starting up with the self-evident basic concept, which is of individuals and the field, and introducing the new concept step by step by placing a little further logical statement. "Character" is a concept, which appears first, and does not exist in common physics. "You can not say that electrons do not have character" is rather a passive statement, however, it is a key statement that forms the basis of Dialogue Principle. If there was found an experimental result, which denies this statement, we would have to admit that this Dialogue Principle is wrong. However, we have stated that electron's character cannot be observed, because of the uncertainty principle, and by assuming the existence of "character", quantum mechanics can be interpreted more naturally. The correctness of quantum mechanics proves the existence of "character". If you translate the word "character" as staying in a different topological space, you can accept it without any sense of incongruity. If the word "character" is interpreted into such a limited meaning, you can discuss it within the framework of the common physics. But we are expecting more of the character with electrons. Such existence of character has not yet proven, but the fact that the essence of quantum mechanics is in probability and statistics is a circumstantial evidence. Applying statistical mathematics to social phenomenon was a necessary method to set numerical value on humans ignoring its character. We think the similarity of the two was generated in order to have characters as a substance for humans and electrons.

Origin of life is created by let's say living electrons

Selected character is called "will". Selection occurs in the manner ruled by statistics in every class. Here, concept of will is generalized. We state the generalized will a "will". Note that "will" is different from consciousness or mind. Consciousness is a concept of memory storage referred to when decision is made. "Will" is the body making a decision. We

see that there is a strong correlation between the statistical selection in human being and probability phenomenon in the microscopic world. We can say that an essence of "will" is a probability phenomenon and probability phenomenon occurs by the "will". Will decision of human beings is done randomly when there is no potential or no incentives. Human often uses word "prefer" or "love", when he makes decision without reasons. We name this free will. In the same sense, wondering circular motion of birds after enough eating is known to be random as same as Brawn motion. Electron motion is the same. Electron's will must be the origin of human's will, and consequently of the DNA's.

REFERENCES

* Japanese version of this paper is published in Soryusiron Kenkyu (素粒子論研究) Yukawa Institute for Theoretical Physics, 94-2 (1996-11)

The Portable Synchrotron MIRRORCLE-6X

Daisuke Hasegawa*, Hironari Yamada, Andrey I. Kleev, Norio Toyosugi,
Taichi Hayashi*, Takanori Yamada*, Isao Tohyama*, and Young-Deok Ro*

Ritsumeikan University, Synchrotron Light Life Science Centre,
* On leave from Photon Production Laboratory. Ltd.

Abstract. The construction of the tabletop synchrotron MIRRORCLE-6X is completed. The electron beam current extracted from the 6-MeV Microtron injector is reached 100 mA with 500 ns pulse duration and 400 Hz repetition rate. The measured main magnetic field distribution of the synchrotron showed good agreement with calculation. We found that the 6-MeV electron beam can be injected from the designed injection port to the synchrotron with 20 mm•mrad acceptance according to the computer simulation using measured magnetic fields.

INTRODUCTION

MIRRORCLE-6X is the first portable synchrotron with 60 cm magnet OD, in which 6 MeV electron beam is stored in the 15 cm radius exact circular orbit. This synchrotron generates a high brilliance x-ray beam comparable to SR by using the inelastic collision of circulating relativistic electrons with a tiny target placed on the central orbit. This novel x-ray source based on a low energy tabletop synchrotron has been proposed by Yamada [1,2] and is demonstrated with its brilliant x-ray production by using MIRRORCLE-20 [3], which is a 20 MeV synchrotron having 15 cm orbit radius, and 1.2 m magnet OD. The observed brilliance of hard x-rays is comparable to SR. This novel source is unique by its few μm x-ray source size, which is determined by target size, by its high energy transfer rate, and by its broadband x-ray spectrum of from few KeV up to the energy of injected electron dominated by the hard components.

At the beginning MIRRORCLE-20 was developed for FIR lasing according to "Photon Storage Ring" (PhSR) theory [4]. The PhSR required the space for placing a concentric barrel-shaped optical resonator inside the chamber surrounding the electron orbit, so that the outer diameter of the main magnet body became 1.2 m. Moreover it was designed to storage 50 MeV electrons at maximum in 15 cm orbit radius using a normal conducting magnet. By decreasing the electron energy to 6 MeV and aiming at only the x-ray

generation, MIRRORCLE-6X is designed to be the 60 cm outer diameter and 28 cm height, which are the one half in the diameter and one quarter in the height of MIRRORCLE-20. The electron energy of 6 MeV is chosen for suppressing neutron generation. The expected brilliance of MIRRORCLE-6X is order of 10^{11} photons, which is the 1000 times more of MIRRORCLE-20. The total flux exceeds that of conventional SR. We expect many application fields with this machine such as, x-ray microscope, protein crystallography, non-destructive testing, fine medical imaging, cancer therapy, x-ray fluorescent analysis of heavy elements, x-ray lithography, and so on. This machine is commercially available from the Photon Production Laboratory Ltd. [5].

The fabrication of MIRRORCLE-6X is completed through the physical designing by computer simulations and mechanical designing. An overview of the completed MIRRORCLE-6X is shown in Fig. 1. We have started commissioning of the Microtron injector in September 2003, and the beam injection to the synchrotron is confirmed in November 2003.

In this paper, MIRRORCLE-6X specifications and the feature of x-ray beam lines are given. Results of the beam test on the 6-MeV Microtron injector and the injection test to the synchrotron are shown. The 1/2-integer resonance injection method utilized in MIRRORCLE-6X is also explained.

CP716, *Portable Synchrotron Light Sources and Advanced Applications,*
edited by H. Yamada, N. Mochizuki-Oda, and M. Sasaki
© 2004 American Institute of Physics 0-7354-0195-0/04/$22.00

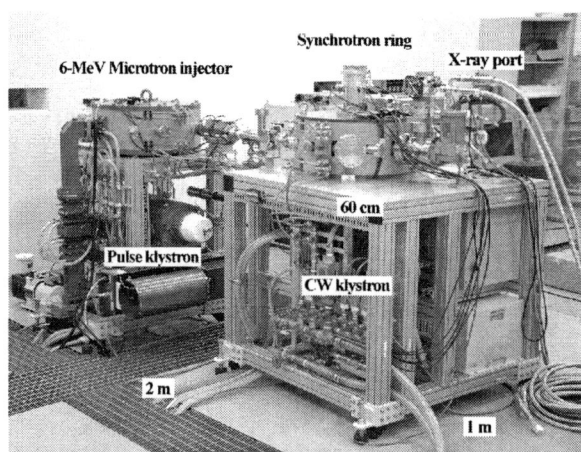

FIGURE 1. An overview of the portable synchrotron "MIRRORCLE-6X".

MIRRORCLE-6X SPECIFICATIONS

MIRRORCLE-6X consists of 6-MeV Microtron injector and Synchrotron storage ring with three x-ray beam ports. MIRRORCLE-6X specifications are listed in table 1. The 6-MeV microtron is the Kapitza type, which has a LaB6 electron emitter inside the single cell-accelerating cavity. The electrons are extracted directory by the RF field. Because of this scheme we are able to apply strong magnetic field, which results in the 60 cm small out diameter magnet. The RF source is a powerful pulse klystron that

TABLE 1. MIRRORCLE-6X specifications.

Injector	6-MeV Microtron
	Peak current: 100 mA
	Pulse width: 1 µs
	Repetition rate: 400 Hz
	RF frequency: 2.45 GHz
Synchrotron	Electron energy: 6 MeV
	Orbit radius: 15 cm
	N-value: 0.72
	Magnet OD: 60 cm
	RF frequency: 2.45 GHz
Radiation Scheme	Target radiation
Radiation Angle	83 mrad
Photon Spectrum	Continuous from 1 KeV to 6 MeV
Time Structure	Variable pulse width: 100 ns – 10 ms
	Repetition: 2.45 GHz
Intensity	1.4 (0.08) Gy/Pulse (200 ns)
Imaging Time	1 Pulse (200 ns)/Frame (576 cm^2)
Average Power	160 (24) Gy/s
Calculated Brilliance	2.5x10^9 Photons/s/mrad^2/mm^2/0.1 %λ
Calculated Total Photons	5.5x10^9 Photons/s/0.1 %λ

utilizes multi-beam with 30 kV low anode voltage. The output power is 5 MW at peak, 5 kW on average, and 45 % efficiency with the 80 cm x 30 cm compact body. Operating frequency is 2.45 GHz.

The synchrotron is made of one piece of cylindrical normal conducting magnet of 60 cm OD. The 1/2-integer resonance is selected to provide stronger vertical focusing power compared with the 2/3-integer resonance used for MIRRORCLE-20, which is suitable to minimize the vertical x-ray source size. To introduce the 1/2-integer resonance injection scheme the n-value of the magnet field is set to 0.72 over 10 cm in radial direction around the central orbit. The expected beam size is less than mm in vertical, and 10 mm in horizontal. The perturbator to trigger the 1/2-integer resonance is one-tern air core magnet, which produces ±0.42 T•mm magnetic field at ∓30 mm from the central orbit.

The synchrotron accommodates three x-ray beam ports as seen in Fig. 2 together with the beam dynamics. Each port is assigned to X-ray crystallography, X-ray microscope, X-ray fluorescent analysis, and X-ray imaging. The X-ray microscope and X-ray crystallography beam line is equipped with a focusing element made of multi-layered 8-set of cylindrical mirrors aligned in concentric.

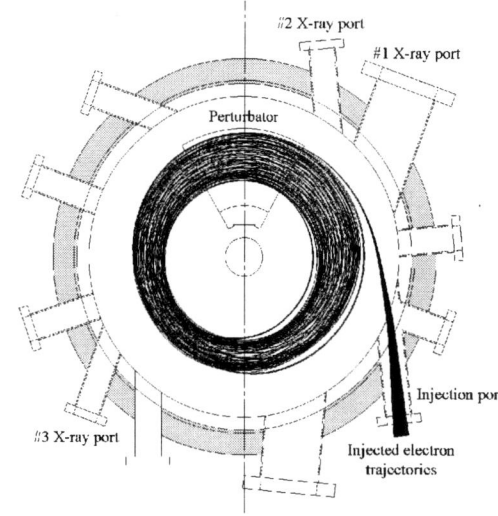

FIGURE 2. Injection trajectories on the median plane of the MIRRORCLE-6X obtained by the computer simulation are shown.

Resonance injection scheme

The resonance injection method is considered to be a time reversal of half integer resonance extraction. Its method is similar to the super-conducting synchrotron

AURORA [6,7,8], and is different from MIRRORCLE-20, which uses the 2/3-integer resonance.

In resonance injection scheme, the perturbator (PB) placed under the main magnetic field kicks the beam electron toward outside from the central orbit and gradually changes its orbit to the central orbit. Electron trajectories with or without PB field are shown in Fig. 3. The injected electron with PB field as shown in Fig. 3(a) is captured in the stable orbit after 20 turns by damping the PB field in a synchronized way, while the injected electron escapes from the stable orbit without PB as shown in Fig. 3(b). Thus the PB dose not disturb the orbiting electrons, the resonance injection method makes a continuous beam injection possible. This is an essential factor for the high brilliance x-ray production.

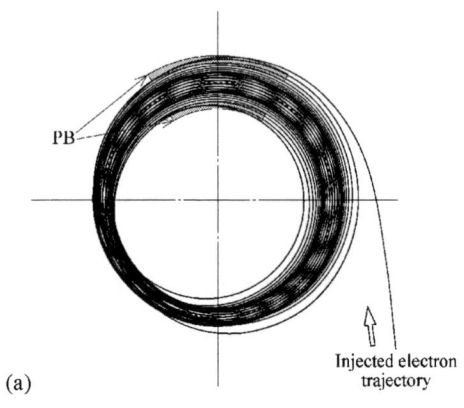

(a)

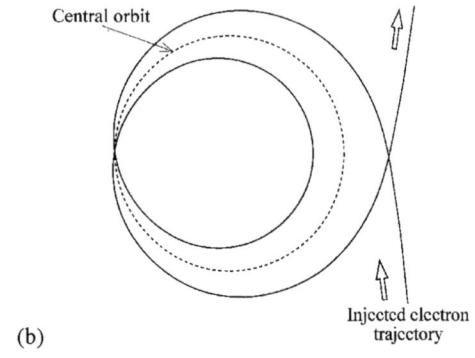

(b)

FIGURE 3. Electron trajectories with (a) or without (b) perturbator field. The dotted line is the central orbit.

Injection trajectory

The main magnetic field distribution of the synchrotron was designed by the code TRIM. The computer simulation of the injection trajectory was made by the code DYNA to decide the position and angle of the injection port. We found that the measured magnetic field distribution is in good agreement with calculated one around the central orbit. However, the measured magnetic field around the fringe has some deviation from the calculated one. To take into account the effect of the fringing field to the injection trajectory, we have carried out the computer simulation by using the measured magnetic fields. We found that the 6-MeV electron beam can be injected from the designed injection port to the synchrotron with 20 mm•mrad acceptance as shown in Fig. 2.

COMMISSIONING OF MIRRORCLE-6X

Commissioning of MIRRORCLE-6X was started with the beam testing of the 6-MeV microtron injector on September 20, 2003. At the first operation, the electron beam current of 5 mA was recorded with the probe placed in the last orbit. After by adjusting the cavity form, and by adjusting the median plane by changing slightly the current of the upper and lower coil, as a result, the electron beam current of more than 100 mA was extracted from the microtron. We use iron extraction channel. The pulse duration was 500 ns as shown in Fig. 4. The fast current transformer (BERGOZ Instrumentation FTC-082-05:1) is used to measure the beam current. The emission of 660 mA from LaB6 single crystal is obtained by heating with accelerated electron beams from the filament at 500 V and 50 mA. Accordingly the capture efficiency of 15 % is reached. The beam current of 100 mA with 400 Hz repetition rate guarantees that the X-ray brilliance of MIRRORCLE-6X is 1000 times higher of MIRRORCLE-20.

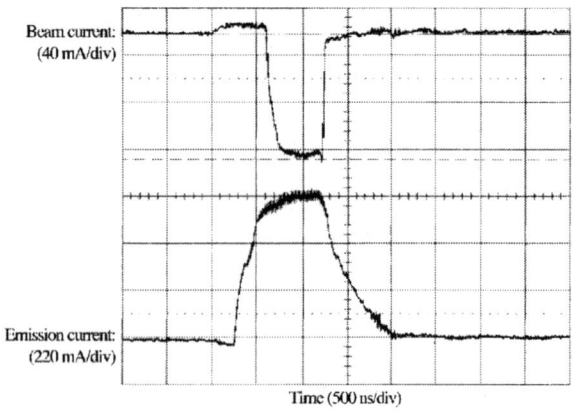

FIGURE 4. The measured beam current (top) and emission current (bottom).

118

The injection of 6-MeV electrons to the synchrotron according to the injection trajectories obtained by the computer simulation was confirmed on Nov. 13, 2003. The electron beam lifetime stored up the synchrotron observed by a photo multiplier was few milliseconds with the target, as shown in Fig. 5. The capture efficiency of the synchrotron is less than 50 %, because the emittance of the electron beam emitted from the microtron is 50 mm•mrad that is larger than the 20 mm•mrad acceptance of the synchrotron. We should increase the capture efficiency by decreasing the microtron emittance, in which the x-ray intensity will be increased and then back ground radiation will be reduced. In this project the back ground radiation from the cavity and the iron extraction channel is the important subject to be minimized. We will minimize these radiations dramatically by introducing electron-focusing elements inside the microtron magnet for medical and industrial use of this machine.

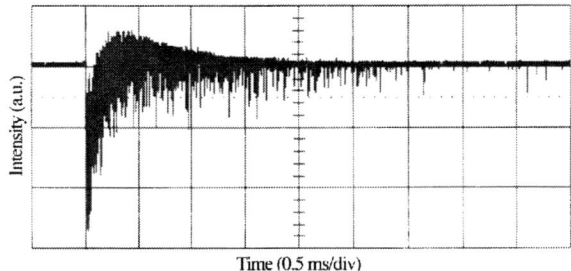

FIGURE 5. The lifetime of the beam is observed by a photo multiplier with the target.

CONCLUSION

The construction of MIRRORCLE-6X, the world smallest synchrotron was completed. The measured magnetic field distribution is in good agreement with calculated one around the central orbit. The electron beam current of 100 mA extracted from the classical microtron was achieved. We confirmed that the 6-MeV electron beam was injected through the designed injection trajectory.

In summarizing, MIRRORCLE-6X will provide satisfactory x-ray flux for variety of applications. The high brilliant x-ray beam generated by only 60 cm OD synchrotron can be utilized on site in medical and industrial applications, which will change the paradigm of X-ray business.

The development and study of MIRRORCLE-6X are supported by a Grant-in-Aid for Science Research from Japan Society of the Promotion of Science.

REFERENCES

1. H. Yamada, *Jpn. J. Appl. Phys.* **35** L182-L185 (1996).

2. H. Yamada, *Nucl. Instr. and Meth. A* **467-468**, 122 (2001).

3. H. Yamada, *Nucl. Instr. and Meth. B* **199**, 509-516 (2003).

4. Photon Production Laboratory Ltd.: 4-2-1(808) Takagai-cho minami, Omihachiman-city, SHIGA, 523-0898, Japan. URL: http://www.photon-production.co.jp

5. H. Yamada, *Adv. Colloid Interface Sci.* **71-72**, 371-392 (1997).

6. T. Takayama, *Nucl. Instr. and Meth. B24/25*, 420-424 (1987).

7. H. Yamada, J. *Vac. Sci. Technol. B* **8** (6) 1628 (1990).

8. N. Takahashi, *Nucl. Instr. and Meth. B24/25* 425-428 (1987).

The Photon Storage Ring

Andrey I. Kleev and Hironari Yamada

*Synchrotron Light Life Science Center, Ritsumeikan University, 1-1-1 Nojihigashi, Kusatsu-City
Shiga 525-8577 Japan*

Abstract. We derive the theory of mode locking phenomena in the novel type of the free electron laser called Photon Storage Ring (PhSR). We found the electron beam parameters for which the mode locking appears. We examine the characteristics of the pulse mode regime and outline the possibility of the femtosecond pulse generation. We also present the current status of the PhSR experimental investigations.

INTRODUCTION

A novel type circular free electron laser, called *Photon Storage Ring* (PhSR), which employs a compact, exactly circular electron storage ring, as in AURORA, MIRRORCLE-20, and MIRRORCLE-6X was proposed by Yamada [1-5]. The differences between PhSR and other free-electron lasers (FEL) are following:

i.   Neither an undulator nor a wiggler are introduced;

ii. Both electron and optical beams are strongly localized in the circular optical cavity;

iii. The whispering gallery or annular modes can be utilized in the PhSR.

The new lasing mechanism without the undulator was described in [1-9]. A schematic configuration of PhSR is shown in Fig. 1. The storage ring is, here, exactly circular. The synchrotron radiation is reflected back tangentially into the circular electron orbit.

Initially, the theory of the PhSR lasing mechanism has been constructed on the assumption that the electromagnetic wave is uniformly filled inside the optical cavity [3, 4], but we have been long time believed that the electromagnetic field must have a pulse structure. The electron beam in the storage ring is a chain of bunches in the order of 10 mm, separated by a distance greater than that of the bunch length. It is clear that the pulse current, corresponding to such a beam, interacts most effectively with the cavity eigenmodes that have the same structure, i.e. with the modes in the form of the short pulses moving through the cavity together with the electron beam.

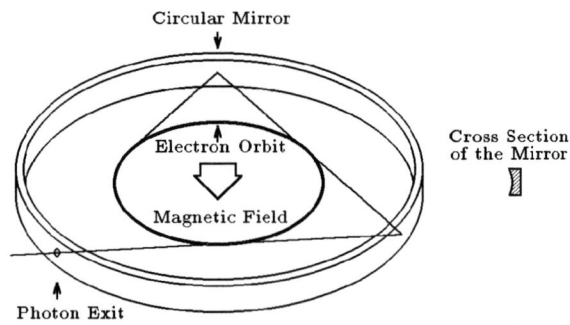

FIGURE 1. The Photon Storage Ring scheme.

THE FORMALIZM OUTLINE

The PhSR formalism is based on the electromagnetic field representation as the sum of radial transmission line eigenmodes $\vec{E}_s^{(\pm)}$:

$$\vec{E}^{(\mathrm{I,II})} = \sum_s \left(C_{+s}^{(\mathrm{I,II})} \vec{E}_s^{(+)} + C_{-s}^{(\mathrm{I,II})} \vec{E}_s^{(-)} \right) \quad (1)$$

where symbols I or II denote the internal (with respect to the electron orbit) and external domain (see Fig. 1);

CP716, *Portable Synchrotron Light Sources and Advanced Applications,*
edited by H. Yamada, N. Mochizuki-Oda, and M. Sasaki
© 2004 American Institute of Physics 0-7354-0195-0/04/$22.00

the symbols $+$ or $-$ denote the forward and backward waves. The matching condition on the beam and regularity conditions lead to the following relations for the unknown coefficients:

$$C_{\pm s}^{(II)} = C_{\pm s}^{(I)} \pm I_s^{\mp}, \ C_{+s}^{(I)} = C_{-s}^{(I)} \qquad (2)$$

where the *interaction integral* $I_s^{\mp}$ is defined by the applying the standard equations for the electromagnetic field, radiated by an arbitrary current in a radial transmission line. After applying the boundary condition on the mirror surface (see details in [6]), we get the following *matrix equation* for the unknown coefficients:

$$\Gamma\left(\mathbf{C} - \mathbf{I}^{(+)}\right) = \mathbf{S}\left(\mathbf{C} + \mathbf{I}^{(-)}\right) \qquad (3)$$

For the given value of the current one can obtain the unknown coefficients from the above equation and find the field in the barrel shaped cavity produced by the given current of the frequency $\omega = ck$ which, in turn, define the value of Γ :

$$kR = v_{M,j}' + i\frac{\ln\Gamma}{2\sin\theta_M}, \ \cos\theta_M = \frac{R_1}{R}, \ M = kR_1 \ (4)$$

where $v_{M,j}'$ are the roots of equation $J_M'(v_{M,j}') = 0$, R_1 is the electron orbit radius, R is the mirror radius (see Fig. 1). If there is no current, $\mathbf{I}^{(\pm)} = 0$ and matrix equation gives the *cold eigenmodes* in the barrel shaped resonator (see also [6]). Using the above matrix equation we can also consider the case when the electric current is excited due to the interaction between the electromagnetic field and the electron beam moving along the circular orbit. In order to take into account this interaction we must obtain the relation between the interaction integral $\mathbf{I}^{(\pm)}$ and the EM field in the cavity defined by the vector $\mathbf{C}$. In the linear approximation the general form of this relation is

$$\mathbf{I}^{(\pm)} = \mathbf{G}^{(\pm)}\mathbf{C} \qquad (5)$$

where the matrix $\mathbf{G}^{(\pm)}$ can be considered as the *generalized matrix of conductivity* because this matrix give us the relation between the EM field and the current, which is induced by this field.

To define the matrix $\mathbf{G}^{(\pm)}$ we must solve a linearized equation of motion. Neglecting the radial motion and taking into account that $\gamma \gg 1$ we can obtain that the equation for the change in energy $\delta\gamma$ is:

$$\frac{\partial\delta\gamma}{\partial t} + \frac{\omega_c}{1-\xi}\frac{\partial\delta\gamma}{\partial\varphi} = \frac{e}{m_e c}E_\varphi, \ \omega_c = -\frac{eH_0}{\gamma m_e c} = ck_c \ (6)$$

where the parameter $\xi \in [\xi_1, \xi_2]$ defines the dimensionless radial displacement of the electron orbit: $\xi = 1 - r/R_1$. We assume that $\xi \ll 1$. The continuity equation for the oscillating current J_φ is:

$$\frac{\partial J_\varphi}{\partial t} + \frac{\omega_c}{1-\xi}\frac{\partial J_\varphi}{\partial\varphi} = \frac{\partial}{\partial t}\left[\rho^{(0)}v_\varphi^{(1)}\right] \qquad (7)$$

where $\rho^{(0)}$ is the electron beam density in the absence of the electromagnetic field (unperturbed density) and the $v_\varphi^{(1)} = -c\delta\gamma/\gamma$ is the azimuth velocity perturbation due to the interaction with the electromagnetic field. We assume that the bunch charge density $\rho^{(0)}$ is an periodical function of both the azimuth angle and time and this function can be written in term of Fourier's series:

$$\rho^{(0)}(r,\varphi,t) = \rho_0 \sum_P \rho_p(\xi)\exp\{pq[(1-\xi)\varphi - \omega_c t]\}(8)$$

where integer q specify the number of electron bunches in the orbit. For example, the "quasi Gaussian" bunches correspond to the following relation for $\rho_p(\xi)$:

$$\rho_p = \frac{\delta_e q}{\sqrt{2\pi}}\exp\left[-\frac{1}{2}\left(q^2 p^2 \delta_e^2 + \xi^2/\xi_0^2\right)\right] \qquad (9)$$

where $\xi = 1 - r/R_1$. For $\delta_e \ll 2\pi/q$ these relations corresponds to the chain of the Gaussian bunches with rms relative azimuthal width δ_e and radial width ξ_0 respectively.

The final matrix equations for the eigenmodes of the resonator with the electron beam has the form:

$$\Gamma\left(\mathbf{C} - \mathbf{G}^{(+)}\mathbf{C}\right) = \mathbf{S}\left(\mathbf{C} + \mathbf{G}^{(-)}\mathbf{C}\right) \qquad (10)$$

THE RESULTS OF CALCULATIONS

In this section we will discuss the most important properties of the solution of Eq. 10. This is the linear eigen value problem and we have used standard procedures to solve it. We found out, that if only the orbit radius is close to some values $R_1^r = R\cos\theta_M^r$, where θ_M^r satisfy the *Yamada's Resonance Condition* [1-5]:

$$\tan\theta_M^r - \theta_M^r = \frac{\pi(n-1)}{ql}, \quad n,l = 1,2,\ldots \quad (11)$$

then the mode locking appears and the modes have the form of short pulses moving in the cavity together with the electron bunches. It is necessary to underline, that the gain in mode locking regime is increased considerably compared with the assumption that the EM field is uniformly filled in the cavity.

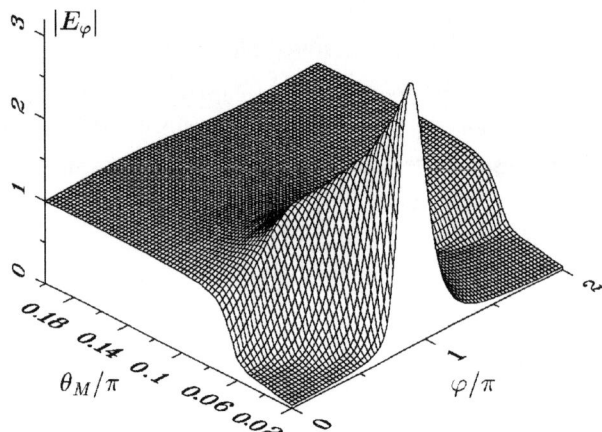

FIGURE 2. The pulse shape evolution under the orbit parameter θ_M changing.

The most interesting and important properties of the pulse modes are demonstrated in the Fig. 2 and Fig.3. We will use the dimensionless co-ordinate $\eta = z/D$ and the following dimensionless current parameter:

$$g_e = \frac{I}{I_A} \frac{\pi k^2 \theta_e^3}{4k_c^3 D \gamma \mu^4}, \text{ where } \mu = \left(\frac{M}{2}\right)^{1/3} \quad (12)$$

and $I_A = mc^3/e \approx 17 \cdot 10^3 \text{A}$. In Fig. 2, the pulse shape evolution due to the orbit parameter θ_M change is shown; in the case under consideration the well-defined pulse mode can exist only when the parameter θ_M is small enough i.e. close to the principal solution ($n = l = 1$) of the *Yamada's Resonance Condition* (11).

In Fig. 3, the pulse mode envelopes for different $\delta\theta_M = \theta_M - \theta_M^r$ values are presented. The width of the EM-pulse can be far less than the electron bunch width, and becomes shorter as θ_M become closer to the resonance value θ_M^r. This specific feature of the PhSR allows us to consider the PhSR as a promising femtosecond pulse source of coherent radiation.

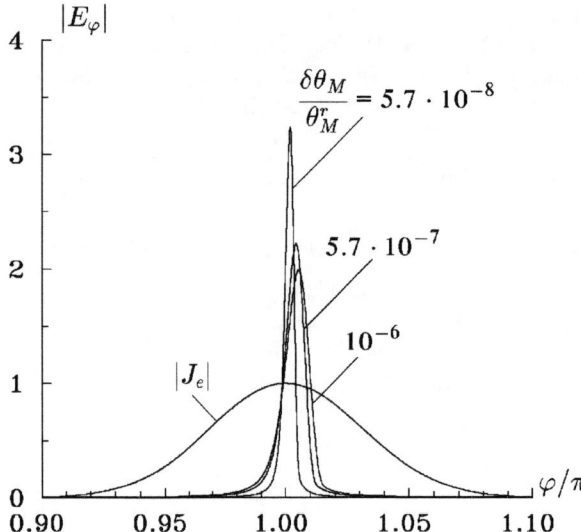

FIGURE 3. The pulse mode envelopes for different $\delta\theta_M/\theta_M^r$ ($\delta\theta_M = \theta_M - \theta_M^r$).

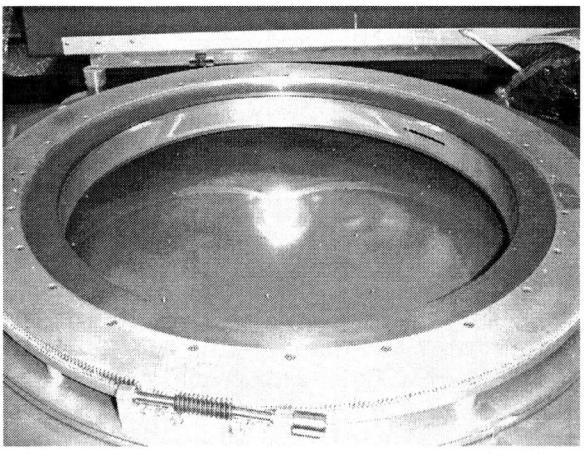

FIGURE 4. The barrel shaped optical cavity.

LASING EXPERIMENT

At present, we have already finished the design and the manufacturing of the barrel-shaped resonator. The experimental investigation of the cold eigenmodes properties has been carried out quite successfully. The Fig. 4 demonstrates the PhSR mirror structure. The experimental setup for the cold mode investigations is shown in the Fig. 5. Now we are planing to install this resonator into the MIRRORCLE-20 storage ring and observe the spontaneous radiation. MIRRORCLE-20 is already converted to machine dedicated for FIR experiments.

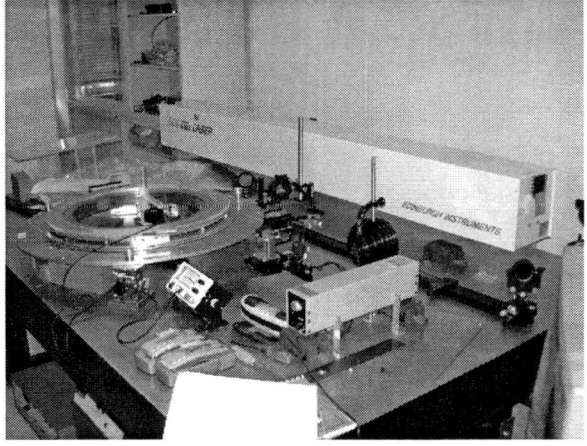

FIGURE 5. The experimental setup.

CONCLUSION

We have demonstrated that the pulse mode is the most probable EM-field in the photon storage ring of the barrel-shaped resonator when the electron beam is bunched. The results have been obtained by the characteristic equations, which we have solved using the quasi-optical approximation. Our approach allowed us to analyze the main characteristics of the pulse modes such as the gain, the pulse shape, and so on. Using this self-consistent approach, we have confirmed the basic conclusions of the previous work about the possibility of short pulse generation in the Photon Storage Ring. We have found some interesting and somewhat obscure properties of the eigenmodes with the electron beam in the barrel-shaped cavity. In this work, we have demonstrated the existence of a new type of eigenmodes called pulse modes. The pulse mode regime gain is larger by $2\pi/q\delta_e$ than the uniform mode gain.

We established that the pulse mode regime only exists when the distance between the electron orbit and the mirror surface is close to the resonance values, which is the same as the regime obtained earlier by Yamada. In general, when the electron orbit is not placed in such a resonant position, only conventional types of the eigenmodes (i.e. modes with uniform azimuth distribution) can exist, even in the case when the electron beam is a chain of bunches. One of the most interesting properties is that the EM pulse width of the pulse mode regime can be made considerably shorter than the electron bunch width. Our simulations have shown that the peak current and the radial bunch size are the main PhSR parameters, which define the lasing possibility.

We have estimated the pulse modes regime to the case of existing tabletop synchrotron [8,10], which has an exact circular electron orbit. It is a 20 MeV electron storage ring having the orbit radius 155.8 mm, the stored current 3 A, the rms bunch length 3 mm, and the horizontal beam size 0.2 mm [9]. To the wavelength 300 μm, we have obtained the pulse mode regime with the gain parameter $1-|\Gamma|\approx1\%$, and the light pulse duration of about 4.8 ps. The gain drops quickly as the wavelength becomes shorter than this value. We will continue the further analysis to optimize the synchrotron system for the shorter wavelength. The question might be a nonlinear effect in the lasing, which we could not include in the present calculation.

REFERENCES

1. Yamada H., *Japan. J. Appl. Phys.* **28**, L1655-L1668 (1989).

2. Yamada H., *Nucl. Instr. and Meth.*. **A304**, 700-702 (1991).

3. Mima K., Shimoda K., Yamada H., *IEEE J. Quantum Electronics* **27**, 2572-2579 (1991).

4. Tsutsui H., Yamada H., Mima K., Shimoda K, *Nucl. Instr. and Meth.* **A 331**, 395-400 (1993).

5. Yamada H., "Stimulated emissions in the exact circular electron storage ring", AIP Conference proceedings 367, New York: American Institute of Physics, 1995, pp.165-180.

6. Kleev A.I., Manenkov A.B. and Yamada H, *Nucl. Instr. and Meth.* **A 358**, 362-364 (1995).

7. Kleev A.I. and Yamada H., *IEEE J. of Quantum Electronics*, **39**, 820-828 (2003).

8. Yamada H., Kitazawa Y., Kanai Y., Tohyama I., Ozaki T., Sakai Y., Sak K., Kleev A.I., Bogomolov G.D., *Nucl. Instr. and Meth.* **A 467-468**, 122-125 (2001).

9. Yamada H., *Advances in Colloid and Interface Science*. **71-72**, 371-392 (1997).

10. H. Yamda, *Novel X-ray Source based on a tabletop synchrotron and its unique features*, Nuclear Instruments and Methods in Physics Research B199, 2003 pp.509-516.

Development of low energy and high brilliance x-ray source using a portable synchrotron MIRRORCLE

Y. Okazaki, N. Toyosugi, H. Yamada
Synchrotron Light Life Science Center, Ritsumeikan University, Noji-Higashi, Kusatsu 525-8577, Japan

Y. Narazaki, T. Takashima, S. Imai
College of Science & Engineering, Ritsumeikan University, Noji-Higashi, Kusatsu 525-8577, Japan

Abstract. We have developed the computer code "RTR1.0" for intensity calculations of the Resonance Transition Radiation produced by high-energy electrons crossing the multi-layer-films target. By using the program "RTR1.0", we have designed the optimum structure of the multi-layer-films target for portable electron synchrotrons "MIRRORCLE".

INTRODUCTION

"MIRRORCLE" is a portable and normal conducting electron synchrotron having the 15 cm exactly circular electron orbit for far infrared ray and hard X-ray productions [1, 2, 3]. The bremsstrahlung mechanism .has been used for the hard-x-ray production [4]. We are now challenging to the soft-x-ray production by using "MIRRORCLE". The Resonance Transition Radiation (RTR) is the one to be used for the soft-x-ray production mechanism. In the past most experiments were carried out by the high energy LINAC, but we use synchrotron for the first time. We expect higher and practical photon numbers. The RTR phenomenon appears when a charged particle with high energy crosses a periodic medium. In this case, the transition radiation, emitted at each interfaces, are interfered and coherent soft-x-rays are produced. The RTR theoretical investigation has been carried out in [4, 5]. The RTR are usually obtained using the charged particles of several GeV [6, 7]. The experimental studies for the electron energy from several tens to a few hundreds MeV was presented in 1980's [8, 9].

As the first step of the soft-x-ray source design, we have developed a computer code "RTR1.0" to investigate the multi-layer-film target. By using this program, we found the candidate materials, thickness and the number of the films for the efficient soft-x-ray emissions. To consider the RTR produced by the low-energy electron beam (for instance, by the 6 or 20 MeV beam of the "MIRRORCLE"), it is necessary to take into account the multiple coulomb scattering, x-ray absorption by the target of different thickness, and so on. These phenomena play important roles in the RTR intensity.

Major applications of the soft X-ray beam in the future will be the X-ray lithography of particularly the condensing projection method.

THEORY

When a charged particle at relativistic energy traverses the interface between two media with different dielectric constant, the transition radiation occurs [7]. Both thickness of material i ($i=1, 2$) should be greater than so-called formation lengths z_i:

$$
\begin{aligned}
z_i &= \frac{2c}{\omega \left[1 - \beta(\varepsilon_i - sin^2\theta)^{1/2}\right]} \\
&\cong \frac{4\chi\beta}{\dfrac{1}{\gamma^2} + \theta^2 + \left(\dfrac{\omega_i}{\omega}\right)^2} ,
\end{aligned}
\tag{1}
$$

where $\beta = v/c$, v is the normalized particle velocity, c the speed of light, $\chi = c/\omega$, $\gamma = (1-\beta^2)^{-1/2}$, $\varepsilon_i = 1 - (\omega_i/\omega)^2$ the permittivity of media, ω_i the plasma frequency of media i, ω the frequency, and θ the photon emission angle with respect to the particle direction. The formation length is the minimum

CP716, *Portable Synchrotron Light Sources and Advanced Applications,*
edited by H. Yamada, N. Mochizuki-Oda, and M. Sasaki

distance over which the electromagnetic wave and a charged particle can exchange energy.

The number of photons emitted per the unit solid angle from a single interface is given in [7] as:

$$F_1 = \frac{\alpha \omega \sin^2 \theta}{16\pi^2 c^2} (z_1 - z_2)^2 , \qquad (2)$$

where $\alpha = e^2/\hbar c \cong 1/137$ is the fine-structure constant.

The RTR differential cross section to the periodic target of M layers with spacing l_1 and thickness l_2 is given in [7, 10] as:

$$\frac{d^2 N(\omega)}{d\omega \, d\Omega} = F_1 \, F_2 \, F_3 , \qquad (3)$$

where $N(\omega)$ is the photon number per unit frequency interval and unit solid angle. The second factor F_2 corresponds to the interference effect between the transition radiations emitted from the two surfaces of the single layer. If the attenuation due to the electron collisions within the layer and photon absorption by the layer are negligible, then,

$$F_2 = 4 \sin^2 \left(\frac{l_2}{z_2} \right) , \qquad (4)$$

The third factor F_3 describes the coherent sum of the contributions from the each layer in the stack:

$$F_3 = \frac{\sin^2 MX}{\sin^2 X} , \qquad (5)$$

where $X=(\lambda_1/z_1)+(l_2/z_2)$.

Taking into account the soft-x-ray absorption by a target as in [9], then,

$$F_3 = \frac{1 + exp(-M\sigma) - 2 exp(-M\sigma/2) cos(2MX)}{1 + exp(-\sigma) - 2 exp(-\sigma/2) cos(2X)} \qquad (6)$$

where $\sigma = \mu_1 l_1 + \mu_2 l_2$ and μ_i are the x-ray absorption coefficients. The resonance conditions for transition radiation are

$$X = r\pi , \qquad (7)$$

$$\frac{l_2}{z_2} = \left(m - \frac{1}{2} \right) \pi , \qquad (8)$$

where r and m are positive integers. For the first resonance $r = m = 1$, the transition radiation has higher intensity.

THE PROGRAM "RTR1.0"

To design the multi-layer target for MIRRORCLE, we have developed the computer code "RTR1.0". By using "RTR1.0" one can calculate both energy and angular distribution of the RTR intensity. "RTR1.0" was made using the programming developer Borland Delphi5 for Windows. The program language is Pascal.

The "RTR1.0" can evaluate RTR intensity in cooperating with the x-ray absorption for the following 21 elements:

Lithium, Beryllium, Magnesium, Aluminum, Silicon, Potassium, Calcium, Copper, Zinc, Gallium, Germanium, Rubidium, Strontium, Silver, Cadmium, Indium, Tin, Cesium, Barium, Gold, Lead.

The calculation flow based on the RTR theory is as follows,

Step 1-1) Select two elements composing the target, calculate the formation lengths $z_{1,2}$ and each thickness for resonance conditions: $m = 1, 2, \cdots, 10$,

Step 1-2) Input each thickness of the two kind layers and the number of layers,

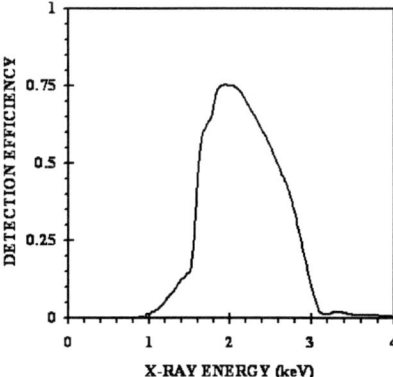

FIGURE 1. The x-ray detection efficiency for the Resist including Br.

Step 2-1) Calculate the RTR angular distribution to the given target,

Step 2-2) Calculate the RTR energy distribution to the given target.

THE TARGET DESIGN

The multi-layer target was designed in the following way:

1)   Select appropriate material and the thickness

We have calculated RTR intensity for 21 kind of materials under the following conditions: 20 or 6 MeV initial electron energies, 1 or 2 keV x-ray transition radiation energies, assume the material 1 to be vacuum, the number of layers M to be 10, and the resonance condition indices r and m to be 1 or 2, respectively.

TABLE 1. The RTR intensity for a multi-layer target in the first resonance condition (E=20(MeV)).

x-ray	1(keV)			2(keV)		
element	l_1(? m)	l_2(? m)	I_{eff} (/eV /sr)	L_1(? m)	l_2(? m)	I_{eff} (/eV /sr)
Li	1.41	1.32	7.31E-06	0.711	0.698	1.01E-05
Be	1.35	0.96	9.85E-05	0.701	0.636	2.76E-04
Mg	1.39	1.23	1.05E-05	0.706	0.683	1.41E-05
Al	1.36	1.07	2.49E-05	0.706	0.659	3.18E-05
Si	1.36	1.06	2.25E-05	0.701	0.654	3.13E-05
K	1.42	1.39	1.35E-07	0.711	0.707	7.37E-07
Ca	1.27	0.66	8.50E-05	0.687	0.548	1.05E-03
Cu	1.39	1.23	2.41E-07	0.706	0.684	1.50E-06
Zn	1.07	0.14	8.87E-04	0.598	0.214	4.30E-03
Ga	1.04	0.11	1.15E-03	0.584	0.172	7.44E-03
Ge	1.36	1.06	5.29E-06	0.701	0.654	1.09E-05
Rb	1.42	1.4	6.38E-08	0.711	0.708	1.53E-07
Sr	1.32	0.83	2.77E-05	0.697	0.603	6.05E-06
Ag	1.4	1.28	1.27E-07	0.711	0.695	1.12E-06
Cd	1.12	0.26	4.38E-05	0.635	0.325	2.12E-03
In	1.38	1.18	4.36E-07	0.706	0.676	8.32E-06
Sn	1.05	0.12	1.91E-04	0.589	0.186	7.80E-03
Cs	1.42	1.4	5.96E-09	0.711	0.709	1.19E-07
Ba	1.34	0.9	3.30E-06	0.697	0.619	7.39E-05
Au	1.4	1.28	1.25E-07	0.711	0.695	6.75E-07
Pb	1.06	0.14	1.54E-04	0.598	0.212	5.51E-03

For the material silicon, we also calculated the RTR intensity for 1.7 keV x-ray. The x-ray mass absorption curve for silicon has a big absorption peak at 1.84 keV and is very small for the energy less than 1.84 keV. The peak intensities calculated for $r = m = 1$ using the given detection efficiency in which the resist including Br (see FIGURE 1) are shown in the TABLE. 1 for 20 MeV electron. Six targets Be, Ca, Zn, Ga, and Pb give relatively large intensity. The structure of the RTR target for "MIRRORCLE" is shown in TABLE 2.

2)   Determine the upper limit of the layer number and optimize the target structure

Due to x-ray absorption by the target, the RTR intensity is saturated when the number of layers M is increased. We have calculated the dependence of RTR intensity versus M for 6 targets picked up in above procedure. These dependencies are shown in Fig. 2. The maximum layer number M_{limit} corresponds to the intersection of two asymptotes (the red solid lines in Fig. 2). The number of layers should be determined so as $M \leq M_{limit}$.

CONCLUSION AND DISCUSSION

We have developed the computer code "RTR1.0" for the intensity calculations of the Resonance Transition Radiation produced by the high-energy electrons crossing the multi-layer target. Using the program "RTR1.0", we could design the optimal structure of the multi-layer target. The Be, Ca, Zn, Ga, and Pb can been selected among 21 elements as the best elements for the target.

We could also determine the target structure, considering the following two points:

1)   We should take into accounts the Bremsstrahlung process. When the charged particle passes through the multi-layer target, the Bremsstrahlung is also produced besides the RTR. We will calculate the Bremsstrahlung cross sections using the simulation code GEANT4 or EGS4, and then we will know the influence of Bremsstrahlung on the circulating electron beam in the storage ring.

2)   We should take into account the random variation in the layer thickness and spacing, we should make the decision over the number of layers.

TABLE 2. The structure of a RTR target for "MIRRORCLE".

Electron energy: E (MeV)	Element	x-ray energy: k (keV)	Resonance condition: m	Intervals: l_1 (μm)	Thickness: l_2 (μm)	M_{limit}
20	Be	2.0	1	0.64	0.70	193
20	Si	1.7	1	0.82	0.75	77
20	Be	2.0	2	1.90	2.10	66
20	Si	1.7	2	2.25	2.47	29
6	Be	2.0	9	1.07	1.08	344
6	Si	1.7	7	0.96	0.97	73

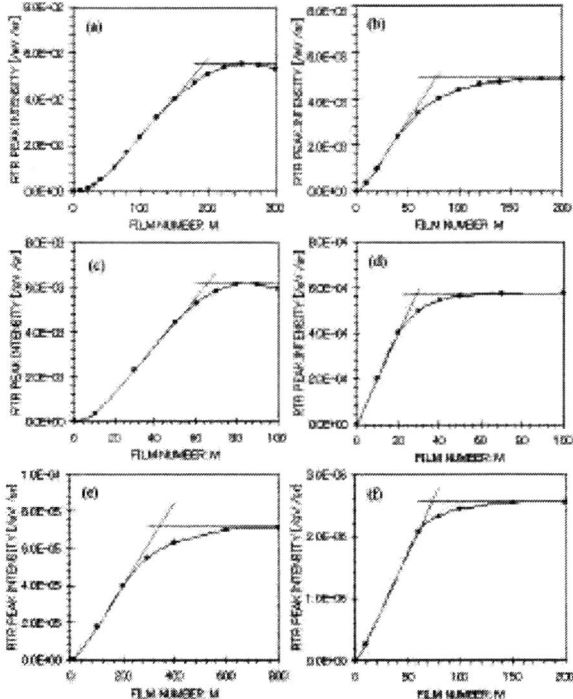

FIGURE 2. The layer number dependence on RTR peak intensity, where (a) the Be target, E=20(MeV), k=2.0(keV), m=1, l_1 =0.64(μm), l_2 =0.70(μm), (b) the Si target, E=20(MeV), k=1.7(keV), m=1, l_1 =0.82(μm), l_2 =0.75(μm), (c) the Be target, E=20(MeV), k=2.0(keV), m=2, l_1 =1.90(μm), l_2 =2.10(μm), (d) the Si target, E=20(MeV), k=1.7(keV), m=2, l_1 =2.25(μm), l_2 =2.47(μm), (e) thte Be target, E=6(MeV), k=2.0(keV), m=9. l_1 =1.07(μm), l_2 =1.08(μm), (f) the Si target, E=6(MeV), k=1.7(keV), m=7, l_1 =0.96(μm), l_2 =0.97(μm).

REFERENCES

1. H. Yamada, *Jpn. J. Appl. Phys.* **35**, L182 (1996).

2. H. Yamada, *J. Synchrotron Rad*, 1326 (2003), Invited paper for SRI97.

3. H. Yamada, *Nuclear Instruments and Methods in Physics Research* B **199**, 509-516 (2003).

4. G. M. Garibyan, *Zh. Eksp. Theor. Fiz.* **33**, 1403 (1958).

5. M. L. Ter-Mikaelian, *Nuclear Physics* **24**, 43-61 (1961).

6. A. I. Alikhanian, S. A. Kankanian, A. G. Oganessian, and A. G. Tamanian, *Physical Review Letters* **30**, 109-111 (1973).

7. M. L. Cherry, G. Hartmann, D. Muller, and T. A. Prince, *Physical Review* D **10**, 3594-3607 (1974).

8. M. J. Moran, B. A. Dahling, P. J. Ebert, M. A. Piestrup, B. L. Berman, J. O. Kephart, *Physical Review Letters* **57**, 1223-1226 (1986).

9. M. A. Piestrup, D. G. Boyers, C. I. Pincus, Qiang Li, G. D. Hallewell, M. J. Moran, D. M. Skopik, R. M. Silzer, X. K. Maruyama, D. D. Snyder, G. B. Rorthbart, *Physical Review* **45**, 1183-1196 (1992).

10. C. W. Fabjan and W. Struczinski, *Physics Letters* **57B**, 483 (1975).

The Development of Hard X-ray Microscopy with MIRRORCLE-6X

T. Hirai[1], T. Tokunaga[1], H. Yamada[2], M. Sasaki[1], D. Hasegawa[3], Y. Ogasaka[4], and H. Yamashita[4]

(1)Faculty of Science and Engineering, Ritsumeikan University
(2) Synchrotron Light Life Science Center, Ritsumeikan University
(3)Photon Production Laboratory Ltd.
(4) Faculty of Science, Nagoya University

Abstract. A laboratory-scale hard X-ray microscope utilizing the portable synchrotron named MIRRORCLE-6X is developed in our laboratory. MIRRORCLE-6X is a X-ray source suitable for hard X-ray microscopy as a result of the X-ray source size of the order of micron, and highly brilliant hard X-rays. Furthermore, when effective focusing elements of hard X-rays are used for MIRRORCLE-6X, this machine could be used for non destructive inspection with high resolution of biological and engineering. We designed hard X-ray optical elements for MIRRORCLE6X and started fabricating X-ray mirrors.

INTRODUCTION

The portable synchrotron MIRRORCLE-6X is a X-ray source suitable for hard X-ray microscopy, because the x-ray produced by MIRRORCLE-6X has the following two characteristics. (1) The effective source spot is determined by the X-ray target which can be the order of micron. (2) Brilliant hard x-rays are produced by MIRRORCLE-6X, because the electrons which bombard targets are re-circulating and reused. Therefore, non-destructive inspection can be carried out with high magnification and high resolution. MIRRORCLE-6X can be utilized for a projection X-ray microscope without using optical elements. Furthermore, when effective focusing elements of hard X-rays are used for MIRRORCLE-6X, it is possible to utilize this machine for nondestructive inspection with high resolution of biological and engineering specimens. Therefore, X-ray optical elements are being developed for a hard X-ray microscope beam line designed for MIRRORCLE-6X.

Recently, high-resolution microscopes have been constructed in synchrotron facilities using a Kirkpartrick-Baez grazning incidence mirror, a Fresnel zone plate, a capillary tube ,and a compound refractive lens. Among many x-ray optical elements, a Wolter type mirror is one of the most suitable for hard X-ray imaging. Due to its small coma and absence of chromatic aberration, it can be used for polychromatic X-rays. In addition, its numerical aperture is not as small as a zone plate, which means that its resolving power is relatively large. Consequently, the zone plate cannot be used to focus X-rays which are characterized by wide radiation-angles. Therefore, we designed a Wolter type multiplex multilayer mirror as a condenser mirror of a hard X-ray microscope (5-10keV).The Wolter type mirror is one of the well-known grazing incidence mirrors for X-ray microscopes [4] and X-ray telescopes [1],[5].

MAGNIFIED X-RAY IMAGING

At first we evaluated spatial resolution of magnified X-ray images obtained using MIRRORCLE-6X. To evaluate spatial resolution , we prepared sample composed by several Pb lines, which are located parallel to each other, within a plane perpendicular to the incident X-ray, and are separated by different thickness air gap. The samples was placed at a distance

CP716, *Portable Synchrotron Light Sources and Advanced Applications,*
edited by H. Yamada, N. Mochizuki-Oda, and M. Sasaki

of 325 mm from the source spot. A diagram of the experimental configuration is shown on Figure 1. X-ray targets were prepared from a metal wire, whose size is the order of micron. A medical X-ray film detector was located at a distance of 1300 mm from the samples. As a result, the spatial resolution was better than 100 μm (5 line/mm) in the 5 times magnified image. Figure 2 shows the 1.5 times-magnified X-ray image of a thylatron, which is made of ceramics and heavy metals. Thin filaments, electrodes and thin metal wires in the thylatron are clearly seen. The good performance, of the X-ray target was demonstrated by obtaining high spatial resolution of magnified X-ray image. When the sample is placed close to the source spot, and the detector is placed at a long distance from the sample, a higher degree of magnification of the image is obtained. Therefore, when the sample is located at focus point by use of X-ray focusing elements for MIRRORCLE-6X, this machine can be utilized for hard X-ray imaging microscopy.

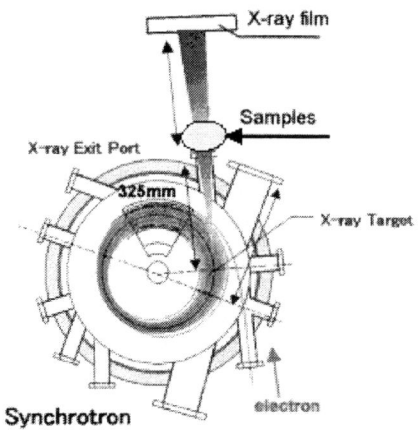

FIGURE 1. Experimental setup for analyzing spatial resolution of the magnified imaging.

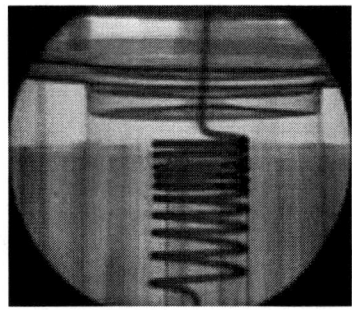

FIGURE 2. Magnified x-ray image of a thylatron. (One half times magnification)

DESIGN OF HARD X-RAY MIRROR

We designed a Wolter type I mirror [3], with grazing incidence optics. It consists of two axially symmetric confocal surfaces of revolution, while each one of them contains a hyperboloid and a paraboloid. One mirror reflects X-rays generated at large radiation angles and transforms them into parallel X-ray beam. When symmetrical Wolter type mirror is located along the optical axis, X-rays with large radiation angles can be focused by 4times-reflections. Figure 3 shows the parameters of one of the mirrors. Furthermore the surface of the mirror is coated a multilayer which consists of Pt and C layers. The multilayer mirror can reflect the hard X-ray with high efficiency, and multiplex mirror increases the focusing efficiency, multiplex wolter type mirror is a set of several Wolter type mirrors which have same optical axis. Figure 4 illustrates the principle of operation of Wolter type multiplex mirror.

The shape of the Wolter type mirror is determined by the following parameters: the incidence angle (4α) and the radius (r_0) of the intersection circle of the two confocal surfaces, the distance from the axial ray focus to the intersection plane of the paraboloid and hyperboloid (z_0), the total length of the mirror (L), and the thicknesses of the mirrors (t) [2]. The thicknesses of the individual layers of the multilayer are optimized to achieve maximum X-ray energy reflected by the mirror.The multiplayer materials of multilayer are selected to be Pt (heavy element) and C (light element). The γ -value which is the ratio between the thickness of a heavy element layer to a light element layer is chosen to be 1/2, because the reflectivity of the first order x-ray has high efficiency, and these materials do not diffuse into each other. Optimized values of the above parameters are given in Table 1.

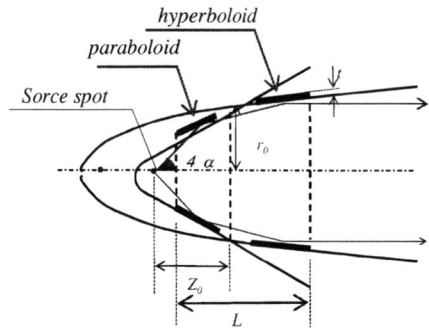

FIGURE 3. Geometrical parameters of one of the mirrors (Wolter Type I mirror)

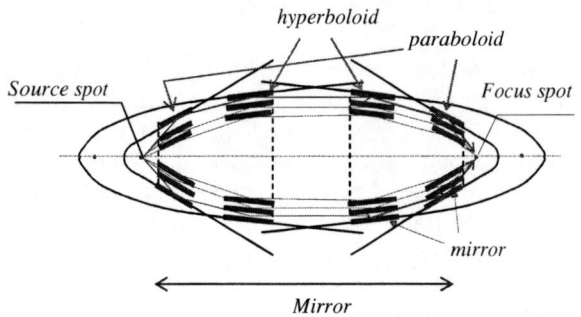

FIGURE 4. Principle of operation of the multiplex Wolter Type I mirror

Table.1 Figure parameter examples of the designed Wolter type mirror

Parameters of the optimized mirror

X-ray energy	10keV(Wave length 0.12nm)
Material of Mirror	Pt/C multilayer
Material of substrate	Ni
Diameter(front)	44.1mm
Diameter(intersection) r_0	51.5mm
Diameter(Rear)	54.5mm
Length of Mirror L	199mm
Length of hyperboloid Lp	89mm
Length of paraboloid Lh	100mm
Grazing incident angle	(hyperboloid Average) αh 0.86deg.
Grazing incident angle	(paraboloid Average) αp 0.77deg.
Forcal length	377.2mm
Distance between mirror and source point	377.2mm
X-ray radiation angle	58.4-55.2mrad (3.35deg.-3.16deg.)

Pt/C Multilayer (γ-value 0.5)

Period length of paraboloid	d_p	4.5nm
Period length of hyperboloid	d_h	5.1nm
Period number of paraboloid	N_p	39
Period number of hyperboloid	N_h	29

Coated multilayer mirror with these parameters can reflect and focus X-rays, with radiation angle between 58mrad and 55mrad. It's total X-ray focusing power is calculated as a product of reflectivity and the area of mirror aperture. Four times-reflections and areas of mirror aperture are presented in Table 3. The calculated mirror total focusing power is 2×10^4 times for ideal mirror surface when 8 multiplex Pt/C multilayer mirrors are used.

Table.2 Four times-reflections and area of mirror aperture

	reflectivity	Area of aperture	
Mirror 1	22%	10.57%	(outside mirror)
Mirror 2	27%	8.56%	
Mirror 3	34%	6.86%	
Mirror 4	41.9%	5.43%	
Mirror 5	45.8%	4.23%	
Mirror 6	33.9%	3.23%	
Mirror 7	39.8%	2.41%	
Mirror 8	55.7%	1.74%	(inside mirror)

FABRICATION OF THE MIRROR

High efficiency grazing incidence mirror is based on total reflection of X-ray on very precise geometrical shape and super smooth surfaces. The most popular method for fabrication of the grazing incidence mirroris the replication method which uses a master mandrel.The fabrication process is visualized in Figure 5. In principle it is possible to fabricate many replicated mirrors using one master mandrel. We selected master mandrel made of oxygen-free copper, because it is practically used for Ni electronic casting. Tolerated surface roughness of the mirror can be determined from Rayleigh's Imaging Criterion. Tolerance between actual surface and ideal surface h must be less than $\lambda/8\sin\theta$, where λ is the wavelength and θ is the incident angle. The tolerated surface roughness for our designed mirror (Table.1) is less than 1.1nm. Master mandrel is fabricated by ELID grinding. In this technology, the axial figure deviation from the designed parameters is less than 1μm, and the surface roughness measured by interferometer is about 2-3nm r.m.s. Figure 6. shows the master mandrel made of oxygen-free copper and its measured surface roughness is presented in Table. 3 .

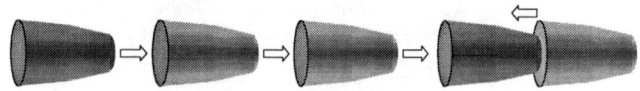

1.Master mandrel 2.Coated multilayer 3.Ni electronic forming 4.Reeplicated multilayer and substrate

FIGURE 5. Method for fabrication of Wolter type multilayer mirror. (Replication method)

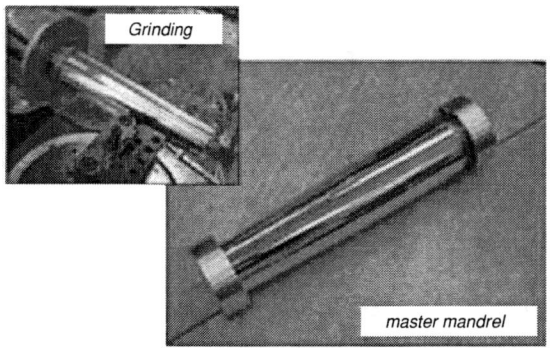

FIGURE.6 ELID grinding master mandrel made of oxygen-free copper. ($50\phi \times 200$ mm)

TABLE.3 Surface roughness of the master mandrel fabricated by ELID grinding. (The measurement area is 140μm$\times$110 μm).

Measurement point	Roughness(r.m.s)
a	2.572 nm
b	2.479nm
c	2.16nm
d	2.219nm
e	2.409nm
f	2.091nm

```
        a  b  c      d  e  f
        X  X  X      X  X  X
```
Measurement point

At the next step, the surface of the master mandrel is coated by Pt/C multiplayer prepared using a DC Magnetron sputtering system (see Fig.7). After that, the mandrel coated by Pt/C multiplayer is coated by Ni plating used as the mirror substrate, the Pt/C multiplayer and them are mould-released simultaneously from the master mandrel As a result, the shape and the roughness of the master mandrel are replicated to the mirrors. We performed successful replication between Ni and oxygen-free copper using Ni electrocasting. However, replication between Pt/C multilayer and oxygen-free copper has not been examined, yet. Therefore we are trying to mould-release Pt/C multilayer from oxygen-free copper by using test plates. Figure 8 shows an oxygen-free copper test plate coated by Pt and Pt/C multilayer, and a test plate stripped from Pt and Pt/C multilayer by Ni electronic casting. The Ni substrate was replicated, but the Pt and Pt/C multilayer were not replicated. Correspondingly, we will try to replicate the Pt/C

multilayer coated by Ni to include one more layer between the oxygen-free copper and multiplayer.

FIGURE.7 DC magnetron sputtering system in Nagoya University.

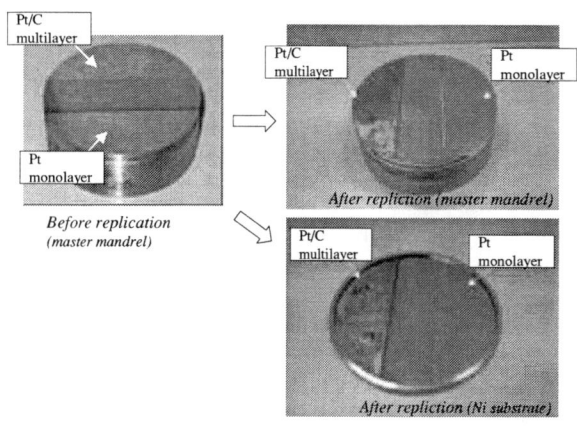

FIGURE.8 Mould-released test by using oxygen-free copper test plate.

REFERENCES

1. K. Tamura, K. Yamashita, H. Kunieda, Y.Tawara, A. Furuzawa, K. Haga, G. S. Lodha , Nakajyo, N. Nakaura, Takashi Okajima, Osamu Tsuda , P. J. Serlemitsos, J. Tueller, R. Petre, Y. Ogasaka, Y.Soong and K. Chan, Proc. SPIE,3113, P. 160-168, 1997.

2. L. P. VanSpeybroeck and R. C. Chase., Appl. Opt 11, 2 1972, p. 440-445

3. H. Wolter, Ann. Phys, 10,1952, P.94-114

4. Aoki, S., Ogata, T.,Sudo,S. & Onuki,T.(1992). Jpn. J. Appl. phys.31, 3477-3480

5. G. S. Lodha, K. Yamasita, H. Kunieda, Y. Tawara, J. Yu, Y. Nanba, and J. M. Bennett, Appl.Opt 37, 22 1998, P170-188.

NOVEL EDGE-ENHANCED X-RAY IMAGING UTILIZING MIRRORCLE

T. Hirai, S. Maki, Y. Sonoda, H. Yamada, M. Sasaki, D. Hasegawa*

Faculty of science and Engineering, Ritsumeikan University
1-1-1 Nojihigashi, Kusatsu, Shiga 525-8577, Japan
* on leave from Photon Production Laboratory. Ltd.

Abstract. Brilliant hard X-rays are produced by the portable synchrotron named MIRRORCLE. The X-ray images taken by MIRRORCLE show enhanced edge effect even with contact imaging. Bodies composed of soft tissues are distinguished in spite of the predominantly hard X-ray components. The imaging mechanism, which is different from the phase contrast, was studied. It was found that the edge effect is partly due to the micron order x-ray source spot size, and MeV region hard X-rays. The characteristic of the medical diagnosis X-ray images and nondestructive inspection X-ray images taken by MIRRORCLE are studied in this paper.

INTRODUCTION

Novel Portable synchrotron named MIRRORCLE [1] provides hard X-rays utilizing stored electron beam with energy only 20 MeV or 6 MeV. Hard X-ray is produced due to the collisions of electrons by the atomic force of a small target placed on the electron orbit, and not by the magnetic force. The X-ray produced by MIRRORCLE has the following characteristics. (1) The effective source spot size is determined by the target. Therefore it can be order of micron, which is much smaller than that of SR source. High brilliance similar to that of SR can be obtained because the electrons are re-circulating and reused. (3) The radiation angle is determined by the kinematics $1/\gamma$ as same as that of SR. MIRRORCLE operates with the low energy electrons and can provide quite wider irradiation field than that of SR. (4) Polychromatic hard X-ray is easily generated, in which X-ray energy is extended up to the electron energy, and the X-ray energy is much higher than that of SR sources. These characteristics are rather useful for the imaging of larger samples compared to SR.

X-ray imaging was carried out using MIRROCLE by a conventional method. High contrast image, which has enhanced boundary of the structures were taken when the sample is set on the imaging device. The X-ray images obtained using MIRROCLE are similar to the "phase contrast" image [2], but the latter use a technique, in which the imaging device and the sample are separated or located within an optical systems. Thus our X-ray imaging is a new type. Therefore the mechanism of this new X-ray imaging is studied. Furthermore, the dependence of the image contrast on the specimen density is investigated to determine its applicability to medical diagnosis X-ray imaging.

EXPERIMENT

Edge enhanced imaging

In the experiment of edge enhanced imaging, the contrast of X-ray images obtained using MIRRORCLE-20 was evaluated. For analyzing image contrast, sample plates made of several materials with different thickness were prepared. The samples were placed at a distance of 1200 mm from the source point and attached to the front side of imaging device such as medical X-ray film, or imaging plate (IP). The irradiation port containing Be window with diameter $\phi = 35$ mm was located at 700 mm away from source point. X-ray target was prepared from lead (Pb) sphere with $\phi = 0.1$ mm. To compare with conventional X-ray

CP716, *Portable Synchrotron Light Sources and Advanced Applications,*
edited by H. Yamada, N. Mochizuki-Oda, and M. Sasaki

imaging, we also obtained images of the same sample utilizing 100 kV X-ray tube. Simulation using EGS4 was carried out to compare the profile of image data recorded on the IP, and to analyze quantitatively the edge effect.

Dependence of the image contrast on the density of the sample

In second experiment of X-ray imaging, the contrast of X-ray images obtained using MIRRORCLE-6X, was evaluated depending on the density of liquid samples. For analyzing density resolution, potassium secondary phosphate aqueous solutions (K_2HPO_4 aq.) with different densities were prepared. K_2HPO_4 aq. has effective atomic number (Z_{eff}) and electron density (ρ_e) which are almost equal to those for bodies such as muscle and fat composed of soft tissues. These solutions were poured in 20 mm thickness acryl box, and X-ray penetration was measured simultaneously using Digital flat panel. The samples were placed at a distance of 1000 mm from the source point and Digital flat panel was placed at a distance of 2000 mm from the source point. The field of view was limited by the port size with $\phi = 35$ mm Capton window, which was 325 mm away from source point. In this case, the X-ray irradiated erea of the flat Panel has ϕ 200 mm. The X-ray target was 27μm Pb foil.

RESULTS AND DISCUSSION

Edge enhanced imaging

In Figure. 1 X-ray images of sample plates on X-ray films are compared. The left images are obtained using MIRRORCLE and the right ones obtained by conventional X-ray tube operated at 100 keV. The X-ray image obtained using by MIRRORCLE shows significant edge effect. The difference between these two kinds of images is quite striking. All of the X-ray images obtained using X-ray tube are absorption-images. It is seen that edge effects in the chemical light elements (g, h, l, m) and thin structures (a) is significant when MIRRORCLE is utilized. X-ray imaging by MIRRORCLE is further investigated to study this effect.

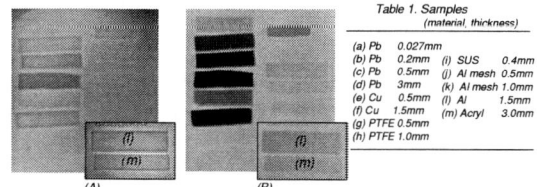

Table 1. Samples (material, thickness)			
(a) Pb	0.027mm		
(b) Pb	0.2mm	(i) SUS	0.4mm
(c) Pb	0.5mm	(j) Al mesh	0.5mm
(d) Pb	3mm	(k) Al mesh	1.0mm
(e) Cu	0.5mm	(l) Al	1.5mm
(f) Cu	1.5mm	(m) Acryl	3.0mm
(g) PTFE	0.5mm		
(h) PTFE	1.0mm		

FIGURE 1. (A) X-ray image obtained using MIRROCLE, and (B) X-ray image obtained using 100 kV X-ray tube. Both X-ray images are recorded on medical X-ray film. X-ray tube used is *Radio Flex 250 EGS 2* produced by RIGAKU.Ltd (Tube voltage 100 kV, current 5 mA) is used.

X-ray edge enhanced imaging of sample plates was carried out with IP in order to evaluate the observed edge pattern. X-ray images of sample plates on IP, using two kinds of X-ray targets, are shown in Figure 2 (A) and 2 (B), and also some profiles are shown in Figure 2 (C). The X-ray targets chosen are spheres with $\phi = 3$ mm in (A) and $\phi = 0.1$ mm in (B).In general sharply focused images occur for smaller target B as expected. The profiles shown in Figure 2(C) indicate that the X-ray flux on the IP is amplified within the plates for target A, but not for target B. The different edge enhancement effect is obtained due to the different X-ray target size. Furthermore, our measurements between two targets are showed are different. We understand that the X-ray spectra must be different between two targets. For the target A the spectrum contains higher energy components. We also proved that there is X-ray amplification within the specimen. The reason for this amplification is the cascade shower due to electron pair creation. We assume that the low energy X-ray components are damped in thicker X-ray target.

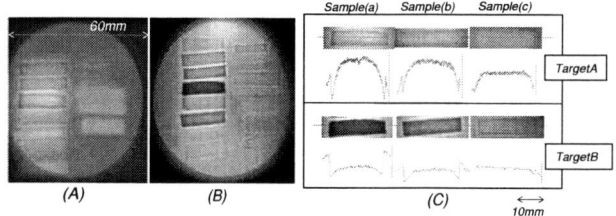

FIGURE 2. The X-ray edge enhanced images of sample plates obtained using IP. The X-ray target chosen is $\phi = 3$ mm in (A) and $\phi = 0.1$ mm in (B). The differences between (A) and (B), are quite striking especially with sample plates a, b and c made of 3mm thick Pb, 1.5mm thick Cu, 1.5mm thick Al, respectively. Differences in the profile are seen most clearly in (C).

Monte Carlo Simulation

Dose distributions on 20μm thick Ag film were calculated by Monte Carlo simulation code EGS4 [3]. The other input parameters for this simulation are the same as for the experimental setup. The Sample is a 3mm thick Pb plate. We chose different X-ray energies. In one case, monochromatic energies of either 100 keV or 3 MeV is assumed as shown in Figure 3(A). In another case, both 100keV and 3MeV are generated simultaneously as seen in Figure 3(B). We learned that at least two colors are necessary to describe the edge effect. High energy X-ray causes cascade shower within the samples and results in amplification of X-ray's number, and low energy X-ray causes absorption in the sample. Edge enhancement is performed by superposition of two X-rays with high and low energies. The experimentally obtained profiles in Figure 2(c) using target B are in good agreement with the simulation using two colors. We believe that the edge enhanced image is obtained due to the wide range spectrum of X-rays from MIRRORCLE including abundant 100 keV region as well as some in the few MeV region .

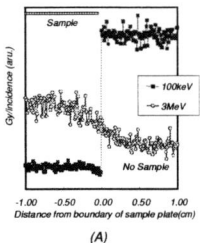
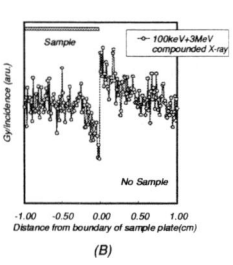

FIGURE3. Dose distributions calculated by the code EGS4 are shown. Monochromatic X-ray of either 100keV or 3MeV is manipulated in (A), and two X-rays including 100keV and 3MeV are generated simultaneously in (B).

The sample's density dependence of the image contrast

X-ray images of K_2HO_4 aq.(11 wt.%) and water on Digital flat panel are shown in Figure 3(A). The left image is for K_2HPO_4 aq. and the right one is for water obtained using the MIRRORCLE. The contrast of X-ray image was investigated by measuring intensity of penetration of X-ray along the blue line drawn in Figure 3(A). The intensity ratio of the X-ray penetration for K_2HPO_4 aq. with different densities and water is shown in Figure 3(B). It was found that difference in densities of < 1% can be distinguished by image contrast with Digital flat pane, when a sample and water with the same thickness and effective Z number are used. Therefore bodies composed of soft tissues can be distinguished utilizing MIRRORCLE.

We expect that tissue having a slightly different density compared to other tissues, such as cancer tissue, can be recognized by MIRRORCLE.

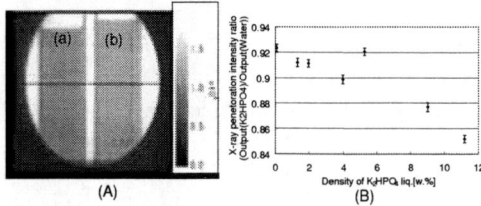

FIGURE 4. (A) Digital flat panel image obtained using by MIRROCLE-6X, and (B) density dependence of the ratio between the intensities of X-ray penetration for solute and water.

CONCLUSION

The potential using MIRRORCLE as a novel X-ray imaging device is demonstrated. The polychromatic X-rays including few MeV components and the small X-ray source spot size lead to novel edge enhanced X-ray imaging. As a result, it is found that images of light chemical elements can be obtained utilizing hard X-rays. Fig. 5 shows X-ray image is a good example of enhanced imaging for Medical diagnosis. We expect that MIRRORCLE can be useful for both cancer therapy and medical diagnostic imaging.

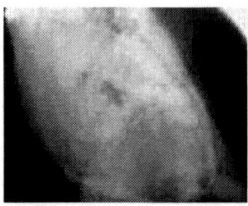

FIGURE5. X-ray image of a chicken obtained utilizing MIRROCLE-6X with X-ray film. The boundaries of soft tissues and bones are clearly seen at the same time due to the enhanced edge effect.

REFERENCES

1. H.Yamada. ; Nucl.Instr.and Meth.in Phys. Res B 199 (2003) 509-516

2. S.W.Willkins, T.E.Gureyev, D.Gao, A.Pogany and A.W.Stevenson.; Nature 384 (1996) 335

3. Nelson, W.R.; Hirayama, H.; Rogers, D.W.O. The EGS4 code system.Stanford, CA: Stanford Linear Accelerator, Stanford University; SLAC-265; 1985

X-Ray Fluorescence Analysis of Heavy Elements with the Portable Synchrotron MIRRORCLE

H. Saisho[*], J. Hirano[†], T. Hirai[†], and H. Yamada[*†]

[*]Synchrotron Light Life Science Center, Ritsumeikan University, Kusatsu, Shiga 525-8577, Japan
[†]Faculty of Science and Engineering, Ritsumeikan University, Kusatsu, Shiga 525-8577, Japan

Abstract. X-Ray fluorescence analysis of five heavy elements (Pb, Pt, W, Sn, and Ag) was performed using the portable synchrotron MIRRORCLE-20. The characteristic K X-rays (Kα and Kβ) were clearly observed for all of the five samples. This verifies that MRRORCLE-20 generates X-rays of energy higher than that of the Pb K absorption edge (88.0 keV). In addition to these characteristic K X-rays, several peaks with short lifetimes were also observed in all the samples.

INTRODUCTION

This paper describes the results of K X-ray fluorescence measurements for heavy elements using radiation from MIRRORCLE-20 (electron energy : 20 MeV, out-diameter : 1.2 m, orbit radius : 0.15 m) [1, 2]. SPring-8 has been the only synchrotron light source in Japan, which enables to analyze heavy elements such as Pb and Pt by the characteristic K X-ray, since it generates X-ray energies higher than those of their K absorption edges. We could, however, for the first time demonstrate the K X-ray analysis of heavy elements by the portable synchrotron. This paper shows the potential of the portable synchrotron as well as the new research field by the X-ray fluorescent analysis of heavy elements.

EXPERIMENTAL

Samples used

Five kinds of samples, Pb (88.0 keV), Pt (78.4 keV), W (69.5 keV), Sn (29.2 keV), and Ag (25.5 keV) were used for X-ray fluorescence spectrum measurements. Values in parentheses denote the energies of K absorption edges. Pb, W, Sn, and Ag were metal sheets, each of 0.50 mm in thickness, and Pt was a 0.03 mm thick metal foil. All the samples

CP716, *Portable Synchrotron Light Sources and Advanced Applications,*
edited by H. Yamada, N. Mochizuki-Oda, and M. Sasaki

purchased from the Nilaco Corporation were cut to squares with about 40 x 40 mm^2.

Detector system

A germanium detector (Canberra Model: GUL0035P) with an active area of 30 mm^2 was employed to measure characteristic X-rays emitted from the samples. The detector type was the ultra low energy one. The observed energy resolutions (FWHM) were 137 eV for the 5.9 keV peak from ^{55}Fe and 447 eV for the 122.0 keV peak from ^{57}Co, respectively. With MIRRORCLE-20 the X-ray beam is widely spread in more than 25mrad. The X-ray exit port of 30 mmϕ is located at a 700 mm distance from the source point. The samples are placed at 400 mm from the port (incidence angle of 45 degrees), thus 1.1 m from the source point as seen in Fig. 1. About 27 mmϕ area of the sample is irradiated. Since the sample - detector spacing was 20 mm (takeoff angle of 45degrees), 14% of emitted X-rays are collected. The detector head is shielded from unnecessary radiation using some Pb blocks to collect X-ray fluorescent signals effectively.

The X-ray flux

Total photons on the sample should be estimated for the X-ray fluorescence analysis. The X-ray total flux on the sample is said to be 1.6 x 10^7 photons/s, 0.1% band when MIRRORCLE-20 is operated at the rate of 40Hz injection and 2 mA injector beam current in this experiment. The counting times were 2,000 s for Pt and 1,200 s for Pb, W, Sn, and Ag. All the measurements were carried out in air atmosphere.

RESULTS AND DISCUSSIONS

The K X-ray fluorescence spectra obtained are shown in Figs. 2-8. All of these spectra show nice peak to background ratios. The peaks of Kα1, Kα2, Kβ1', and Kβ2' for Pb are very clearly observed in Fig. 2 (According to Siegbahn's notation, Kβ1'=Kβ1+Kβ3+Kβ5, and Kβ2'=Kβ2+Kβ4.). Pt Kα1 and Pt Kα2 are also clearly observed in Fig. 3. The presence of Pb Kα1 is from the Pb blocks used for shielding, which interferes with Pt Kβ1'. Figure 4 exhibits four peaks of W Kα and Kβ distinctly. In the case of Sn, Kα is, however, composed of only one peak as shown in Fig. 5. This is because the

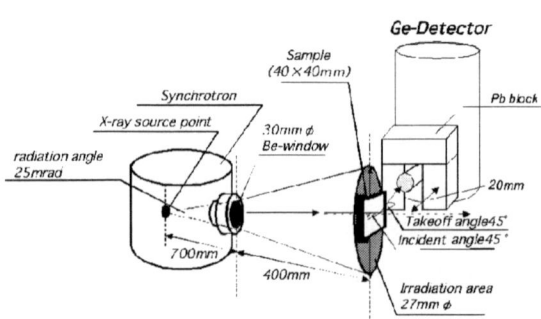

FIGURE 1. Schematic drawing of experimental setup.

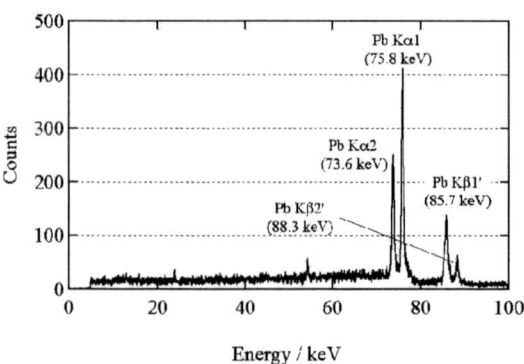

FIGURE 2. X-Ray fluoresence spectrum of Pb.

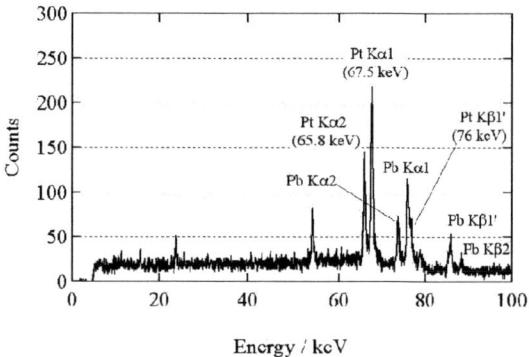

FIGURE 3. X-Ray fluoresence spectrum of Pt.

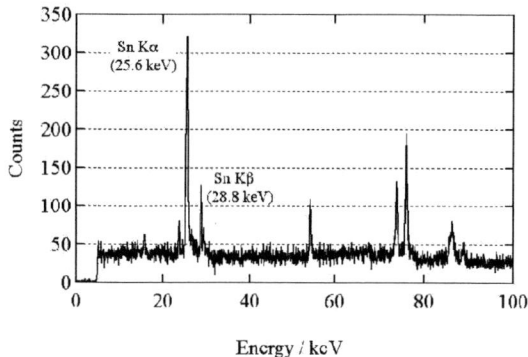

FIGURE 5. X-Ray fluoresence spectrum of Sn.

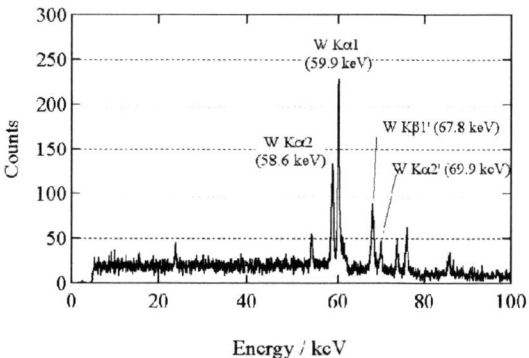

FIGURE 4. X-Ray fluoresence spectrum of W.

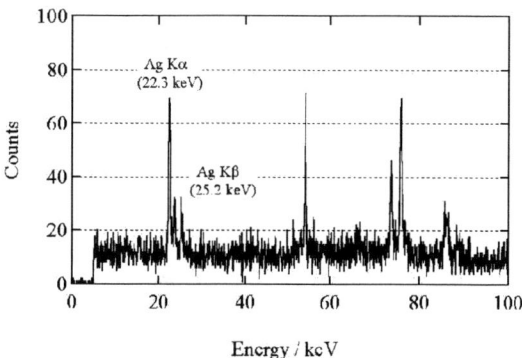

FIGURE 6. X-Ray fluoresence spectrum of Ag.

energy difference between Kα1 and Kα2 is very close. The present X-ray detector is poor in the energy resolution to distinguish two peaks. The peak of Sn Kβ is also seen as one peak. Figure 6 shows the K X-ray fluorescence spectrum of Ag. In this spectrum, both Ag Kα and Kβ are seen to be one peak as in the case of Sn. All the peak energies assigned to characteristic K X-rays show good agreement with the previously reported energy values [3].

The five spectra above-mentioned have two additional peaks at 23 keV and 53 keV, which are not assigned to the characteristic K X-rays. Furthermore, we observe some other peaks in the higher energy range from 130 to 200 keV (see Fig. 7). The peak profiles from 130 to 200 keV in the W metal sample of

Fig. 7 are similar to those of other four metal samples. In order to search the origin of these peaks we measured the spectrum after beam stop. Figure 8 is the spectrum from the W metal sample taken for 300 s and 2 min after stopped beam. The other four metal samples also show a very similar spectrum to the W metal sample taken 2 min after the beam stop. The spectral intensities up to about 130 keV were nearly equal to the background level taken without beam. The peaks below 100 keV disappear completely, while those above 130 keV remain but decay rapidly. The peaks from 130 to 140 keV decay but remain in the same spectrum profile, but the new peak at the vicinity of 200 keV was observed instead of the disappearance from the spectrum in the energy range from about 160 to 200 keV.

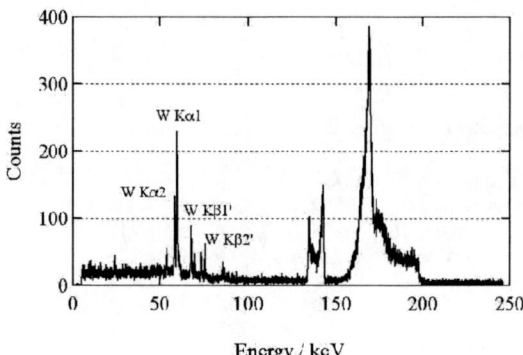

FIGURE 7. X-Ray fluorescence spectrum of W (energy range up to 250 keV).

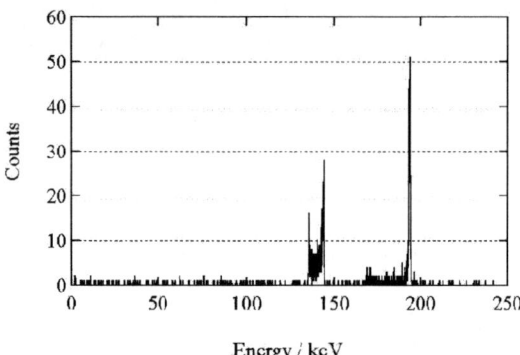

FIGURE 8. X-Ray fluorescence spectrum of W (taken for 300 s after 2min of the beam stop).

Those non-characteristic K X-rays including 23 keV and 53 keV in Figs. 7 and 8 have very short lifetimes. Those could be resulted from radioactive nuclides produced by photonuclear reactions like (γ, γ') or (γ, n) with high-energy photons. We have studied the photonuclear reactions by possible targets such as N_2, O_2, and Ar in air components, and Cu and Fe used for MIRRORCLE, as well as Pb, Pt, W, Sn, and Ag. We have, however, not yet succeeded in assigning those unknown peaks.

CONCLUSIONS

We could definitely demonstrate that the portable synchrotron MIRRORCLE-20 has a great deal of potential for K X-ray fluorescence analysis of heavy elements such as Pb, Pt, W, and rare earth metals. The K X-ray fluorescence measurements lead to a further increase in sensitivity and accuracy of heavy element analysis since the K X-ray with higher intensity and energy than those of the L X-ray of the same atom minimizes the interference from a sample matrix.

Studies on chemical states (coordination number, oxidation number, number of unpaired electrons, and so on) based on the changes in fine features of K X-ray lines (such as shifting in energy, distortion of line shape, and intensity changes) of heavy elements will probably become feasible using the portable synchrotron MIRRORCLE and provide us useful information on life and materials science.

REFERENCES

1. H. Yamada, *Nucl. Instrum. Methods*, **B199,** 509-516 (2003).
2. H. Yamada, *J. Synchrotron Rad*, **5,** 1326-1331 (1998).
3. R. B. Firestone, C. M. Baglin, S. Y. F. Chu (Ed.), *Table of Isotopes (Eighth Edition)*, John Wiley & Sons, New York, 1999

Problem of radiation safety in the diagnosis using MIRRORCLE

Y.Suetsugu, H.Yamada, T.Takuya, D.Hasegawa[1], T. Hirai, and M. Inoue

Ritsumeikan University Faculty of Science and Engineering, [1]Photon Production Laboratory,Ltd.

Abstract. We have evaluated the absorption dose by simulation in a phantom irradiated by the X-ray from the 6MeV electron synchrotron MIRRORCLE-6X. Comparison with the dose by the X-ray from a 100keV X-ray tube was made. It shows that the averaged dose integrated over the phantom to make one flame of the X-ray image in the case of MIRRORCLE-6X is much smaller than that in the case of the X-ray. We have also measured the dose using MIRRORCLE-6X under the various conditions. We could confirmed this results experimentally by observing dose at the surface, although it was slightly higher than expected.

1.Introduction

MIRRORCLE-6X is a portable electron synchrotron that is developed at Ritsumeikan University. It has a potential to provide the X-ray for both cancer therapy and diagnosis simultaneously. A circular microtron is used as an injector that accelerates electrons up to 6MeV. The electron beam from the microtron is injected into MIRRORCLE-6X by means of the 1/2 resonance and stored at an exact circular orbit. The electron orbit radius is 15cm. The X-ray is produced by Bremsstrahlung with very small targets placed in the electron orbit. MIRRORCLE-6X generates high energy and high flux X-rays with low emittance at high efficiency. Tabletop X-ray source must be extremely useful in medical applications.

Because the X-ray from MIRRORCLE-6X has much higher energy than a usual X-ray tube, we expect that it will make lower effect on human body in case of imaging than the usual X-ray tube. Our purpose is to study the effect on human body during the imaging by MIRRORCLE-6X.

2.Radiation safety in a clinical imaging

2.1 Calculated dose distribution by simulation code EGS4

In the simulation by the code EGS4, we assume 1Gy

X-ray absorptions in the Ag-film behind the phantom. We compare the dose distribution induced by "MIRRORCLE-6X" with that by ordinary X-ray tube. Fig.1 shows the arrangement in this simulation. X-rays are also generated by the codes EGS4 as the spectrem is shown in Fig. 2, where the 6 MeV elecvtron beam and 1mm thick Pb target is assumed.

Fig.3 shows the result of the simulation. The total absorption dose over the phantom in case of the X-ray tube is 42.5Gy. In case of MIRRORCLE-6X with a 1mm-Pb target the total dose is 3.15Gy, and the dose with a 1mm-Al target is 2.66Gy. "MIRRORCLE-6X" produces smaller effect than the 100keV X-ray tube by factor 1/15. Large amount of X-rays from the conventional X-ray tube is absorbed at the surface of the body.

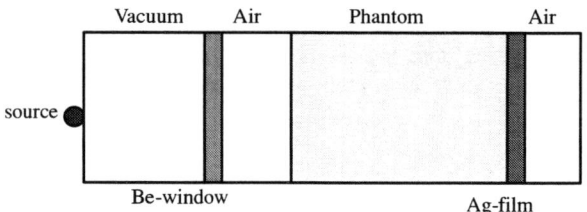

Fig.1 Arrangement in the calculated simulation

CP716, *Portable Synchrotron Light Sources and Advanced Applications,*
edited by H. Yamada, N. Mochizuki-Oda, and M. Sasaki

a

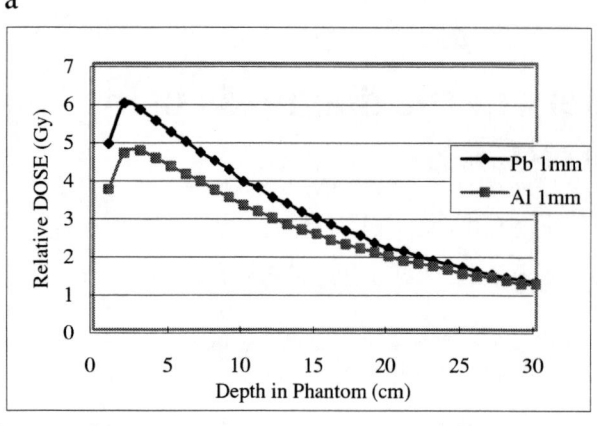

b

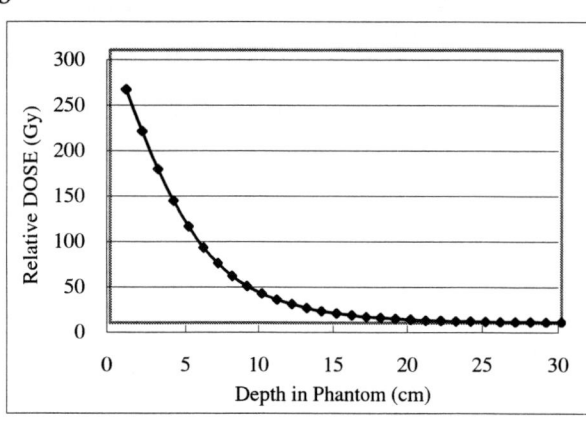

Fig.3 Dose distribution by X-rays simulated with EGS4
a, "MIRRORCLE-6X"   b, 100keV X-ray tube

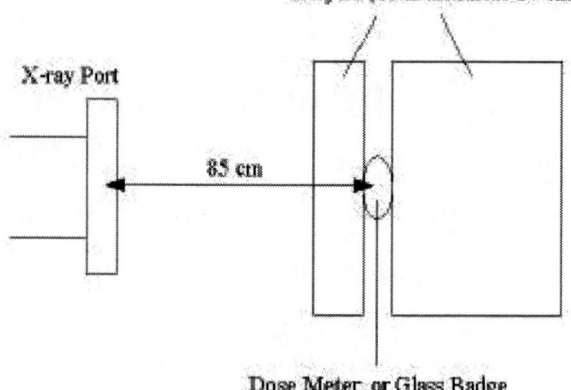

Fig.4 Layout of the DOSE measurement

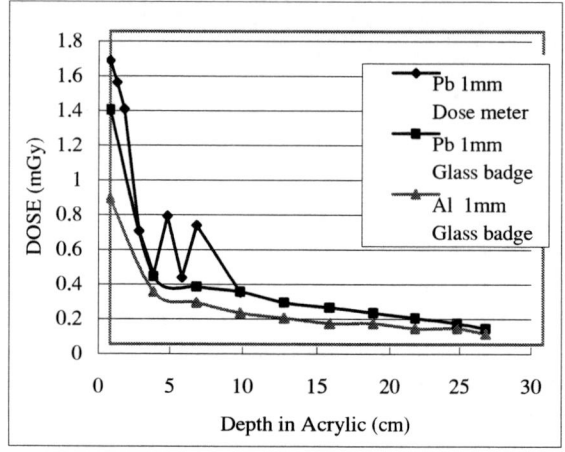

Fig.5 Dose distribution induced by X-rays from
MIRRORCLE-6X

2.2 DOSE measurement

Fig.4 shows the layout of the dose measurement. A dose meter and a glass badge were placed at around 120cm distance from the source point of MIRRORCLE-6X. We measured the dose by changing the front acrylic thickness from 0cm to 26cm. We adjusted the thickness of the backward acrylic so as to be 27cm in total of the acrylic thickness. Fig.5 shows the observed results of the dose distribution in the phantom. We found that the large amount of X-ray flux is absorbed at near the surface. This implies that the low energy X-ray components are dominant than that we have expected from the simulation. These must be due to the background radiations that come from other than the X-ray target, which are generated by the electrons not captured in the injection process.

3.Conclusion

As the result of the simulation, we found that the total absorption dose over the phantom is much lower in the case of MIRRORCLE-6X than in the case of the 100keV X-ray tube by factor 1/15. But for MIRRORCLE-6X, the result of our experiment suggests the absorption dose is fairly high at the surface of the phantom compared with the estimated dose by the simulation as shown in Fig.5. It seems that the actual X-ray from MIRRORCLE-6X has low energy component much more than the assumed X-ray in the simulation. The significant background component of the low-energy X-ray might come from other than the target.

The low-energy component is not desirable because it increases the surface dose and decreases clearness in the imaging. Therefore we will investigate to reduce the low-energy background photons. And as the X-ray spectrum can be changed by changing the target species, we will also study to find an ideal target to decrease the undesirable dose.

Protein Crystallography Beam Line at MIRRORCLE

M. Sasaki, T. Hirai, and H. Yamada

Synchrotron Light Life Science Center, Ritsumeikan Univ.

Abstract.

A Protein crystallography beam line using MIRRORCLE-6X, which is a compact synchrotron for hard x-ray production, is under construction. Since MIRRORCLE-6X provides hard x-ray with high brightness, it is suitable for x-ray source of accurate protein crystallography research in a small laboratory.

The divergent x-ray emitted from a target is focused by a condensing mirror. The mirrors used for the condensing mirror are a multi layer mirror of Wolter type and a rotating ellipsoidal mirror. Photon energies dispersed by them are set to the K-edge of Zn and of Se, respectively. The flux at a focusing point is expected to be $10^3 \sim 10^4$ times larger than that without the condensing mirror. The focusing size is less than 100 µm of diameter. The x-ray after the mirror is transformed to parallel beam using a Fresnel zone plate and irradiated on a protein crystal. The diffraction figure is detected by the an imaging-plate detector.

INTRODUCTION

Protein crystallography is one of the most important methods for determining atomic structures of biopolymers. Recent life science research on the structural functions of biopolymers, which are based on the atomic structures investigated by the protein crystallography, has been actively carried out.

Since the size of protein crystal is too small (~ 100 µm), the diffraction intensity is also small, and therefore, monochromatic x-ray with small beam size and high brightness is necessary for crystallography. At present, synchrotron radiation in huge laboratories, such as SPring-8 (Japan Synchrotron Radiation Research Institute in Hyogo, Japan) [1] and KEK-PF (High Energy Accelerator Research Organization – Photon Factory in Tsukuba, Japan) [2], has been frequently used. By using synchrotron radiation, accurate atomic structures of protein have been investigated. In recent years, the MAD (Multiwavelength Anomalous Diffraction) structure analysis method has been adopted for protein crystals including anomalous diffraction atoms [3], instead of the MIR (Multiple Isomorphous Replacement) method. Since the protein structures can be determined in the MAD method from diffraction data only for one crystal using x-rays with several different wavelengths, atomic structures of proteins have been rapidly analyzed one

after another. Correspondingly, protein crystallography using synchrotron radiation has played an important role for the advancement of life science research.

However, since beam lines for protein crystallography are too large (~ 50 m for BL41XU at SPring-8 [1]), they are not convenient for applications to life science. Therefore, a small beam line, which users can use easily in their laboratory, has been demanded. Development of such beam lines is expected to accelerate the progress of life science. Correspondingly, a protein crystallography beam line using MIRRORCLE-6X, which is a compact synchrotron developed by Yamada[4], is under construction.

DESIGN OF PROTEIN CRYSTALLOGRAPHY BEAM LINE USING "MIRRORCLE-6X"

MIRRORCLE-6X is a compact synchrotron x-ray source (1 x 2 m), which consists of a microtron, a beam transport line, and a storage ring. Electron beam accelerated up to 6 MeV by the microtron moves along

CP716, *Portable Synchrotron Light Sources and Advanced Applications,*
edited by H. Yamada, N. Mochizuki-Oda, and M. Sasaki
© 2004 American Institute of Physics 0-7354-0195-0/04/$22.00

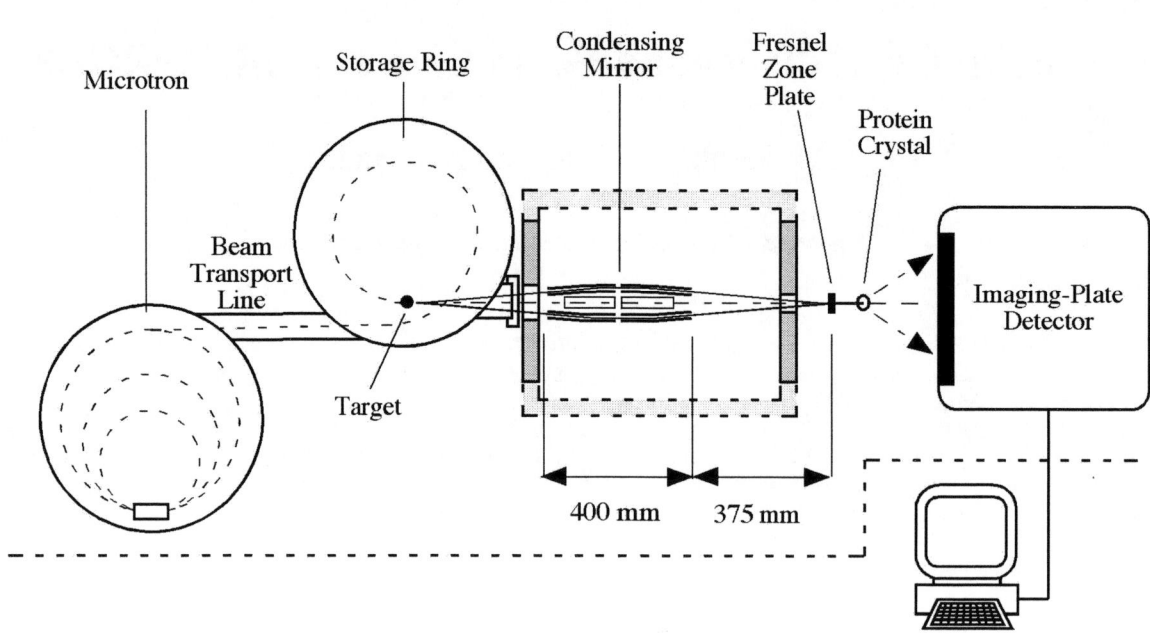

FIGURE 1. Designed protein crystallography beam line.

a circular orbit with a radius of 15 cm in the storage ring and bombards a small target with diameter ~μm located on the orbit. The hard x-ray, which is the bremsstrahlung, is relativistically emitted from this target in a forward direction. The generated hard x-ray has high brightness (~10^{11} ph/s/mm^2/mrad^2/0.1%λ), and its divergent angle is 83 mrad. The x-ray energy spectrum ranges continuously up to 6 MeV. In addition, MIRRORCLE-6X can stably supply a pulse x-ray. From these characteristics, it is expected that diffraction images with high resolution can be observed. Therefore, MIRRORCLE-6X is suitable for the x-ray source of precise protein crystallography [5].

The designed protein crystallography beam line using MIRRORCLE-6X is shown in FIGURE 1. This beam line consists of MIRRORCLE-6X, a condensing mirror, a Fresnel zone plate, a goniometer for a protein crystal, and an imaging-plate detector. Since this measuring system is quite small (~ 5 m) even including MIRRORCLE-6X, it can be built in a small laboratory.

The divergent x-ray emitted from the source point is focused by the condensing mirror. The mirrors used for this condensing mirror are a multiple layer mirror of Wolter type I and a rotating ellipsoidal mirror, and are being developed [6]. Their properties are listed in TABLE 1. The Wolter mirror contains two sets of 8 rotating parabolic and hyperbolic mirrors. Photon

energy dispersed by the Wolter mirror is set to the K-edge of Zn (9.7 ± 0.3 keV). The focusing efficiency, which is a ratio of photon flux at the focusing point to that without the mirror, is expected to be ~ 10^4 times. The focusing size is less than 100 μm.

The rotating ellipsoidal mirror is used for protein crystallography by the MAD method. The dispersive

TABLE 1. Properties of the condensing mirrors [6].

	Wolter Mirror	**Rotating Ellipsoidal Mirror**
Size	55 (φ) x 400 (L) [mm]	60 (φ) x 400 (L) [mm]
Number of Mirrors	2 sets of 8 mirrors (in symmetry)	1 mirror in symmetry
Shape	Parabolic and Hyperbolic	Ellipse
Reflective Material	Pt and C (Max. 39 Layers)	W and C (Multi Layers)
Dispersive Energy	9.7 ± 0.3 [keV]	12.7 ± 0.3 [keV]
Focusing Efficiency	x ~ 10^4	x ~ 10^3
Focusing Size	< 100 μm	< 100 μm

FIGURE 2. Imaging-plate detector, *"R-AXIS VII"*

photon energy is set to the K-edge of Se (12.7 ± 0.3 keV). The focusing efficiency and focusing size of this mirror are expected to be ~ 10^3 times and be less than 100 μm, respectively.

The focused x-ray beam is transformed into parallel x-ray beam by the Fresnel zone plate in order to observe diffraction images with high resolution. To achieve energy resolution sufficient for utilization of the MAD method, a spectrometer is set at the back of the Fresnel zone plate. The parallel x-ray beam is irradiated into a protein crystal on a goniometer, and the diffraction image is obtained by an imaging-plate detector, namely the *R-AXIS VII* made by *RIGAKU Corporation* and shown in FIGURE 2.

In order to estimate the protein crystallography beam line using MIRRORCLE-6X, diffraction experiments will be carried out using a well-known crystal such as Au. From diffraction intensity

determined by Laue method using white x-ray and monochromatic x-ray dispersed by a spectrometer through a pinhole collimator, it will be verified that the x-ray intensity obtained from MIRRORCLE-6X is sufficient for performing protein crystallography. This information will influence the development of the condensing mirrors of this measuring system.

REFERENCES

[1] M. Kawamoto, Y. Kawano, and N. Kamiya, *Nucl. Instrum. and Meth.* **A467-468** No.2, pp.1375-1379 (2001)

[2] N. Watanabe, A. Nakagawa, S. Adachi, and N. Sakabe, *Rev. Sci. Instrum.* **66**, pp.1824-1826 (1995)

[3] M. A. Walsh, G. Evans, R. Sanishvili, I. Dementieva, and A. Joachimiak, *Acta Cryst.* **D55**, pp.1726-1732 (1999)

[4] H. Yamada, *"Features of the Portable Synchrotron Named MIRRORCLE"*, in this proceedings.

[5] H. Yamada, *Nucl. Instrum. and Meth.* **B199**, pp.509-516 (2003)

[6] T. Hirai, et al., *"Development of hard x-ray microscopy with MIRRORCLE-6X"*, in this proceedings.

High-Energy X-ray Microprobe by Multilayer Zone Plate and Microscopy at SPring-8

Shigeharu Tamura[1,4], Masato Yasumoto[2,4], Nagao Kamijo[3,4], Yoshio Suzuki[4],
Mitsuhiro Awaji[4], Akihisa Takeuchi[4], Hidekazu Takano[5] and Kentaro Uesugi[4]

[1]AIST Kansai, Ikeda, Osaka 563-8577, Japan
[2]AIST Tsukuba, Tsukuba, Ibaraki 305-8568, Japan
[3]Kansai Medical University, Hirakata, Osaka 573-1136, Japan
[4]SPring-8/JASRI Mikazuki, Sayo, Hyogo 679-5198,Japan
[5]Nihon University, Setagaya, Tokyo 156-8550, Japan

Abstract. Microfocusing experiments of high-brilliance, high-energy X-ray by using a multilayer Fresnel zone plates were performed at SPring-8. It is proved that the multilayer FZPs can be used as focusing elements with high spatial resolution in a wide range X-ray wavelength domain up to 100 keV. A Cu/Al FZP with the thickness of 40 micron has attained the spatial resolution of 0.7 ~ 1.8 micron in a wide range X-ray wavelength domain of 18.6 ~ 113 keV. Three types of microscopic image of an Au mesh with 1500 lines per inch were taken by a X-ray microscopy experiment by using the multilayer FZP at 82 keV: a scanning microscopic transmission image, a scanning microscopic fluorescent one and an imaging microscopic one with the spatial resolution of ~ 0.7 micron. High-energy X-ray microprobe by using the multilayer FZP will be powerful tool for non-destructive various analyses of thick materials (bulk, IC, etc) with submicron spatial resolution.

INTRODUCTION

The advent the third-generation SR facilities such as SPring-8, have enabled us to utilize X-rays with high-brilliance, high-energy X-ray for the first time. One of advantages of the high-energy X-ray is the highly penetrating energetic X-ray photons. It is remarkable that the path lengths of more than many millimetres are possible for various materials at high-energy X-ray [1]. Representative applications of such an X-ray are (1) trace heavy-element detection by K-shell spectroscopy [2], (2) stress or strain measurement of thick materials [3,4], (3) non-destructive analyses of thick materials (bulk, IC, etc), etc.

The microprobe (microbeam) of such a high-brilliance, high-energy X-ray, therefore, may be a promising tool for various field of research. During the past decade, X-ray microprobes with submicron spot size have been formed by various optics (8 ~ 30 keV) : Fresnel zone plates (FZP), Bragg Fresnel lenses (BFL), Kirkpatrick-Baez (K-B) mirrors, compound refractive lenses (CRL), etc [5]. Though some of them

can focus X-rays of 50 ~ 100 keV, it is difficult to realize submicron microprobe [2, 6].

A multilayer (sputtered-sliced) FZP [7-9] is one of promising focusing optics for high energy X-ray region over 50 keV, because a large "aspect ratio" (the ratio of the FZP thickness to zone width) can easily be realized. We have developed various types of multilayer FZP and tested at SPring-8 (BL20XU) [5,10], and recently a submicron (500 nm) microprobe has attained at 100 keV [11]. In this article, our recent experimental results on the fabrication of the multilayer FZP and on the X-ray microscopy are described.

ZONE PLATE FABRICATION

The multilayer FZP is a circular diffraction grating composed of alternate transparent (low-Z material) zones and opaque (high-Z one) zones and acts as a focusing lens with high spatial resolution for use in high energy (short wavelength) X-ray domain at the

CP716, *Portable Synchrotron Light Sources and Advanced Applications,*
edited by H. Yamada, N. Mochizuki-Oda, and M. Sasaki
© 2004 American Institute of Physics 0-7354-0195-0/04/$22.00

synchrotron radiation (SR) high-brilliance X-ray beamlines. The FZP was fabricated by a sputtered-sliced process[5]. This process consists of two parts: a coating process and a mechanical manufacture one. Cu/Al multilayer coatings have been deposited onto rotating Au wire substrate (diameter: 50 micron) by a DC magnetron sputtering apparatus with two DC sputtering guns. The coatings were fabricated under various deposition conditions; the coating rates were 0.2 ~ 1 nm/s, the Ar gas pressures were 0.2 ~ 1.33 Pa and the substrate rotating speeds were 20 ~ 100 rpm. Among various conditions, the coating fabricated under the low coating rate of 0.2 nm/s, the low Ar gas pressures of 0.2 Pa and the medium rotating speed of 20 ~ 50 rpm has formed comparatively smooth zones (multilayer interface)(FIGURE 1). The substrate temperature was not controlled. An overcoat protective layer (Cu: 3 µm) was also deposited.

After the deposition process, the multilayer sample was fixed into a low melting point alloy (Sn: 60 %, Pb: 40 %, melting point 180°C), sliced into a plate of 1 mm thickness perpendicular to the wire axis by using a band saw micro-cutting. Next, this polished surface was fixed on the graphite plate (20 mm × 20 mm × 1 mm) by using resin glue (Loctite 420) and polished until the desired thickness was attained.

The design parameters of the FZP are summarized in TABLE 1 (in the TABLE 1 , another experimental results are also described).

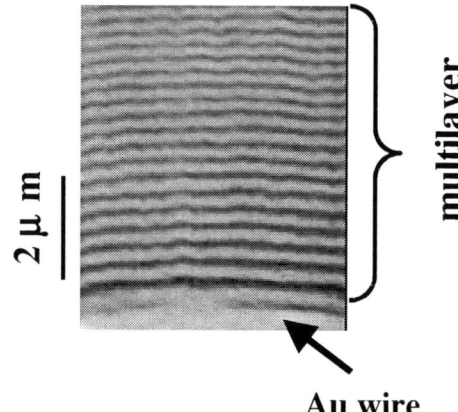

FIGURE 1. Scanning electron micrograph of Cu/Al concentric multilayer. Black and white zones are Al and Cu layers, respectively.

TABLE 1. Parameters and experimental results of multilayer Fresnel zone plates.

Samples (note)	Thickness, outermost zone width (micron)	Focused beam size (micron)	Working X-ray energy domain (keV)
Cu/Al-1 (ref .10, FIGURE 2)	20 , 0.25	0.3	8 – 12.4
Cu/Al-2 (FIGURE 2)	36 , 0.25	1.1 – 1.3	25 – 82.7
Cu/Al-3 (FIGURE 3)	40 , 0.16	0.7 – 1.8	18.2 – 113

MICROBEAM FORMING

The focusing test of the FZP was performed mainly at the BL20XU undulator beamline of SPring-8. Representative experimental set-up is described in ref [10,11]. Scanning electron micrographs (SEM) of concentric multilayers for FZPs used are shown in FIGURE 2 and FIGURE 3.

The domains of X-ray energy which these Cu/Al multilayer FZPs can work were examined. The experimental results are shown in Table.1.

It is proved that the multilayer FZPs can be used as focusing elements with high spatial resolution in a wide range X-ray wavelength domain.

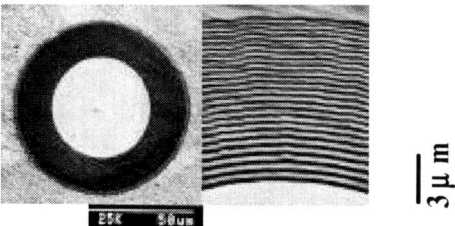

FIGURE 2. Scanning electron micrographs of Cu/Al concentric multilayer for FZP on Au wire substrate: full view (left) and close-up view (right). Black and white zones are Al and Cu layers, respectively.

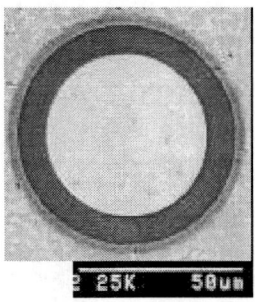

FIGURE 3. Scanning electron micrograph of Cu/Al concentric multilayer for FZP on Au wire substrate.

X-RAY MICROSCOPY AT 82 keV

The X-ray microscopy experiment by using the multilayer FZP was performed at 82 keV. Representative experimental set-up is described in ref [12]. Three types of image of an Au mesh with 1500 lines per inch (a wire diameter of 5.6 micron, a nominal aperture of 11 micron and a thickness of 3.8 micron) were taken: a scanning microscopic transmission image [12], a scanning microscopic fluorescent one and an imaging microscopic one (FIGURE 5).

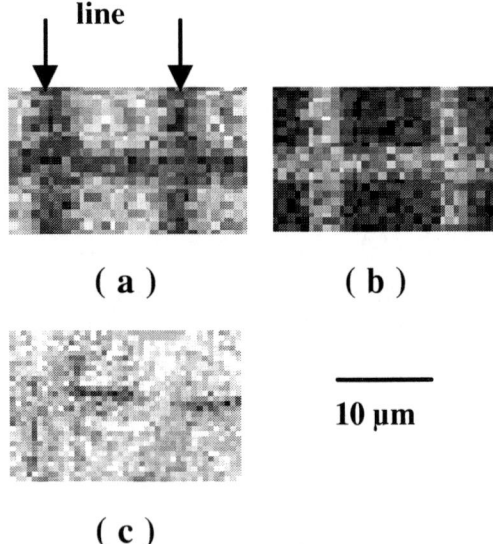

FIGURE 4. Various types of X-ray microscopic image of an Au mesh with 1500 lines per inch: (a) scanning microscopic transmission image , (b) a scanning microscopic fluorescent image and (c) imaging microscopic image. The magnification of imaging microscopy is 100 (sample to FZP: 1.64 m, FZP to image plane: 164 m).

The spatial resolution of each microscopic image is estimated about 0.7 micron. The attenuation lengths to 66% attenuation for X-ray at 100 keV were calculated by using the NIST database [13] : 0.11 mm for Au, 2.8 mm for Cu and 24 mm for Al. High-energy X-ray microprobe by using the multilayer FZP will, therefore, be powerful tool for non-destructive various analyses of thick materials (bulk, IC, etc) with submicron spatial resolution.

The details of the experiments of the scanning microscopic fluorescent microscopy and and the imaging one will be reported elsewhere.

CONCLUSIONS

The Cu/Al multilayer FZPs have been developed by the DC sputtering deposition and their focusing characteristics have been evaluated at high energy X-ray region over 50 keV. Such a high-energy X-ray microprobe will contribute to various types of advanced studies in a wide range of fields ranging from materials science to life science.

Since the thickness of the FZP in ref [11] has been 190 micron (aspect ratio : 1000) and has worked well at 100 keV, the multilayer FZP of the form similar to zone fiber (FIGURE 5) will also be effective as a focusing element at higher energy region of ~ 1 MeV.

FIGURE 5. Schematic diagram of multilayer Fresnel zone plate for high-energy X-ray above 100 keV.

The FZP will also be a useful optical element to portable synchrotrons such as MILLORCLE-6X.

Acknowledgments

The synchrotron radiation experiments have been performed at SPring-8 with the approval of the Japan Synchrotron Radiation Research Institute (JASRI)(Proposal No.2001B0517-NM-np and No.2002B0372-NM-np).

REFERENCES

1. Withers, P. J, Preuss, M., Webster, P. J., Hughes, D. J. and Korsunsky, A. M., *Material Science Forum*, **404-407**, 1-12(2002).

2. Shastri, S. D., Maser, J. M., Lai, B. and Tys, J., *Optics Communications,* **197,** 9-14(2001).

2. Korsunsky, A. M. and Wells, K. E., *Material Science Forum*, **321-324**, 218-223(2000).

3. Lienert, U.. Poulsen, H. F., Martins, R. V. and Kvick, Å.,*Material Science Forum*, **347-349,** 95-100(2000).

5. Tamura, S., Yasumoto, M., Kamijo, N., Suzuki, Y., Awaji,M., Takeuchi, A.,Takano, H. and Handa, K., *Journal of Synchrotron Radiation,* **9**, 154-159(2002).

6. Grigoriev, M., Shabelnikov, L., Yunkin, V., Snigirev, A., Snigireva, I., Michiel, M. Di, Kuznetsov, S., Hoffmann, M. and Voges, E., Proceedings of SPIE **4501**, 185-192(2001).

7. Rudolph, D., Niemann, B. and Schmahl, G, Proceedings of SPIE **316**, 103-105(1981).

8. Saitoh, K., Inagawa, K., Kohra, K., Hayashi, C., Iida, A. and Kato, N, *Jpn. J. Appl. Phys.,* **27**, L2131-L2133(1988).

9. Bionta, R. M., Ables, E., Clamp, O., Edwards, O. D., Gabriele, P. C., Ott, L. L., Skulina, K. M. and Viada, T., *Opt. Eng.* **29**, 576-580(1990).

10. Kamijo, N., Suzuki, Y., Awaji, M., Takeuchi, A., Takano, H., Ninomiya, T, Tamura, S. and Yasumoto, M., *Journal of Synchrotron Radiation,* **9**, 182-186(2002).

11. Kamijo, N., Suzuki, Y., Takano, H., Tamura, S., Takeuchi, A., Yasumoto, M. and Awaji, M., *Review of Scientific Instruments*, **74**, 5101-5104(2003).

12. Awaji, M, Suzuki, Y., Takeuchi, A., Takano, H., Kamijo, N., Yasumoto, M., Terada, Y. and Tamura, S, *Review of Scientific Instruments*, **74**, 4948-4949(2003).

13. http://physics.nist.gov/PhysRefData/FFast/Text/ cover.html

Recent Progress of X-Ray Microscopy at Ritsumeikan University Synchrotron Radiation Center

K. Takemoto[1], M. Kimura[1], K. Kojima[2], T. Matsumoto[2], B. Niemann[3],
M. Hettwer[3], D. Rudolph[3], E. Anderson[4], D. Attwood[4], D.P. Kern[5], H. Iwasaki[2]
and H. Kihara[1]

[1] Department of Physics, Kansai Medical University, Osaka, Japan
[2] Department of Photonics, Faculty of Science and Engineering, Ritsumeikan University, Shiga, Japan
[3] Institut für Röntgenphysik, Universität Göttingen, Göttingen, Germany
[4] Center for X-ray Optics, Lawrence Berkeley National Laboratory, California, USA
[5] Institut für Angewandte Physik, Universität Tübingen, Tübingen, Germany

Abstract. X-ray microscopy enables high-resolution analysis of thick specimens such as several microns in aqueous and atmospheric pressure environments in fields of biological and material sciences. Most commonly, zone plates (ZPs) are used as optical elements in combination with the use of synchrotron radiation (SR) as an x-ray source. It is realized at the x-ray microscopy station at BL-12 of Ritsumeikan University SR center. The highest spatial resolution is 50 nm. The beam line is open to researchers from universities, government research centers and companies and has been applied in biology, medicine, and material science. The present paper describes a recent progress and recent results of BL12.

INTRODUCTION

X-ray microscopy enables high-resolution analysis of thick specimens such as several microns in aqueous environments in fields of biological and material sciences. Since 1996, we have been operating a transmission X-ray microscope beamline, BL12, at SR center of Ritsumeikan University ([575MeV, 300mA], Kusatsu, Japan) [1]. It was transferred from UVSOR (Okazaki, Japan), where it was initially installed [2]. Zone plates (ZPs) are used as optical elements. The achieved resolution upto now is 50 nm, judging from the edge analysis [3]. Polystyrene latex spheres of 0.23 μm diameter and fine structures of 0.1 μm size in diatom cells could clearly be observed [3]. The X-ray microscope has been upgraded to get a higher performance.

The SR center is supported by research projects on nanotechnology and materials of Ministry of Education,

Culture, Sports, Science and Technology in Japan till 2006. The X-ray microscope beam line is open to researchers from universities, government research centers and companies with this support.

In this paper, we report a recent progress of X-ray microscopy beam line and some results supported by nanotechnology researches network projects.

X—RAY MICROSOPE BEAM LINE AT RITSUMEIKAN UNIVERSITY SR CENTER

Instrument Description

A high resolution soft X-ray microscope station has been installed at BL12 of the Ritsumeikan synchrotron radiation center. At the center, a super-conducting compact storage ring AURORA, designed and

CP716, *Portable Synchrotron Light Sources and Advanced Applications,*
edited by H. Yamada, N. Mochizuki-Oda, and M. Sasaki
© 2004 American Institute of Physics 0-7354-0195-0/04/$22.00

manufactured by Sumitomo Heavy Industries [3-5], was installed in 1996. Since then the ring has successfully been operated. Its total operation time is more than 1200 hours per year. It is usually operated at an energy of 575 MeV with an initial beam current of 300 mA. The critical wavelength of the radiation is 1.47 nm.

Figures 1 and 2 show the lay-out and optical configuration of the soft X-ray microscope station [6,7]. The station is of the same type as the Göttingen X-ray microscope at BESSY I [8]. The optical stage is composed of a condenser ZP (CZP: the Göttingen KZP 7 type CZP [9]: diameter: 9 mm, outermost zone width: 53.7nm, number of the zones: 41890) a pinhole (diameter: 20 μm), a specimen stage, an objective ZP (OZP: IBM/LBNL OZP [10]: diameter: 50 μm,

outermost zone width: 45 nm, number of the zones: 277) and a backside-illuminated CCD camera (SITe SI502A: 512 × 512 pixel, each 24 μm × 24 μm). The CZP chamber and an optical microscope are placed on a moving stage. They are switched with a pneumatic cylinder. For object finding and pre-focusing, the light microscope is placed on a moving stage. The highest spatial resolution is 50 nm [1, 6, 7].

The microscope covers the wavelength range including the water window region (0.28 ~ 0.53 keV). Samples are placed under atmospheric pressure. For wet samples, a special wet cell is prepared, which consists of two thin polyimide films of 300 nm thickness supported by thick ones. Wet samples are placed between the two thin polyimide films and is sealed with silicone grease [11].

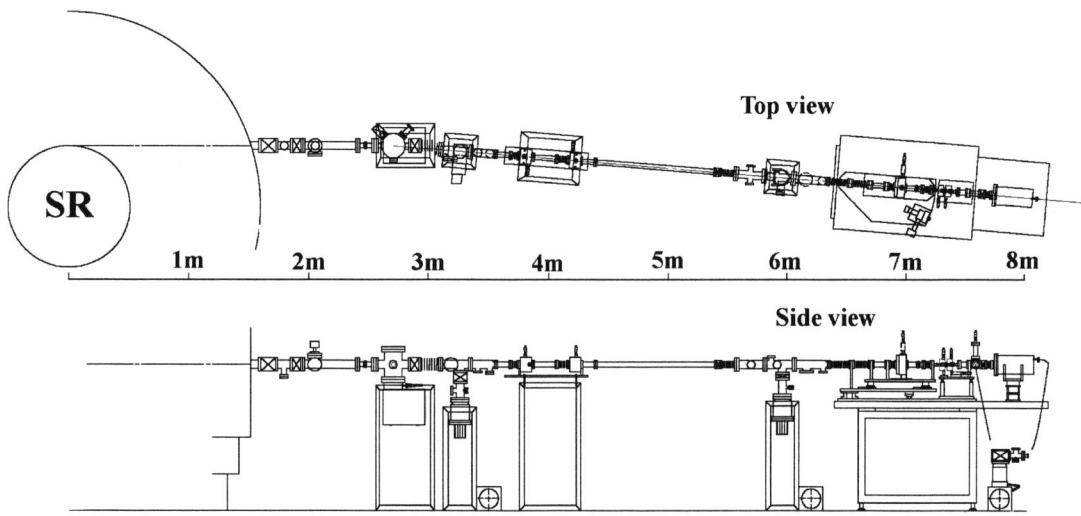

FIGURE 1. Lay-out of the soft X-ray microscope station BL12.

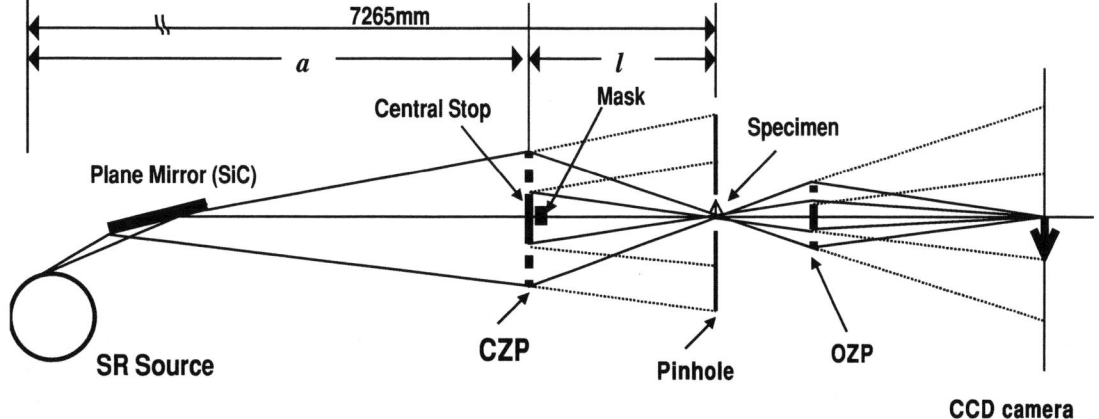

FIGURE 2. Optical configuration of the X-ray microscopy at BL-12.

New CZP and OZP Camber

CZP is used to monochromatize and collimate white light from SR bending source, and OZP is used to magnify the image. Thus, CZP can be considered as circular diffraction gratings with radially increasing line densities.

ZP is characterized by three parameters. For instance, if we use λ, Δr, and n as basic parameters, the other ZP parameters are given by

$$D = 4n\Delta r, \quad f = 4n(\Delta r)^2 / \lambda,$$

where n is the zone number, Δr is the outermost zone width, λ is the wavelength, and f is the first-order focal length and D is the diameter of ZP [12-14]. f can also be written as

$$f = r_1^2 / \lambda,$$

where r_1 is the first radius of ZP. Thus, when ZP is used as an optical element of the monochromator, the observed wavelength λ is determined by the focal length, f.

In current status of X-ray microscopy at BL-12, ZPs are only put on the position calculated geometrically by manual control. In order to install an automatically wavelength sweep system, new CZP and OZP cambers were designed. Figure 3 is planes of the new CZP and OZP cambers. A wavelength in water window region is used for imaging now. Covering the region, relative variation of the CZP position is about 100 mm and the OZP position is about 0.42 mm at the first order diffraction.

(a)

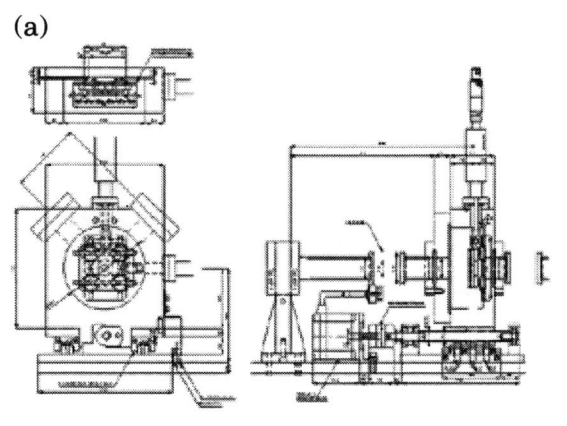

(b)

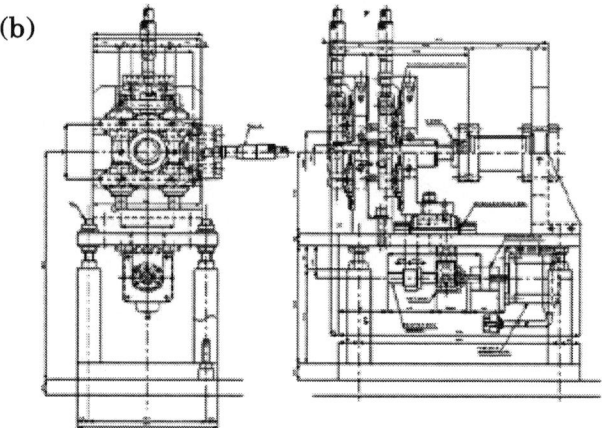

FIGURE 3. New plane of a CZP (a) and an OZP (b) chamber.

APPLICATIONS SUPPORTED BY NANOTECHNOLOGY RESEARCHES NETWORK PROJECT

At BL12, X-ray microcopy beam line, six proposals were accepted and supported by nanotechnology researchers network project [15]. In those, several works continue in 2003. The CH_3-SiO_2 organo-silica micro-particle produced by a sol-gel method using a W/O emulsion is porous material so that it has an internal density distribution. Using a gas absorption technique and SEM observation, a specific surface area and a mean particle diameter were obtained [16]. Each result showed that remarkable structure change happened on their surface below 400 degrees C. However, each method could not solve the structure change without destroying their structure.

X-rays are used as a probe for the non-destructive tests of a substance, since transmittance is high. Figure 4 shows the X-ray micrographs of CH_3-SiO_2 organo-silica micro-particles. Image contrast depends on an electron density. The diameter of each particle is so large that an immediate sharp variation of contrast near outline is expected when it is an ideal solid particle. The variation reflects the micro-structure change near surface. In order to estimate it, a contrast profile across the diameter direction was obtained and a width at 20 – 80 % intensity of it was estimated. Compare to the values obtained from the particles which processed at each temperature, remarkable peak was observed at 200-300 degrees C. The results imply that generating micro-pores. The observed results agreed with the result of the gas absorption technique. The results imply that generating micro-pores near surface depends on the temperature of heat treatment. The reaction was promoted at 200 to 300 degrees C.

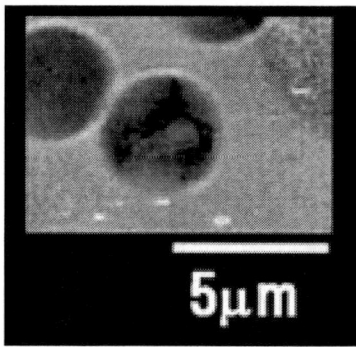

FIGURE 4. X-ray Micrograph of CH_3-SiO_2 organo-silica micro-particles.

SUMMARY

A transmission X-ray microscope with zone plates has been operating at BL-12 of SR Center of Ritsumeikan University. The beam line is open to researchers. In 2003, with the newly introduced CZP and OZP chambers, an automatically wavelength sweep system was designed and produced. This X ray microscope has been applied in biology, medicine, and material science.

ACKNOWLEDGMENT

We are grateful to Ritsumeikan SR center staff for their support.

REFERENCES

1. A. Hirai, K. Takemoto, K. Nishino, B. Niemann, M. Hettwer, D. Rudolph, E. Anderson, D. Attwood, D. P. Kern, Y. Nakayama, and H. Kihara, *Jpn. J. Appl. Phys.* **38**, 274-278 (1999).

2. K. Takemoto, A. Hirai, B. Niemann, K. Nishino, M. Hettwer, D. Rudolph, E. Anderson, D. Attwood, D.P. Kern, Y. Nakayama and H. Kihara, in *X-ray Microscopy*, edited by W. Meyer-Ilse, T. Warwick and D. Attwood, AIP Proc. of XRM99, 2000, pp.446-451.

3. N. Takahashi, *Nucl. Instrum. Methods*, **B24/25**, 425-428 (1987).

4. H. Yamada and SHI Accelerator Research Group, *Rev. Sci. Instrum.*, **60**, 1786-1789 (1989).

5. H. Yamada, Proc. Asian Forum on Synchrotron Radiation, edited by T. Ohta, S. Suga, and S. Kikuta, Ionics Pub., Tokyo, 1994, pp.227.

6. A. Hirai, K. Takemoto, K. Nishino, N. Watanabe, E. Anderson, D. Attwood, D. Kern, M. Hettwer, D. Rudolph, S. Aoki, Y. Nakayama and H. Kihara, *J. Synchrotron Rad.* **5**, 1102- 1104 (1998).

7. A. Hirai, N. Watanabe, K. Takemoto, N. Nishino, E. Anderson, D. Attwood, D. Kern, S. Shimizu, H. Nagata, S. Aoki, Y. Nakayama and H. Kihara, in *X-Ray Microscopy and Spectromicroscopy*, edited by. J. Thieme, G. Schmahl, D. Rudolph and E. Umbach, Proc. of VRM96, Springer-Verlag, Heidelberg, 1998, pp. I-123-I-172.

8. P. Guttmann, G. Schneider, J. Thieme, C. David, M. Diehl, R. Medenwaldt, B. Niemann, D. Rudolph, and G. Schmahl, in *Soft X-Ray Microscopy*, edited by. C. J. Jacobsen and J. E. Trebes, proc. SPIE 1741, 1992, pp. 52-61.

9. G. Schmahl, D. Rudolph, B. Niemann, P. Guttmann, J. Thieme, G. Schneider, C. David, M. Diehl and T. Wilhein, *Optik*, **93**, 95-102 (1993).

10. E. Anderson and D. Kern, in *X-Ray Microscopy III*, edited by A.G. Michette, G.R. Morrison and C.J. Buckley, Springer-Verlag, Berlin, 1992, pp.75-78.

11. K. Takemoto, N. Watanabe, A. Hirai, Y. Nakayama, and H. Kihara, in *X-ray Microscopy and Spectromicroscopy* edited by. J. Thieme, G. Schmahl, D. Rudolph, and E. Umbach, Proc. of XRM96, Springer-Verlag, Heidelberg, 1998, pp.I-129 - I-134.

12. A. G. Michette, *Optical Systems for Soft X-Rays*, Plenum, London, 1986.

13. G. R. Morrison, "Diffractive X-Ray Optics, Chapter 8", in *X-Ray Science and Technology*, edited by in A. G. Michette and C. J. Buckley, Inst. Phys., London, 1993.

14. D. T. Attwood, *Soft X-Rays and Extreme Ultraviolet Radiation: Principles and Applications*, Cambridge Univ. Press, Cambridge, 1999.

15. Technical Report 2002 in the SR CENTER of Ritsumeikan University, 2003.

16. T. Matsumoto, Y. Takayama, N. Wada, H. Onoda, K. Kojima, H. Yamada and H. Wakabayashi, *J. Mater. Chem.*, **13**, 1764-1770 (2003).

The X-ray Microscopy Project at Saga LS

M. Yasumoto[1], K. Takemoto[2], E. Ishiguro[3], H. Kihara[2], T. Tomimasu[4], N. Kamijo[2] and Y. Chikaura[5]

1 Photonics Research Institute, National Institute of Advanced Industrial Science and Technology (AIST), Tsukuba, 305-8568, Japan
2 Department of Physics, Kansai Medical University, Osaka, 573-1136, Japan
3 College of Education, University of the Ryukyus, Okinawa, 903-0213, Japan
4 Saga Synchrotron Light Source, Saga, 840-8570, Japan
5 Faculty of Engineering, Kyushu Institute of Technology, Fukuoka, 804-8550, Japan

Abstract. A new high resolution X-ray microscopy project is proposed at Saga synchrotron light source, a third generation synchrotron light facility in Kyusyu Island, Japan. Two microscopy beamlines are planned. One is a scanning microscope in the water window region and the other is a full-field imaging microscope in the multi-keV X-ray energy region. To demonstrate the feasibility of the project, an optical layout of the scanning microscope was designed. The beamline mainly consists of a 3.5 cm periodical undulator, a varied line-spacing plane grating monochoromator (600 lines/mm) and an end-station including a zone plate. The calculated X-ray properties focused on the sample position are as follows: the spot size is ~70 nm, the monochromaticity is ~2000, and the photon flux is $10^9 \sim 10^{10}$ photons/s.

INTRODUCTION

Saga Synchrotron Light Source (Saga LS), the first synchrotron light facility in Kyushu Island, is under construction in Tosu city by Saga prefecture government. The ring will be operational in fiscal year of 2004. At Saga LS, a high resolution X-ray microscopy project is proposed in which two beamlines are planned. One is a scanning X-ray microscope (SXM) in the X-ray energy range of 0.28 ~ 0.53 keV (*water window*), and the other is a full-field imaging X-ray microscope in the X-ray energy range of 2 ~ 7 keV (*multi keV*). This paper reports the project and discusses the feasibility of the water window microscopy beamline at Saga LS.

SAGA SYNCHROTRON LIGHT SOURCE (SAGA LS)

The Saga LS is being built to promote the interaction for research and development in Kyushu

and throughout Asia. It is part of the Northern Kyushu Science Cities' Development Plan (ASIAS Kyushu) for creating a network of academic and research facilities. The facility is the first SR facility to have been established by a local government. Therefore, the objectives of this facility are to focus on applied research for the industrial uses of SR and employ the research results to create new industries and enhance regional industry.

Saga LS has a compact third-generation storage ring of 1.4 GeV electron energy. It is the smallest third-generation ring in the world [1]. The 1.4 GeV storage ring is a double bend achromat type lattice and has a circumference of 75.6 m. There are 8 straight sections of about 3 m long for 6 insertion devices, and 20 front-ends in all. The normal stored electron current is 300 mA. Table 1 indicates the electron parameters of the storage ring, and table 2 indicates the source parameters at the straight section.

The storage ring operates at an extremely low emittance mode of 25.1 nm rad [2]. The 7.5 T super-conducting wiggler and the undulators are placed at

CP716, *Portable Synchrotron Light Sources and Advanced Applications*,
edited by H. Yamada, N. Mochizuki-Oda, and M. Sasaki
© 2004 American Institute of Physics 0-7354-0195-0/04/$22.00

the straight sections. The synchrotron radiation (SR) from the 7.5 T wiggler covers a broad range, from infrared to hard X-ray [1]. However, undulator parameters were designed for the microscopy beamline. The SR from the undulator covers the water window energy region in the first harmonic [3]. The undulator has a 3.5 cm period length, 64 periods and a permanent-magnet type with a K variable of between 0.01~1.5. Moreover, the higher harmonic SR from the undulator covers up to 2 keV. Figure 1 shows the photon flux and brilliance of the undulator SR, and the bend magnet SR as a function of the X-ray photon energy.

TABLE 1. Electron storage ring parameters of the Saga-LS

Injection / Maximum electron energy	250 MeV / 1.4 GeV
Storage current	300 mA
Beam life time	> 5 hours
Natural emittance	25.1 nm rad (10% coupling)
Ring circumference	75.6 m
Critical energy	1.9 keV

TABLE 2. Source parameters at straight section.

β (H)	7.657 m
β (V)	6.929 m
Beam size (H)	581 μm
Beam size (V)	126 μm
Beam spread (H)	54.6 μrad
Beam spread (V)	18.1 μrad

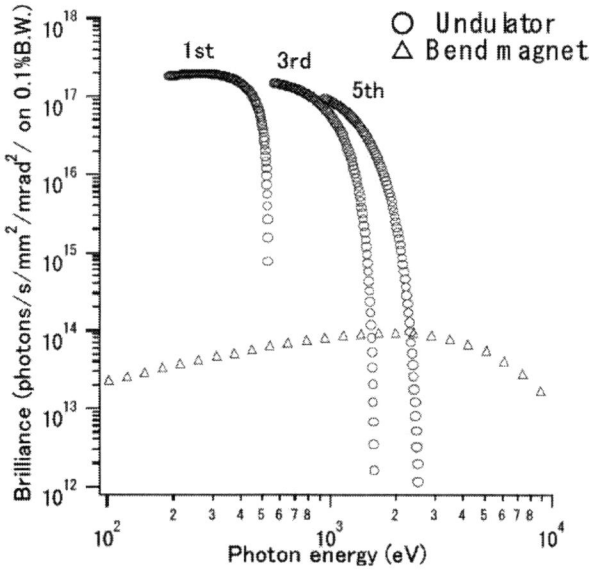 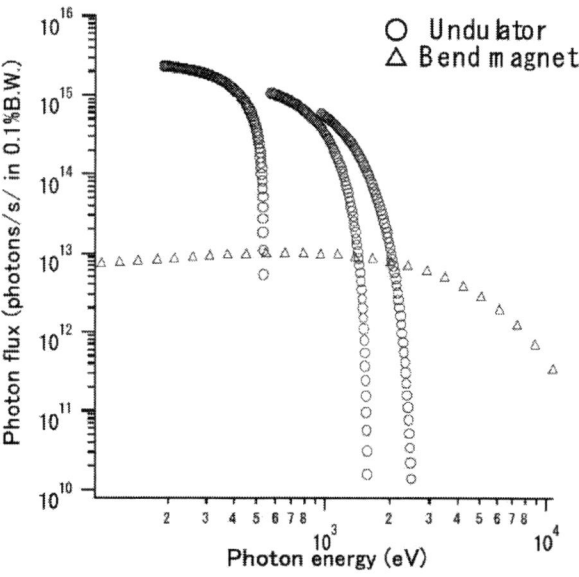

FIGURE 1. Calculated photon flux and brilliance of SRs from undulator and the bend magnet, when the beam current is 300A and the HV coupling is 10%.

X-RAY MICROSCOPY BEAMLINE

Scanning X-ray Microscope

The SXM allows investigation with high resolution of space and energy, permitting quantitative analysis and elemental mapping. We designed an optical setup for the SXM that is located at the undulator beamline.

The optical configuration of the beamline is as simple as possible to obtain the high flux and the necessary high-energy resolution shown in Figure 2. The SXM mainly comprises two water-cooled spherical mirrors (M1 and M2), a plane grating (G), two exit slits (S and S') and an end-station including focusing optics of a zone plate (ZP). M1 and M2 are used for focusing the source image onto the exit slits. The exit slit is 50 μm wide. The parameters of designed ZP are as follows; the outer diameter is 100 mm, the number of zones is 1250, and the width of the outermost zone is 40nm.

The plane grating is a varied line-spacing plane grating with variable line density, so that the plane grating combined with the spherical mirror calls for a varied line-spacing plane grating monochromator (VLSPGM) [4]. The VLSPGM has no entrance slit. When the variable line density is 600 lines/mm, the energy resolving power is about 2000 in water window region [6]. The focal spot size depends on the distance from the ZP. The spot diagrams as a function of the distance from the ZP are shown in Figure 3. The minimum spot size of the focused X-ray is calculated at about 70 nm in this beamline configuration, when the distance from the ZP is 2.6665 mm.

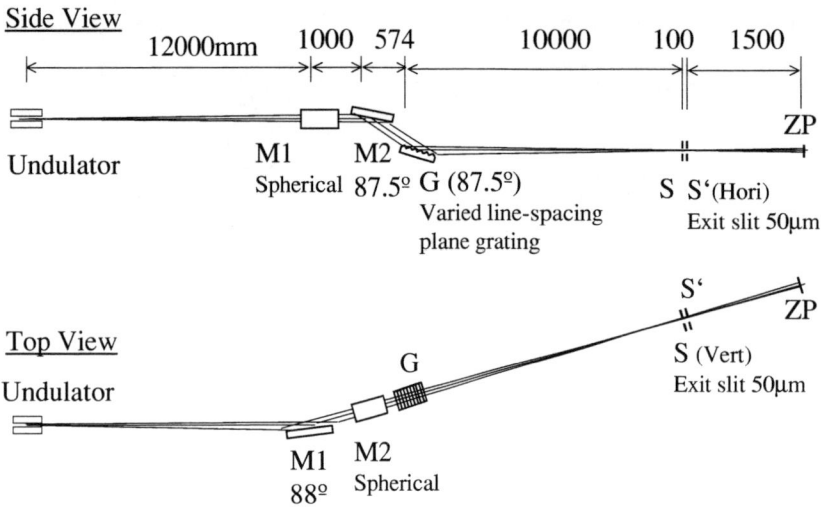

FIGURE 2. Schematic optical layout of the scanning X-ray microscopy beamline.

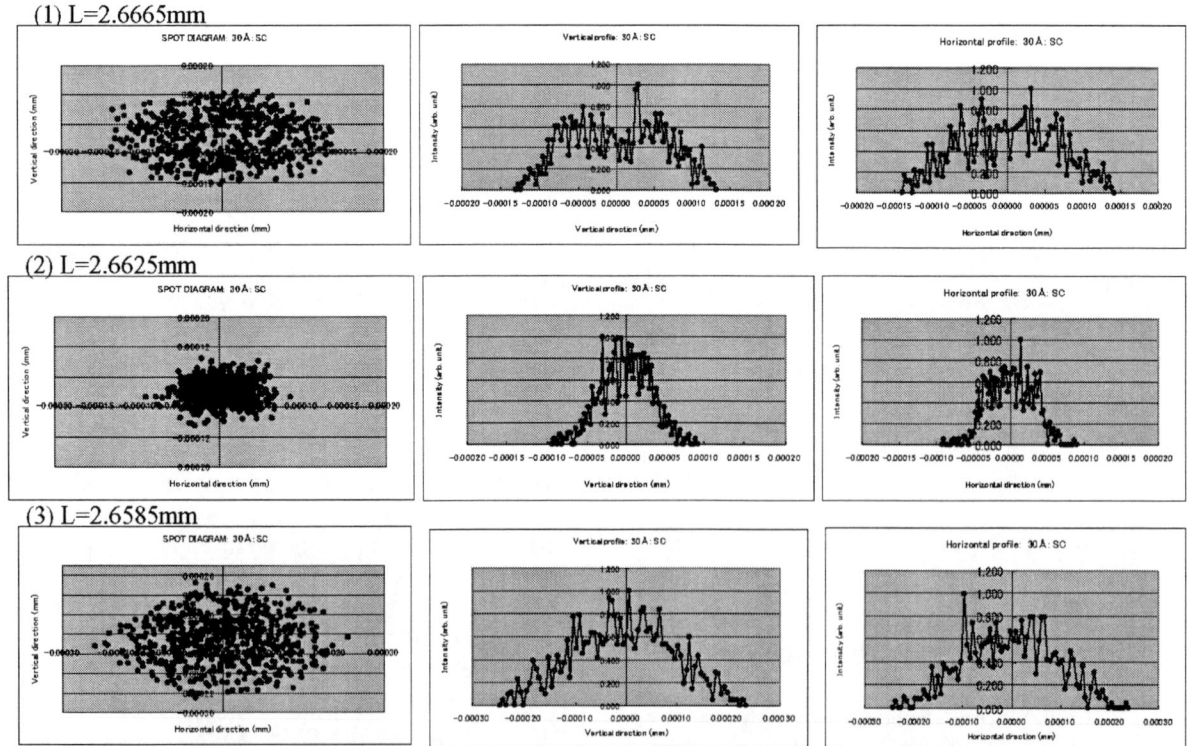

FIGURE 3. Calculated X-ray spot images according to the distance from the ZP, when the X-ray energy is 413 eV and the exit slit of the monochromator is 50 μm wide. L is a distance from the ZP to the focal plane.

Full-field Imaging X-ray Microscope

The full-field imaging X-ray microscope in the multi keV allows the direct investigation of the transmitted image. There are K absorption edges of medium light elements such as Na, Mg, P, S, K and Ca in the energy range of 2 ~ 7 keV (Fig. 4). The energy range is already open for the user at the ESRF [5]. This beamline uses the SR from the 7.5 T super-conducting wiggler. The spectral flux of the wiggler is 1×10^{13} (photons/s/0.1% b.w.) in the multi-keV region. The beamline is composed of mirrors, a double crystal monochromator, a condenser ZP, an objective ZP and an X-ray CCD camera. The spatial resolution of the microscope is expected to decrease to sub-100 nm.

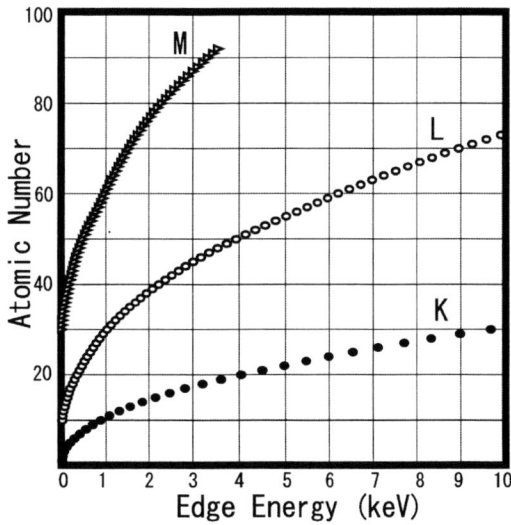

FIGURE 4. Absorption edges in the multi-keV X-ray region. K is a K absorption edge, L is an L edge, and M is a M edge.

CONCLUSION

A new microscopy project at Saga LS is reported, and discussed the feasibility of the microscopy beamline. Two microscopy beamlines were planned in the soft X-ray energy region and the multi-keV energy region. An optical setup for the scanning microscopy beamline in the water window region was designed. The beamline is a simple configuration composed of an undulator (3.5 cm-period-length, 64 periods, K = 0.01~1.5), a varied line-spacing plane grating monochoromator of 600 lines/mm, and a zone plate of 40 nm outermost zone width. In this beam configuration, an expected micro-beam size at sample position are as follows; the spot size is ~70 nm, the energy resolution (E/δE) is ~2000 and the photon flux is $10^9 ~ 10^{10}$ (photons/s) at 0.413 keV with an exit slit of 50 μm.

REFERENCES

1. T. Tomimasu, S. Koda, Y. Iwasaki, H. Toyokawa and M. Yasumoto, IEEE Proc. of PAC'03, 902-904 (2003).

2. T. Tomimasu, M. Yasumoto, N. Koga, Y. Hashiguchi, Y. Ochiai, and M. Ishibashi., *Nuc.l Instrum. Meth.* **475**, 454.45, (2001).

3. T. Tomimasu, H. Ohgaki, H. Toyokawa, M. Yasumoto, Y. Iwasaki, Y. Yamatsu, N. Koga, Y. Hashiguchi, and Y. Ochiai, Proc. of APAC2001, 340-342 (2001).

4. Y. Tamenori, H. Ohashi, E. Ishiguro, T. Ishikawa, *Rev. Sci. Instru.* **73**, 1588-1590 (2002).

5. J. Susini, R. Barrett, B. Kaulich, S. Oestreich, and M. Salomé, in *X-ray microscopy*, edited by W. Meyer-IIse, T. Warwick and D. Attwood, AIP Proc. of XRM99, 2000, pp19 - 26.

6. M. Yasumoto, E. Ishiguro, K. Takemoto, T. Tomimasu, H. Kihara, N. Kamijo, T. Tsurushima, A. Takahara, K. Hara, Y. Chikaura, in *X-ray microscopy*, edited by J. Susini et al., J. de. Phys. IV Proc. XRM02, EDP Science, Paris, 2003, pp63 - 66.

X-Ray Talbot Interferometry for Medical Phase Imaging

Atsushi Momose[1], Shinya Kawamoto[2], and Ichiro Koyama[1]

[1]Department of Advanced Materials Science, Graduate School of Frontier Sciences, The University of Tokyo,
5-1-5 Kashiwanoha, Kashiwa, Chiba 277-8651, Japan
[2]Department of Applied Physics, School of Engineering, The University of Tokyo,
7-3-1 Hongo, Bunkyo-ku, Tokyo 113-8656, Japan

Abstract. X-ray Talbot Interferometry (XTI) is demonstrated as a novel and simple X-ray phase-sensitive imaging method. Differential phase contrast is generated with XTI and quantitative determination of the differential phase is attained in combination with a phase-shifting technique, further enabling phase tomography. Because XTI does not use crystals but transmission gratings, divergent X-rays with a broad energy band are available. This is an advantage for medical applications especially using a compact X-ray source.

INTRODUCTION

The advantage of using X-ray phase information is well recognized for high-sensitive X-ray imaging particularly for biological soft tissues. Currently a variety of methods for X-ray phase-sensitive imaging are proposed and demonstrated. Medical applications should be attractive targets of the phase-sensitive imaging methods. However, prospects are not straightforward because huge synchrotron facilities are used and indispensable in many studies. From a clinical point of view, the technique of phase-contrast formation and a compact X-ray source should be compatible.

One restriction in the phase-sensitive methods, such as diffraction enhanced imaging (DEI) [1,2] and Bonse-Hart X-ray interferometry [3-5], is that crystal optics must be employed. This implies that a highly monochromatic and highly collimated beam is required and therefore we have to submit ourselves to filtering out most of X-rays from a source to extract such a beam.

The propagation-based method [6,7] does not require crystal optics in principle. If a spatially coherent beam is available, this method can produce a contrast outlining structural boundaries in an object, but a high-resolution imaging system is preferred to resolve the outline contrast.

We propose X-ray Talbot Interferometry (XTI) as a candidate of phase-sensitive imaging method that also omits the use of crystal optics and allows us to use a diverging X-ray beam of a broad energy band, provided that the partial spatial coherency of the beam is assured to some extent. One advantage characteristic of XTI is that the contrast is not localized at structural boundaries as much as the propagation-based method. Therefore, high-resolution X-ray image detector is not always needed. In addition, it should be pointed out that the differential phase can be obtained quantitatively [8] with XTI by introducing a phase-shifting technique. Because the phase shift caused by an object is obtained by integrating the image of the differential phase, phase tomography is also attained by XTI [9].

We report demonstrative experiments of XTI and phase tomography with it, which were carried out with synchrotron radiation at SPring-8. Finally, based on the experimental results, we discuss the possibility of medical phase imaging with XTI using a compact X-ray source.

X-RAY TALBOT INTERFEROMETER

As shown in Fig. 1, a Talbot interferometer employs two transmission gratings. When the first grating G1 is illuminated coherently, an image, which

CP716, *Portable Synchrotron Light Sources and Advanced Applications,*
edited by H. Yamada, N. Mochizuki-Oda, and M. Sasaki

has a spatial intensity modulation corresponding to the transmittance function of G1, is generated at a specific distance Z_T downstream from G1 (Talbot effect [10]). The Z_T satisfies

$$Z_\mathrm{T} = \begin{cases} md^2/\lambda & \text{for amplitde gatings,} \\ (m+1/2)d^2/\lambda & \text{for phase gratings,} \end{cases} \quad (1)$$

where m is an integer and d and λ are the period of G1 and X-ray wavelength.

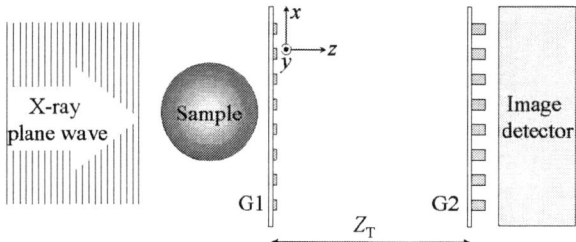

FIGURE 1. Setup of an X-ray Talbot interferometer.

The image is deformed when a sample is placed in front of G1. The refraction by the sample is responsible for the image deformation. Talbot interferometry visualizes the deformation by placing the second grating G2 on the deformed image. The image contrast generated by the Talbot interferometer is understood as a superposition of the deformed image and the transmission function of G2.

Here we use plane wave approximation and assume that G2 is an amplitude grating with the period of d and aligned in parallel with G1. Then, the deformed image is converted to an image $I(x, y)$ just behind G2 given by

$$I(x, y) = a_0 b_0$$
$$+ 2 \sum_{k>0} a_k b_k \cos\left[\frac{2\pi k}{d}\left\{Z_\mathrm{T}\varphi(x, y) + \xi\right\}\right], \quad (2)$$

where ξ is the relative displacement of G2 against G1 in the x direction. $\varphi(x, y)$ is the refractive angular deviation of X-rays by the sample given by

$$\varphi(x, y) = \frac{\lambda}{2\pi}\frac{\partial\Phi(x, y)}{\partial x}, \quad (3)$$

where $\Phi(x, y)$ is the phase shift caused by the sample. a_k and b_k are the kth Fourier coefficients of the self-image of G1 and the intensity transmission function of G2. Thus, a contrast is generated when G2 has a power of amplitude modulation. In principle, the contrast reaches 100% if ideal gratings are fabricated.

By acquiring multiple images changing ξ with a step of d/M (M is an integer over 2), $\varphi(x, y)$ can be determined, if $a_k b_{k'}$ is negligible for k' satisfying $k' \geq M - 1$, as below [11]:

$$\varphi(x, y) = \frac{d}{2\pi Z_\mathrm{T}} \arg\left[\sum_{p=1}^{M} I_p(x, y)\exp\left(-\frac{2\pi ip}{M}\right)\right], \quad (4)$$

where arg[ ] implies the extraction of the argument of a complex value, and $I_p(x, y)$ is the image obtained when $\xi = pd/M$.

The phase shift $\Phi(x, y)$ is given by a projection form:

$$\Phi(x, y) = -\frac{2\pi}{\lambda}\int\delta(x, y, z)dz, \quad (5)$$

where $\delta(x, y, z)$ is the refractive index decrement from unity of the sample. Thus, the measurement of $\Phi(x, y)$ is important for structural understanding of a sample. If the phase shift is measured in multiple projection directions, $\delta(x, y, z)$ is reconstructed with the technique of computed tomography [4].

As mentioned, XTI with the fringe-scanning method enables the measurement of $\varphi(x, y)$, the integration of which yields $\Phi(x, y)$. Although the measurement of $\Phi(x, y)$ is also attained with the propagation-based method, multiple image acquisition is required by changing the distance between a sample and an image detector. The XTI is advantageous in that the sample-detector distance can be fixed.

PHASE IMAGING

Evidence of X-ray Talbot effect

The Talbot effect in the hard X-ray region was first reported from the European Synchrotron Radiation Facility (ESRF) [12]. We performed an experiment to confirm an X-ray Talbot effect at an undulator beamline BL20XU of SPring-8 where X-rays were available 245 m downstream from the source point through a double-crystal monochromator. We prepared a phase grating fabricated by forming gold stripes with a pitch of 8 µm on a glass plate 150 µm in thickness by means of optical lithography. The height of stripes was 1.25 µm, which was determined to produce a π/2 phase difference for 0.1-nm X-rays used in the experiments so that the contrast in an image observed at the distance given by eq. (1) was maximum.

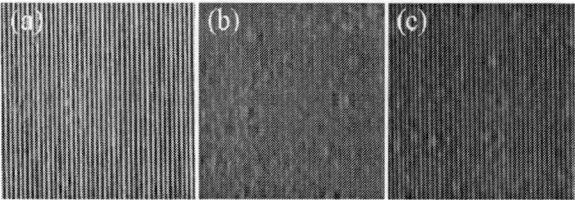

FIGURE 2. Images observed at (a) 32 cm, (b) 64 cm, and (c) 96 cm downstream from the phase grating.

Figure 2 shows images observed downstream from the grating with a CCD image detector (Hamamatsu Photonics K.K., C4880-50) coupled with optical lens and a phosphor screen. Effective pixel size was 0.54 μm. Figures 2(a) and 2(c) were obtained under the conditions of $m = 0$ and 1; that is, the distances between the grating and the detector were 32 cm and 96 cm, respectively. Stripes with the same spatial frequency as that of the phase grating were seen. However, the contrast decreased at intermediate position (64 cm downstream from the grating) as shown in Fig. 2(b).

Images obtained by XTI

We prepared an 8-μm-pitch gold grating for G2 whose pattern height was 8 μm, which was the marginal value with normal optical lithography. Although a pattern with a higher aspect ratio was preferred in order to obtain higher contrast, 92% of 0.1-nm X-rays could be blocked by the gold strips.

An X-ray Talbot interferometer was assembled by aligning the phase grating and the amplitude grating described above with a separation of 32 cm. A plastic sphere 1.2 mm in diameter was placed in front of G1 as a phase object. Figure 3(a) shows a resultant image detected behind G2. Since, in this observation, one does not need to resolve the contrast of stripes as in Fig. 2, the effective pixel size was increased to 6.33 μm by changing the coupling lens. The contrast corresponding to the phase gradient caused by the sphere and bubbles in it were seen.

For the measurement of $\varphi(x, y)$, G2 was displaced with a step of $d/5$ in the x direction and five images were acquired. By processing the images with eq. (4), Fig. 3(b) was obtained. Figure 3(c) shows an image of a phase map $\Phi(x, y)$ obtained by integrating Fig. 3(b). The sample was rotated on the y axis shown in Fig. 1,

and the measurement of a phase map was repeated at every angular position of the sample rotation. A three-dimensional rendering view of the reconstructed tomographic data is shown in Fig. 3(d). A quadrant of the data has been cropped to show bubbles resolved inside the sphere. The noise at a plastic region of the tomogram was 5×10^{-9} (standard deviation), which corresponds to the density deviation of 4 mg/cm^3.

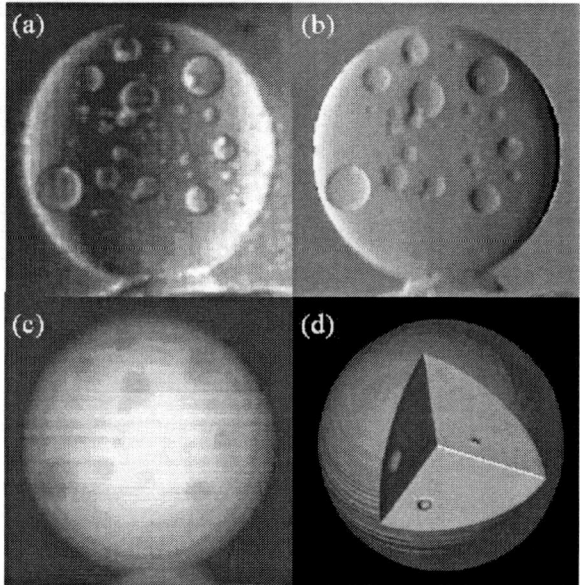

FIGURE 3. Images obtained for a plastic sphere 1.2 mm in diameter: (a) a raw image generated by the X-ray Talbot interferometer, (b) an image mapping the differential phase measured by the phase-shifting technique, (c) a phase map obtained by integrating (b), and (d) a three-dimensional rendering view of the data reconstructed by phase tomography. Structures seen in the sphere are air bubbles.

PROSPECT OF MEDICAL APPLICATIONS

As demonstrated with the experiments at SPring-8, XTI is feasible as a phase-sensitive imaging method. However, for medical applications, three problems are remaining to be solved; that is, (1) using high-energy X-rays, (2) expanding the field of view, and (3) ensuring the compatibility with compact X-ray sources.

In clinical applications, high-energy X-rays that can penetrate a body should be used. Therefore, the fabrication of an amplitude grating becomes more difficult because an aspect ratio higher than that of the grating used in the present experiment is needed. Although optical lithography seems to be unavailable for that purpose, we consider that the X-ray LIGA

(Lithographie, Galvanoformung, Abformung) process [13] would be an alternative approach to the fabrication of such a grating. Because the grating pattern required for XTI is simple, the fabrication of a grating with a large effective area would be feasible by the LIGA process leading to the expansion of the field of view.

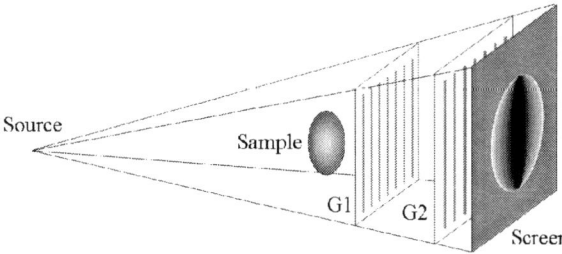

FIGURE 4. An X-ray Talbot interferometer using a diverging X-ray beam. The ratio of the source-G1 distance to the period of G1 is the same as the ratio of the source-G2 distance to the period of G2.

In discussing the compatibility with a compact X-ray source, it should be noted that an X-ray Talbot interferometer can be constructed for spherical-wave X-rays as illustrated in Fig. 4. This advantage of the XTI is attractive because a compact X-ray source, which normally emits divergent X-rays, can be used without crucial loss of X-rays.

CONCLUSIONS

XTI was presented as a novel and simple phase-sensitive X-ray imaging method, three-dimensional imaging with which was also attained. Although the demonstrative experiment was performed with plane-wave monochromatic X-rays at SPring-8, spherical-wave X-rays with a broad energy band are available for XTI in principle. This implies that XTI has high compatibility with compact X-ray sources. It should be added to the advantages of XTI that an image detector with the spatial resolution as high as that used in the propagation-based method is not always required. XTI is also advantageous in that a contrast reaching 100% is generated if ideal gratings are fabricated. Thus, XTI would be feasible as a phase-sensitive X-ray imaging method for future medical applications.

ACKNOWLEDGMENTS

We would like to thank Dr. Y. Suzuki for the discussions and technical support for the experiments at SPring-8, which were carried out under the approvals of the SPring-8 committee, 2003A0458-NML2-np and 2003A0459-NML2-np.

REFERENCES

1. Davis, J., Gao, Gureyev, T.E., Stevenson, A.W., and Wilkins, S.W., *Nature* **373**, 595-598 (1995).

2. Chapman, D., Thomlinson, W., Johnston, R.E., Washburn, D., Pisano, E., Gmür, N., Zhong, Z., Menk, R., Arfelli, F., and Sayers, D., *Phys. Med. Biol.* **42**, 2015-2025 (1997).

3. Bonse, U. and Hart, M., *Appl. Phys. Lett.* **6**, 155-156 (1965).

4. Momose, A., *Nucl Instrum Meth A* **352**, 622-628 (1995).

5. Yoneyama, A., Momose, A., Koyama, I., Seya, E., Takeda, T., Itai, Y., Hirano, K., and Hyodo, K., *J. Synchrotron Rad.* **9**, 277-281 (2002).

6. Wilkins, S.W., Gureyev, T.E., Gao, D., Pogany, A., and Stevenson, A.W., *Nature* **384**, 335-338 (1996).

7. Cloetens, P., Ludwig, W., Baruchel, J., Van Dyck, D., Van Landuyt, J., Guigay, J.P., and Schlenker, M., *Appl. Phys. Lett.* **75**, 2912-2914 (1999).

8. Momose, A, Kawamoto, S., Koyama, I., Hamaishi, Y., Takai, K., and Suzuki, Y.., *Jpn. J. Appl. Phys.* **42**, L866-L868 (2003).

9. Momose, A., Koyama, I., Hamaishi, Y., Kawamoto, S., Takai, K., Uesugi, K., and Suzuki, Y., in *Proceedings of Eighth International Conference on Synchrotron Radiation Instrumentation 2003*, AIP Conference Proceedings, New York: American Institute of Physics, 2004, in press.

10. Talbot, H., *Philos. Mag.* **9**, 401-407 (1836).

11. Stetson, K.A. and Brohinsky, R, *Appl. Opt.* **24**, 3631-3637 (1985).

12. Cloetens, P., Guigay, J.P., De Martino, C., and Baruchel, J., *Opt. Lett.* **22**, 1059-1061 (1997).

13. Becker, E.W., Ehrfeld, W., Hagmann, P., Maner, A., and Münchmeyer, D., *Microelectron. Eng.* **4**, 35-56 (1986).

Development of Dual-Energy X-ray CT using Synchrotron Radiation

M. Torikoshi, T. Tsunoo, M. Sasaki[*], M. Endo, Y. Noda, Y. Ohno[1], T. Kohno[1], M. Natsuhori[2], T. Kakizaki[2], N. Yamada[2], N. Itoh[2], K. Hyodo[3], K. Uesugi[4] and N. Yagi[4]

National Institute of Radiological Sciences, 4-9-1 Anagawa, Inage-ku, Chiba-shi 263-8555, Japan

[1]*Tokyo Institute of Technology, 4259 Nagatsuta-cho, Midori-ku, Yokohama-shi, 226-8502, Japan*

[2]*Kitasato University, School of Veterinary Medicine & Animal Sciences, 35-1 Higashi 23-Bancho, Towada-Shi, Aomori 034-8628, Japan*

[3]*High Energy Accelerator Research Organization, 1-1 Oho, Tsukuba-shi, Ibaragi 305-0801, Japan*

[4]*Japan Synchrotron Radiation Research Institute, 1-1 Koto, Mikazuki-cho, Sayo-gun, Hyogo 679-5198, Japan*

Abstract. Monochromatic x-ray CT at two different energies provides information about electron density without ambiguity due to the beam hardening effect. This information makes the treatment planning for heavy-ion radiotherapy more precise. We have started a feasibility study on the dual-energy x-ray CT by using synchrotron radiation. We developed a linear scanning CT system in order to evaluate what precision in the measurement was achieved. The experiments were carried out using monochromatic x-rays of 40, 70 and 80 keV. Comparison of measured electron densities with the theoretical values proved that these values were in agreement in 0.9 % on average. We have developed a 2D-CT system with a two-dimensional scintillator array to take images in a short time. At present, it has been proved that the electron density is measured in the precision of about 1 % with the 2D-CT system. Effective atomic numbers are obtained as well from the dual-energy x-ray CT. The CT images are reconstructed based on each of the electron density and the effective atomic number. They present different features of an object, and the contrasts in the images differ from that in a conventional CT image. The feature of the images may help distinguish more clearly tissues or organs with aid of different contrast from the conventional CT images.

INTRODUCTION

The information about an electron density in a human body plays an important role in treatment planning for radiotherapy. Especially, for proton and heavy ion radiotherapy, the electron density is indispensable to predict the range of the particle beams and the dose distribution in a body. In order to measure the electron density, we proposed the method of dual-energy x-ray CT using synchrotron radiation (SR)[1].

As the first step, we developed a one-dimensional CT system to establish the technique for the quantitative CT[2]. In the study, the electron densities of several samples were directly measured by dual-energy x-ray CT using monochromatic x-rays produced from SR at the facility of Japan Synchrotron Radiation Research Institute (SPring-8) and Accumulator Ring (AR) of High Energy Accelerator Research Organization. The values were compared with the theoretical values or reference values measured in a complementary method.

---

[*] Present; Synchrotron Light Life Science Center, Ritsumeikan University

CP716, *Portable Synchrotron Light Sources and Advanced Applications*,
edited by H. Yamada, N. Mochizuki-Oda, and M. Sasaki

As the next step, a more practical CT system was required to scan a wide area in a short time. CT systems were constructed using a two-dimensional camera: an image-intensifier and a CCD camera. Since both of the cameras did not have enough wide dynamic range in photon intensity to take images of thick objects, the objects were constrained to be small[3]. We developed the two-dimensional CT scanning system with a detector of scintillator array that had the wider dynamic range. The read-out has been developed to achieve a fast frame-rate up to more than a hundred frames per second. In this paper, we described as follows: the principle for derivation of the electron density from the dual-energy x-ray CT, the brief review of the dual-energy x-ray CT with 1D-CT system, and preliminary results of the dual-energy x-ray CT performed with the 2D-CT system.

BASIC CONCEPT OF DUAL-ENERGY X-RAY CT

An x-ray linear attenuation coefficient of a material at an energy k is described in terms of a photoelectric absorption cross-section, a coherent scattering cross-section and an incoherent scattering cross-section. D. F. Jackson and D. J. Hawkes[4] proposed formulae for each cross-section to easily use in medical applications. We used the formulae and simplified them as follows,

$$\mu = \rho_e \left[Z^4 F(k,Z) + G(k,Z) \right],$$

where ρ_e is the electron density, k is photon energy and Z is an atomic number of a material. $\rho_e Z^4 F(k,Z)$ denotes a photoelectric absorption term and $\rho_e G(k,Z)$ includes the coherent scattering term and the incoherent scattering term. Measuring the linear attenuation coefficients with two x-rays of energy k_1 and k_2, we obtain simultaneous equations with respect to the unknown variables of ρ_e and Z. Assuming that neither the term of $F(k,Z)$ nor the term of $G(k,Z)$ strongly depends on Z, we solve the simultaneous equations with respect to the Z^4 at first; that is,

$$Z^4 = \frac{\mu(k_2)G(k_1,Z) - \mu(k_1)G(k_2,Z)}{\mu(k_1)F(k_2,Z) - \mu(k_2)F(k_1,Z)}.$$

This equation can be solved iteratively. We called this parameter Z an effective atomic number. Once we obtain the Z, we can derive the electron density using the Z from follows,

$$\rho_e = \frac{\mu(k_1)F(k_2,Z) - \mu(k_2)F(k_1,Z)}{F(k_2,Z)G(k_1,Z) - F(k_1,Z)G(k_2,Z)}.$$

In the calculation of the coherent scattering term, we took oxygen as the standard elements as mentioned by D. F. Jackson and D. J. Hawkes[4]. The oxygen cross-sections were quoted from calculations done by Hubbell *et al*[5]. This method is available for low Z elements, but not for high Z elements that have the absorption edges in the interesting energy region.

BRIEF REVIEW OF DUAL-ENERGY X-RAY CT WITH 1D-CT SYSTEM

A translation-rotation CT scanning system (1D-CT) consisted of rotating and sliding tables, a plastic scintillator as an x-ray detector and an ion-chamber. The output current of the detector was experimentally proved to be proportional to the x-ray intensity in the range from 10^4 to 10^9 ph/mm²/s. In the dual-energy x-ray CT, the projection data were taken by being exposed to low energy x-rays and high energy x-rays alternately. Experiments were carried out at the beam-lines of BL20B2 of SPring-8 and NE5A of AR. In BL20B2, 40 and 70 keV x-rays were used, and in NE5A 40 and 80 keV were used. Several samples were used: samples equivalent to a soft tissue, an adipose tissue, a cartilage bone, a compact bone and a lung, graphite, solutions of dipotassium hydrogen-phosphate K_2HPO_4, and water.

The electron densities of the samples were obtained by averaging pixel values within ROI's defined in CT images based on the electron density[2]. They were compared with the theoretical values or the reference values. The measured values were proved to be in agreement with the theoretical or reference values in 0.9 % on average. They are summarized in Figure1.

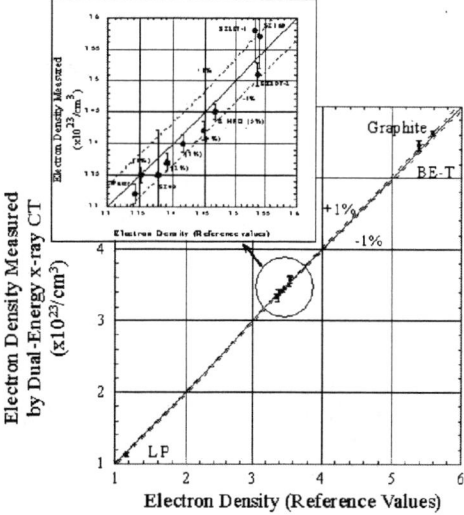

FIGURE 1. The electron densities measured with the dual-energy x-ray CT are plotted with respect to the theoretical value and reference values. This was quoted from Ref. 2.

DUAL-ENERGY X-RAY CT WITH 2D-CT SYSTEM

The CT scanning system consists of a two-dimensional x-ray detector of a 256×96 pixel-array of Gd_2O_2S scintillator. The pixel size is 0.89 mm (H) and 1.03 mm (V). The output current was integrated and digitized into 16-bit. Samples were rotated $180°$ in a $0.4°$ step. The exposure time was adjusted to make the maximum pixel value lower than 55,000 without a sample. Since the detector system was able to take more than a hundred projection images per second, it took about 3 minutes to scan $180°$. The experiments were carried out at BL20B2 of SPring-8 using 40 and 70 keV x-rays.

Solutions of K_2HPO_4, ethanol and water were used to evaluate the precision of the electron density measured with 2D-CT system. Samples of brain, kidney, liver, spleen and pancreas of a pig (6-month old) and a rat (about 20 cm long) were also used to make images of biological samples. The tissues were extracted soon after slaughter and were dipped in agar to immobilize. The agar was antisepticised by adding sodium-azide. The rat was contained in a vessel of polyethlene terephthalete (PET) without the agar.

Preliminary Results and Discussion

According to the results of liquid samples, the electron density was measured in precision of about ± 1 %. This seems same level as that of the 1D-CT. Some typical images are shown below. The images of a chest of the rat are in Figure 2(A) and (B). The images of brain are shown in Figure 3(A), (B) and (C).

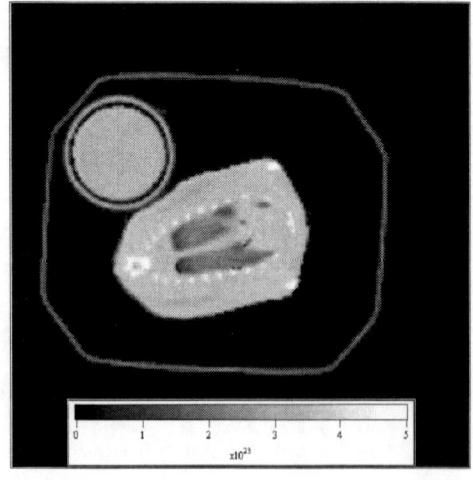

FIGURE 2(A). The electron density image of the rat's chest. The lung are clearly presented.

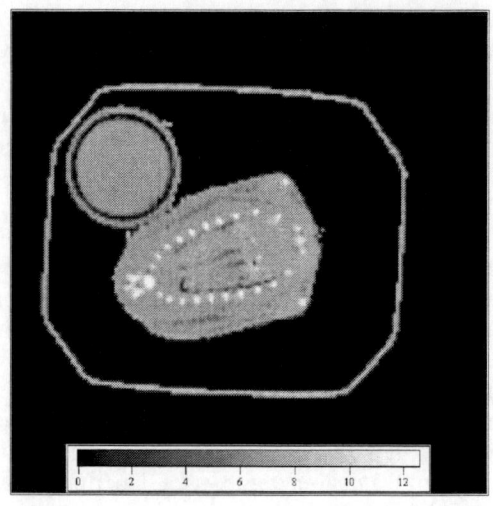

FIGURE 2(B). The effective atomic number image of the rat's chest. The lung is no longer distinguished clearly from other tissues in the image.

The images of (A) and (B) of each figure denote the electron density image and the effective atomic number image, respectively. 3(C) is the image based on the linear attenuation coefficient for 40 keV x-ray.

Rings in the images are the cross section of vessels for water samples. The water sample was contained in each sample vessel to verify the precision of electron density measured in the experiment. In the case of rat sample, the electron density of water is in agreement with the theoretical value in 0.4 %, while that of the brain sample agrees in 0.6 %.

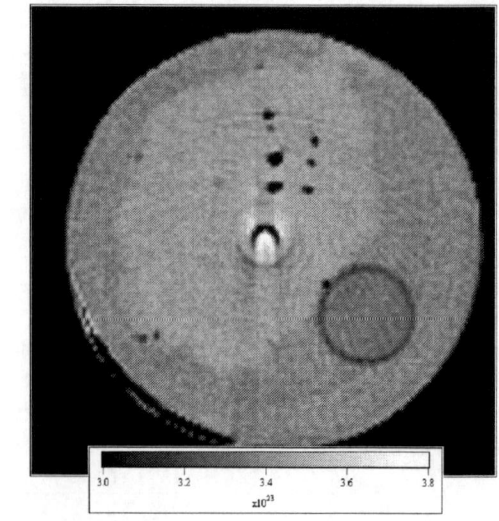

FIGURE 3(A). The electron density image of the brain sample of a pig.

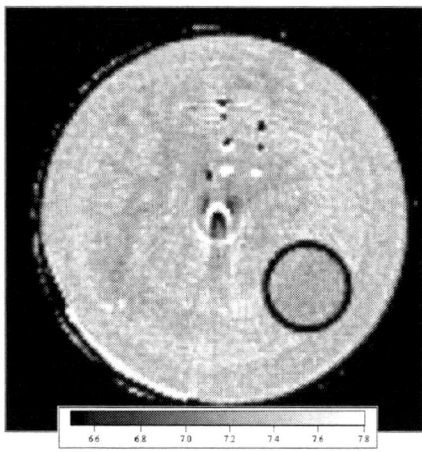

FIGURE 3(B). The effective atomic number image of brain of a pig.

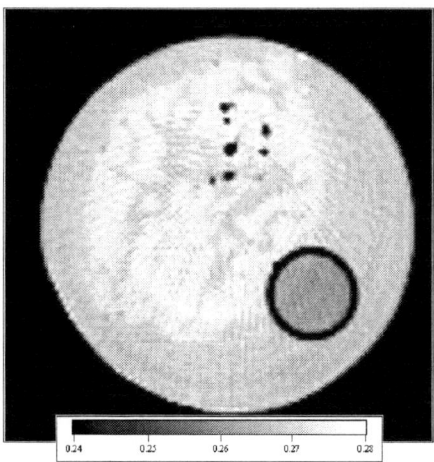

FIGURE 3(C). The image of the brain sample of a pig. It is based on the linear attenuation coefficient for 40 keV x-ray.

The lung of the rat clearly appears in the electron density image of Figure 2(A), while in the image of 2(B) it is difficult to distinguish the lung from other tissues. Since the lung is a lump of alveoli and bronchial tubes, its average electron density is much lower than the adjacent tissues. On the other hand, since alveoli and bronchial tubes consist of cells, their elemental materials should be approximately same as those of other tissues. It decreases contrast between the lung and adjacent tissues in the effective atomic number image.

Window levels of all of the images in Figure 3 were expanded to enhance the subtle variance in the electron density and effective atomic number. In this process, the artifacts were also enhanced. Figure 3(C) is the image based on the linear attenuation coefficient for 40 keV x-ray. In this image, the sulci clearly appear. It is interesting that the sulci are not visible at all in the electron density image. In the effective atomic number image, however, only the sulci appear with high contrast from other part. The effective atomic number of parenchyma of the brain is very close to that of the agar.

SUMMARY

The precision of the electron density measured by the dual-energy x-ray CT achieved to ±1 % with the 2D-CT as well as 1D-CT systems.

The images of the electron density and effective atomic number give the qualitatively different information from the information on linear attenuation coefficients or CT-numbers.

ACKNOWLEDGMENTS

Mr. Y. Fujii and Mr. Y. Toji of Nishi Harima Meat Inspection Office kindly offered us the tissue samples and helped us to prepare them. We deeply appreciate both of them and all of the staffs of the office.

REFERENCES

1. Torikoshi M., Tsunoo T., Endo M., Noda K., Kumada M., Yamada S., Soga F. and Hyodo K., J. BioMed. Opt. 6, 371-377(2001).

2. Torikoshi M., Tsunoo T., Sasaki M., Endo M., Noda Y., Ohno Y., Kohno T., Hyodo K., Uesugi K., and Yagi N., Phys. Med. Biol. 48, 673-685(2003)

3. Tsunoo T., Torikoshi M., Sasaki M., Endo M., Yagi N., Uesugi K., IEEE Trans. Nucl. Sci. Vol. 50, 1678-1682(2003).

4. Jackson D. F., and Hawkes D. J., Phys. Rep. 70, 169-233(1981)

5. Hubbell J. H. et al., J. Phys. Chem. Ref. Data 4, 471-538(1975)

Analysis of Pheochromocytoma (PC12) Membrane Potential under the Exposure to Millimeter-wave Radiation

M. Mizuno[*], A. Hirata[†], K. Kawase[*], C. Otani[*], and T. Nagatsuma[†]

[*] RIKEN, 2-1 Hirosawa, Wako, 351-0198, Japan
[†] NTT Microsystem Integration Laboratories,3-1 Morinosato Wakamiya, Atsugi,Kanagawa, 243-0198, Japan

Abstract. Non-thermal effects of millimeter wave (MMW) on Pheochromocytoma (PC12) were studied by potential measurement with a voltage sensitive dye (DiBAC$_4$(3)). Cells were irradiated at fixed frequencies of 30, 40, 60, 76GHz as well as sweeping frequency between 10 and 100 GHz by an MMW generator based on a uni-traveling-carrier photodiode (UTC-PD), the most widely tunable MMW source. However there were no significant changes in membrane potential between MMW-irradiated and control cells. The results suggest that MMW irradiation in the range from 10 to 100GHz appears to be safe for ordinary PC12 cells under non-thermal conditions.

INTRODUCTION

Health related effects of millimeter-wave (MMW) have attracted the attention of many research groups. The question of whether low power MMW can affect biological systems is still not clear, and the theoretical consideration of such an interaction in the cell is under way. The biological effects of low power MMW were first discovered in the late 1960s. After that they were studied using a wide variety of objects ranging from biomolecules to bacteria and tissues of more complex organisms. Based on the biological experiments, MMW began to be used in medical applications. However, this treatment is not applied in the Western medicine because the nature of MMW exposure on the animal and human organism is not well understood.

H. Fröhlich [1] suggested effects of MMW radiation might occur through a resonance-type interaction, since some of the biomolecules and structural elements of the cells have their own theoretically calculated resonant frequency within the range of 10^{10}-10^{11} Hz. Several experiments showing narrow resonant frequency dependence of biological effects of MMW seem to support this hypothesis [2]. However, other physical mechanisms may also be responsible. The relation between this hypothesis and the medical effects of MMW is not clear.

MMW is already used in industrial and alternative medical applications, therefore the effects of MMW exposure need scientific clarification.

MATERIALS AND METHODS

Cell Culture

PC12 were continuously cultured in RPMI-1640 medium with 5% FBS and 10% HS at 37°C in 5% CO_2. DiBAC$_4$(3) (3.87 µmol/l) was added into the vessel and allowed 60 minutes to permeate the cell. DiBAC$_4$(3) is bis-(1,3-dibutylbarbituric acid)-trimethine oxonol, a voltage sensitive fluorescent dye, for the measurement of membrane potentials [3]. It has been used to evaluate antibiotics and medicines [4]. We can observe depolarization and hyperpolarization on membrane potentials by using this dye. During the measurement, the pH in the medium was stabilized at 7.3 ± 0.1 by CO_2 control, and the temperature was maintained at 37 ± 0.1°C using a temperature controller.

CP716, *Portable Synchrotron Light Sources and Advanced Applications,*
edited by H. Yamada, N. Mochizuki-Oda, and M. Sasaki
© 2004 American Institute of Physics 0-7354-0195-0/04/$22.00

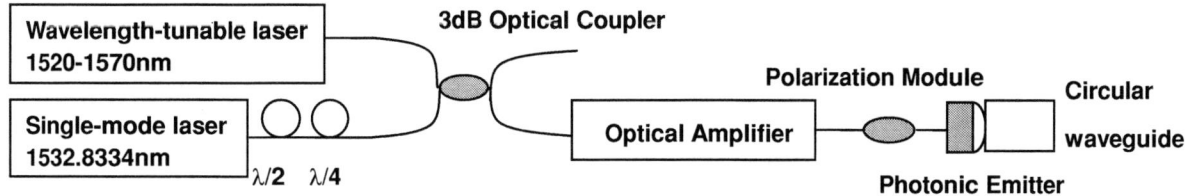

FIGURE 1. Setup of the millimeter wave generator using a UTC-PD

MMW Exposure

For research on biological effects of MMW we used an MMW generator based on a uni-traveling-carrier photodiode (UTC-PD). This generator provides the widest tunability in MMW obtainable from a single component. The schematic of the experimental setup is shown in Figure 1. The MMW power at 10-100GHz was generated by the two-mode beating of a wavelength-tunable laser and a single-mode laser, and input to the photonic emitter via an optical fiber. The photonic emitter consists of an optical lens, a UTC-PD chip, a wideband antenna and Si lens. The optical signals were injected from the backside of the UTC-PD chip that converts the optical signal into an electric signal [5]. Then, the MMW signal was radiated into a circular waveguide by a wideband antenna, and directed to the cells.

PC12 cells in standard tissue culture plants were irradiated from the top of the dish. MMW energy is absorbed in the medium. The power of MMW arriving to the cells was ~0.1µW at 30 GHz on calculated results [Figure 2]. The cells were irradiated with fixed frequencies as well as sweeping frequency from 10 to 100 GHz.

RESULTS AND DISCUSSION

PC12 cells were irradiated with low power at 30, 40, 60 and 76 GHz for 1hour, during which time the membrane potential change was measured using a MRC1000 confocal laser microscope system (BIORAD). The average brightness value of over 6000 pixels in fluorescence image was displayed on the graph [Figure 3]. The brightness depends on the mem-

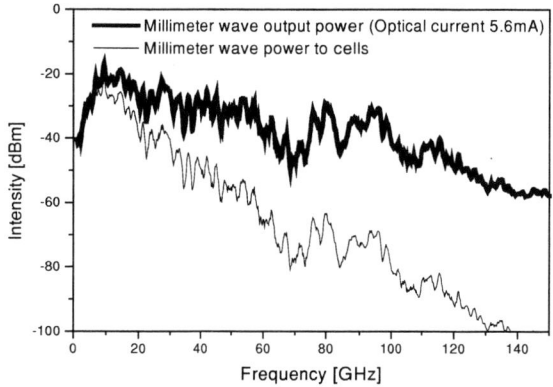

FIGURE 2. MMW radiation power spectra. Thick line is the MMW source power. Thin line is the MMW power to cells after absorbed by water.

brane potential and an increase of the brightness means a depolarization of the membrane potential. We compared membrane potential change in MMW-irradiated cells to that of non-irradiated (control) cells, and there was no significant difference.

In another series of experiments, cells were irradiated with sweeping frequency from 100 down to 10GHz. The sweeping rate was 0.2 GHz/4sec. Using the background as a reference (the areas without cells), there were no significant changes between the MMW-irradiated cells and the control cells.

In summary, we investigated the effects of low-power MMW irradiation on the PC12 membrane potential. The membrane potential change was measured using a voltage sensitive fluorescent dye, however there were no significant brightness changes by MMW irradiation at various frequencies.

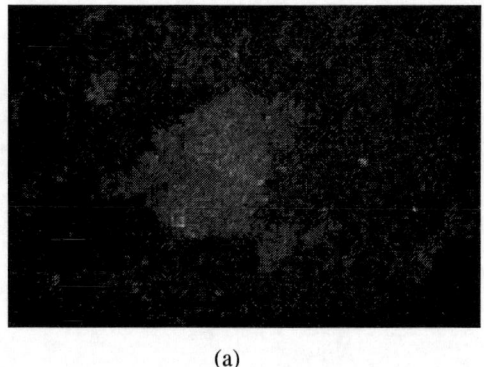

(a)

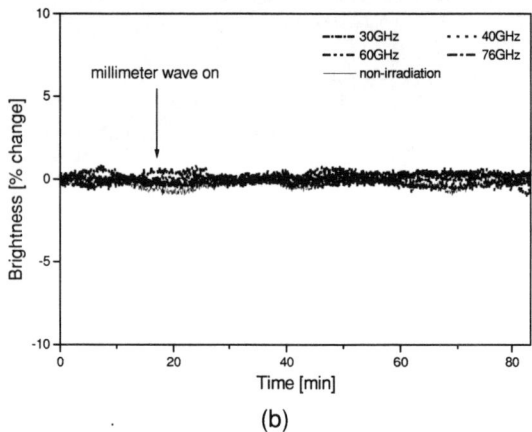

(b)

FIGURE 3. The effect of fixed frequency MMW on PC12. (a) Fluorescence image of membrane potential on PC12 (768×512pixels(Pixel Size: 1.650μm)). (b) Membrane potential change of MMW-irradiated PC12 cells.

The results suggest that MMW irradiation in the 10-100GHz range appears to be safe for ordinary PC12 cells under non-thermal conditions. In the future, we intend to perform experiments at higher MMW power.

REFERENCES

1. Fröhlich H, "Further evidence for coherent excitations in biological systems." *Phys Letters,* 1985, 110A, p480.
2. Pakhomov A, Prol H, Mathur S, Akyel Y, Campbell CBG, "Search for frequency-specific effects of millimeter-wave radiation on isolated nerve function." *Bioelectromagnetics,* 1997, 18, p324.
3. T. Brauner, D. F. Hulser, R. J. Strasser, "Comparative measurements of membrane potentials with microelectrodes and voltage-sensitive dyes." *Biochim. Biophys. Acta.,* 1984, 771, p208.
4. Yamada A, Gaja N, Ohya S, Muraki K, Narita H, Ohwada T, Imaizumi Y, "Usefulness and limitation of DiBAC4(3), a voltage-sensitive fluorescent dye, for conductance Ca^{++}-activated K^+ cannels in HEK293 cells." *Jpn. J. Pharmacol.* 2001, 86, p3.
5. T. Ishibashi, N. Shimizu, S. Kodama, H. Ito, T. Nagatsuma, and T. Furuta, "Uni-traveling-carrier photodiodes." *Tech. Dig. Ultrafast Electron. and Optoelectron.* 1997, p166.

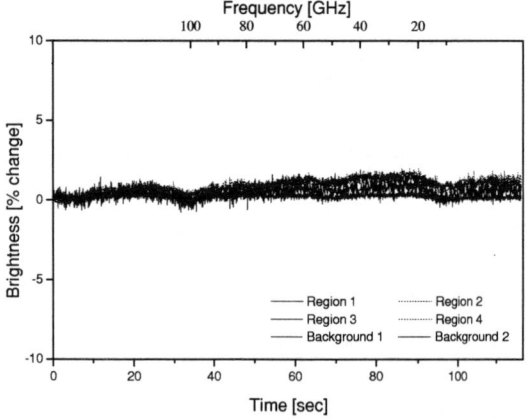

FIGURE 4. Membrane potential change on PC12 irradiated by sweeping frequency from 100 down to 10 GHz.

FIR Beam Line for MIRRORCLE-20

A. Moon[a,b], Y. Nakamura[a], T. Toma[a], H. Yamada[a,b]

[a] Faculty of Science & Engineering, Ritsumeikan University
[b] Synchrotron Light Life Science Center, Ritsumeikan University

Abstract. The recent progress in the portable synchrotron MIRRORCLE-20 for brilliant FIR or IR source for the life Science research is reported. The new MIRRORCLE-20 FIR beam line is consist of the special optics which collect all radiation, emitted from the exactly circular 156mm radius electron orbit, by using both circular mirror and quasi-ellipsoidal mirror (magic mirror). The optics designed is somewhat different from the conventional SR beam line optics. In case the whole radiation is collected and the e-beam lifetime reaches 1 minute, 10W power in total are available. The beam line was tuned to maximize the light extraction.

INTRODUCTION

The IR or FIR radiation plays important role in the life science research. Unfortunately, there are no powerful light sources for this region, except some lasers. The free electron laser has enough power and a broadband tunability. Nevertheless, the monochromatic FEL radiation is not suitable for fundamental investigations in which the broadband spectrum is required. For these investigations, the radiation with broad spectrum is more convenient than monochromatic radiation. The blackbody radiation has broad spectrum but the power is too small. The conventional synchrotron radiation also has broad spectrum but the power is not sufficient for some applications. The MIRRORCLE-20 [1,2] has a large capability as the high power FIR or IR synchrotron radiation source. Particularly, it is possible to collect all radiation emitted from the exactly circular electron orbit, since the orbit radius is only 156mm[3]. We have developed special optics to collect all radiation in MIRRORCLE-20 and assembled the beam line for the brilliant FIR or IR radiation.

Magic mirror

The mirror for collecting the synchrotron radiation from the considerably large arc of the electron orbit is known as the magic mirror [4]. This quasi-ellipsoidal mirror has two "generalized focal point" which are considered as focusing point and radiation point respectively. The shape of the magic mirror is

obtained by solving equation (1) and (2), as shown in Fig.1.

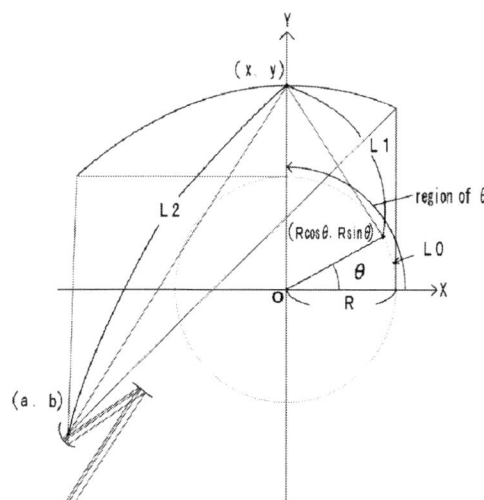

Fig.1. Geometric note of the magic mirror.

$$y = \frac{\cos\theta}{\sin\theta}x + \frac{R}{\sin\theta} \quad (1)$$

$$d = L0 + L1 + L2$$

$$= R\theta + \sqrt{(a-x)^2 + (b-y)^2}$$

$$+ \sqrt{(R\cos\theta - x)^2 + (R\sin\theta - y)^2} \quad (2)$$

CP716, *Portable Synchrotron Light Sources and Advanced Applications,*
edited by H. Yamada, N. Mochizuki-Oda, and M. Sasaki
© 2004 American Institute of Physics 0-7354-0195-0/04/$22.00

The length d and the focusing point (a,b) are supposed to be 900mm and (-310mm, -201.5mm) respectively (the origin is placed to the center of the e-beam orbit). The focused radiation is converted to parallel rays by parabolic mirror and transported to users through the beam line, purged by the dry N_2.

Circular mirror

The magic mirror itself cannot collect all radiation emitted from the exactly circular orbit. The experimental magic mirror covers $\pi/2$ at maximum, but the beam divergence growth. To avoid these disadvantages, we have used both magic mirror and circular mirror to collect all radiation. In the Fig.2 the optical layout for collecting all synchrotron radiation from the exact circular orbit is shown. The entrance hole for the electron beam in the circular mirror is used as the output port for the IR radiation, which, in turn, is focused by the magic mirror on the plane mirror and are converted to parallel rays by parabolic mirror.

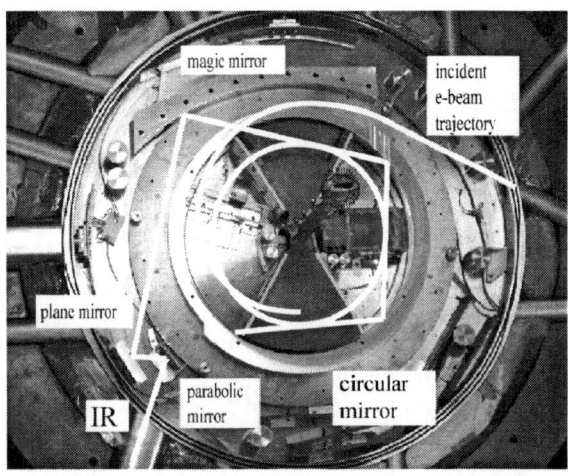

Fig.2. The optical layout for collecting all synchrotron radiation from the exact circular orbit.

Note, that the circular mirror can be used not only for collecting radiation but also as the optical resonator. For the specific wavelength, interaction between e-beam and radiation is expected to be 100 ~ 10000 times larger than the interaction in general case. The interaction has the same mechanism as for the free electron laser. This phenomenon is used in the novel type of the free electron laser called the photon storage ring [5-8]). In the Fig.3 the IR spectrum in the MIRRORCLE-20, calculated for the 19.5MeV e-beam, is shown [2]. The dashed line shows SR power, integrated over 2π and the solid line shows the same dependence including the interference effect.

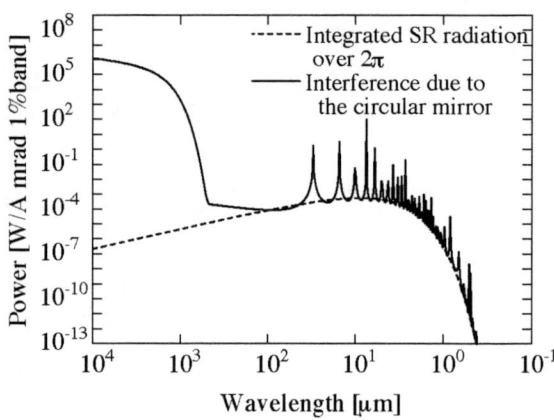

Fig.3. The IR spectrum calculated for 19.5MeV e-beam. The MIRRORCLE-20 covers up to 1μm ($\lambda c = 8$μm).

IR measurements

The IR power was measured by the cooled MCT detector. The detector assembly including amplifier, and DC power supply were stored in the double shielded metal box to decrease EM noises. In Fig.4 the experimental setup is shown schematically. In Fig.5 the typical IR signal at the MIRRORCLE-20 exit is shown. Using only magic mirror, we get about 100mV IR signal peak intensity. With both magic and circular mirror, we have measured about 300mV peak signal.

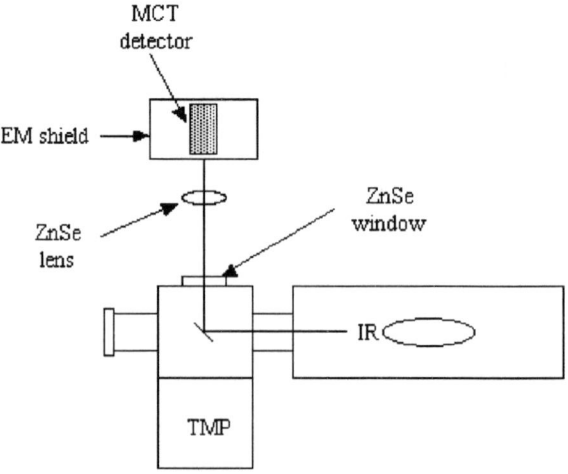

Fig.4. The measurements setup.

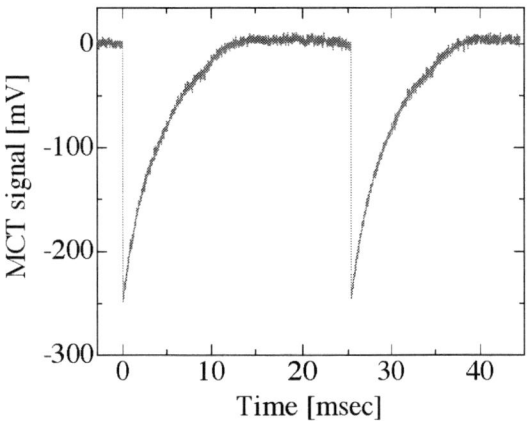

Fig.5. The typical IR signal measured by the cooled MCT detector.

The e-beam lifetime

Improving vacuum is one of the important technical issues to increase the e-beam lifetime, which, in turn, result in the increased IR average power. The e-beam lifetime is determined by the scattering with the residual gas, so by the vacuum level. In the Fig.6 the e-beam lifetime versus vacuum level of the MIRRORCLE-20 is shown. The typical lifetime is around 10msec and is insufficient for the most of applications. The lifetime decreases due to the scattering by the residual gas, which leads to the beam trap by the created ions.

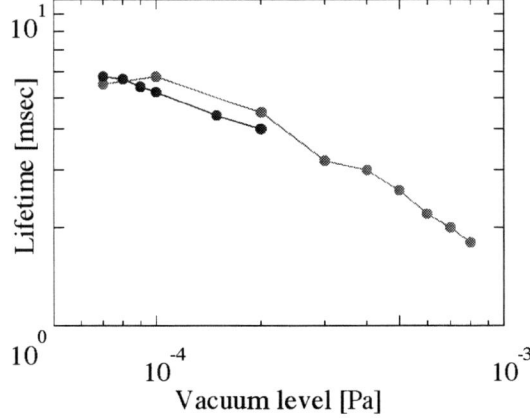

Fig.6. The e-beam lifetime versus vacuum level

The lifetime become 6 msec fore 10^{-4} Pa vacuum level, but this is limited by the IR detector amplifier. We believe that the real lifetime is longer than this

value. In the typical operation conditions the vacuum level reaches 10^{-5} Pa and so the lifetime is the 10msec level. We are going to improve the vacuum by installing the cryogenic pump and the ion trap.

The e-beam instability

In some operation conditions the beam instability appears. In the Fig.7 the typical beam instability signal is shown. After 10msec the IR signal suddenly drops. But just before the e-beam loss, the IR signal (i.e. e-beam lifetime) increases slightly. We assume that some kind of e-beam instability took place when the beam size decreases due to the radiation damping. The origin of this instability and concomitant phenomena are not so clear and obvious: the appropriate careful investigations will be the goal of our future work.

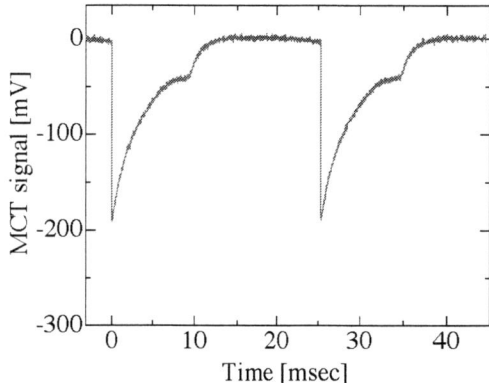

Fig.7. Typical signal of an e-beam instability.

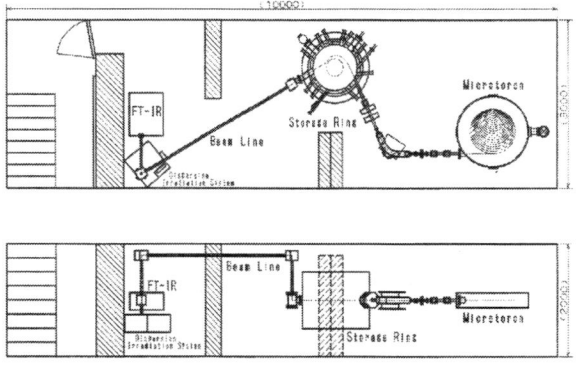

Fig.8. The MIRRORCLE-20 experimental setup. The physical arrangement of equipments and the IR beam line are shown.

The IR beam line

In Fig.8 both the physical arrangement of the equipments and the IR beam line of the MIRRORCLE-20 are shown. The FIR or IR radiation is transported to the FT-IR or the grating monochromator through the 5m beam line, purged by the dry N_2.

CONCLUSION

To collect the whole radiation from the MIRRORCLE-20, the optical system, consist of the both quasi-ellipsoidal mirror (magic mirror) and circular mirror was developed. Using the magic mirror together with the circular one, we have measured about 300mV peak signal by cooled MCT detector placed at one-meter distance from the exit port. The parallel rays from parabolic mirror are transported to the FT-IR or the grating monochromator and can be utilized by users. We are going to increase the beam lifetime for the high power IR extraction. The investigation of the e-beam instability will be our next target.

REFERENCES

1. Yamada H., *Nucl. Instr. and Meth.* **B199**, 509-516 (2003).

2. Yamada H., *J. of Synchrotron Radiation*, 1326-1331 (1998).

3. Yamada H., present status The 28th International Conference on Infrared and Milimete Waves (2003).

4. Lopez-Delgado R. and Szwarc H., *Opt. Commun.* **19**, 286 (1976).

5. Yamada H., *Jpn. J. Appl. Phys.*, **28**, 1655 (1989).

6. Mima K. Shimoda K. Yamada H., *IEEE J. of Quantum Electronics*, **27**, 2572-2579 (1991).

7. Tsutsui H., Yamada H., Mima K., Shimoda K., *Nucl. Instr. and Meth.* **A331**, 395-400 (1993).

8. Kleev A.I. and Yamada H., *IEEE J. of Quantum Electronics*, **39**, 820-828 (2003).

Dynamical Study of Water Structure by Infrared Synchrotron Light

Nobuhiro Miura[*], Hironari Yamada[*#], Ahsa Moon[#], Kishi Nishikawa[#]

*Synchrotron Light Life Science Center, Ritsumeikan University, Noji-higashi 1-1-1, Kusatsu, Shiga 525-8577, Japan
#Department of Photonics, Faculty of Science and Engineering, Ritsumeikan University, Noji-higashi 1-1-1, Kusatsu, Shiga 525-8577, Japan

Abstract. Portable synchrotron, MIRRORCLE 20 in Ritsumeikan University is prepared as the infrared and far infrared synchrotron light source for the molecular dynamics study. The FTIR spectrometer and the dispersive-type IR irradiation system using grating monochrometer are installed in the beam line. Sample cells with a temperature controller is prepared for the FTIR measurements of water-rich samples like the aqueous solution, which encounters difficulty in a conventional FTIR system. The irradiation system provides monochromatic IR-rays in the wave length between 2µm and 25µm. Dielectric relaxation spectroscopy (DRS) with network analyzer in the frequency range of 45MHz to 20GHz is prepared for the measurements of the sample irradiated by monochromatic IR-rays. Research plans with the analysis system for dynamical water structure is described in this report.

INTRODUCTION

Water maintains a molecular arrangement like the ice lattice although the network of hydrogen bonding is warped by micro-Brownian motion of molecules [1]. The arrangement in liquid state is normally called "water structure" or "liquid structure". In most of the aqueous solutions, water molecule interacts with the solute through hydrogen bonding and constructs the water structure [2].

We prepare the analysis system for molecular dynamics utilizing high intensity IR synchrotron light from MIRRORCLE 20 [3, 4]. The system includes the FTIR spectrometer and the dispersive-type IR irradiation system. Our purpose is to investigate the liquid structure in the water and the aqueous solutions with the infrared synchrotron light. In this paper, we describe our research objects with the water analysis system.

In water, absorption peaks of around 3500 cm^{-1}, 1600 cm^{-1} and 700 cm^{-1}, due to OH stretching, bending and lattice vibrations, respectively, have been observed by the infrared spectroscopy [1]. For the study of the water structure, it is important to know the condition of the intermolecular vibration below 1000cm^{-1}. Recent computer simulation studies make possible to obtain the far-infrared absorption curve in the water and the aqueous solution [5,6]. However, there are only a few reports about the experimental results [7] since light sources for the spectroscopy are limited.

The energy transfer within and between molecules in the solutions is the fundamental and important process in molecular dynamics and has been studied actively for the past decade. When OH-stretching vibration of alcohol was excited by the mid-infrared laser pulses, the movement of the vibrational energy toward CH_2 or CH groups was observed [8]. A computer simulation in a protein showed that the energy of the activated normal mode was transferred to a few selective normal modes [9]. Bacteriorhodopsin photochemical intermediates were found at a photoexcitation by the nanosecond time-resolved infrared spectroscopy [10]. When a tryptophan residue in the enzyme protein, subtilisin, was excited by a pulse UV laser, the energy loss due to the influence of the solvent was observed by a fluorescent measurements [11]. We now investigate the effect on the water structure by the activation of the solute. In the IR irradiation system, it is possible to take monochromatic IR from the infrared synchrotron light in MIRRORCLE 20 and to irradiate a sample for selective activation of a molecular vibration.

CP716, *Portable Synchrotron Light Sources and Advanced Applications,*
edited by H. Yamada, N. Mochizuki-Oda, and M. Sasaki
© 2004 American Institute of Physics 0-7354-0195-0/04/$22.00

WATER ANALYSIS SYSTEM

Portable synchrotron "MIRRORCLE 20" is the infrared and far-infrared light source, and it is equipped with the optical system to concentrate the infrared synchrotron radiation from the whole electron orbit. Critical wavelength is 10μm, and infrared light with range from 2μm to 300μm is available for the spectroscopy or the irradiation experiments. Schematic configuration of the analysis system is shown in Figure 1. The IR spectroscopy by Fourier Transform Infrared Spectrometer is used to observe the molecular vibration. Monochromatic IR can activate the vibration selectively under the irradiation system. Physical property measurements can be performed in the sample section of mono-irradiation.

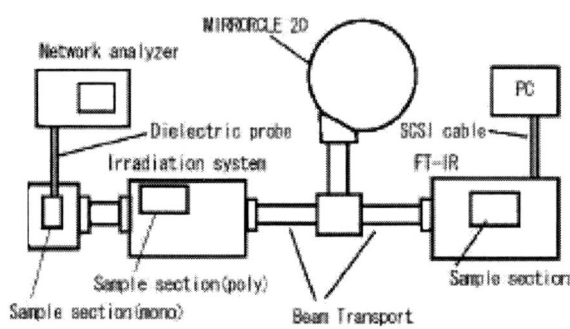

FIGURE 1. Schematic configuration of the water analysis system.

Spectroscopy with the high intensity synchrotron light can be useful for the high accuracy experiments for the samples of large absorption. FT/IR-600 (JASCO Corp.) is used for the spectroscopy. Available wave number is in the range between 7800cm^{-1} and 50cm^{-1}. The FTIR main body including the interferometer, the sample section, and the detectors is in vacuum chamber. We prepared three liquid sample cells made of ZnSe and Si for the mid-infrared transmission, and polyethylene for the far-infrared transmission. The sample holder is equipped with the Peltier elements and the temperature controller.

We performed the FTIR measurements for the saccharose aqueous solution. Figure 2 is the obtained mid-IR. Absorption peaks due to OH-stretching around 3500cm^{-1} and due to bending mode around 1600cm^{-1} were clearly observed even though absorption of the water vapor and the carbon dioxide was overlapped. The gas contribution in the spectrum

can be removed due to decompression of the FT-IR main body by a vacuum pomp.

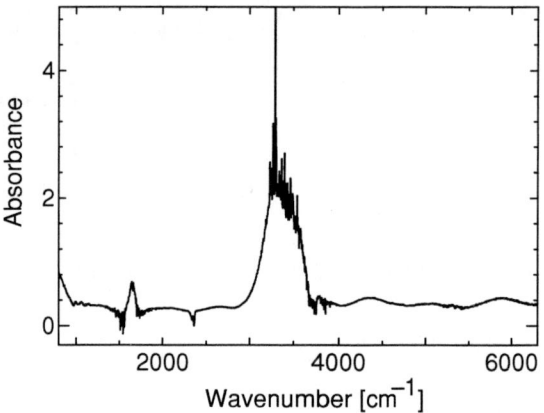

FIGURE 2. Mid-IR spectrum of a saccharose aqueous solution.

The dispersive-type IR irradiation system has the grating monochromators for 2μm to 25μm. As shown in Figure 3, the dielectric relaxation spectroscopy (DRS) with Network Analyzer (Agilent, PNA series E8362B) and the dielectric probe (Agilent, 85070D) were prepared for the physical property measurements under the irradiation.

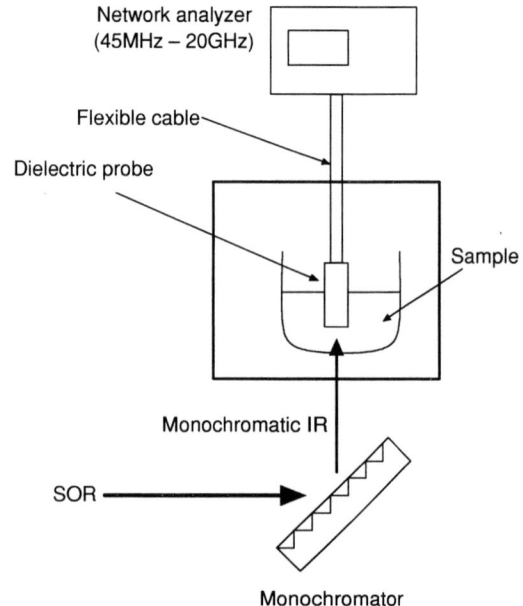

FIGURE 3. Schematic configuration of the microwave dielectric relaxation spectroscopy (DRS) in the monochromatic IR irradiation.

For the example of the dielectric spectroscopy, we measured the complex permittivity in water, air and ethanol at room temperature and plotted them against frequency in Figure 4. The relaxation curves for water and ethanol were well explained by the solid lines calculated by Debye equation [12] and Cole-Cole equation [13], respectively. These relaxation processes are due to reorientation of polar molecules [14].

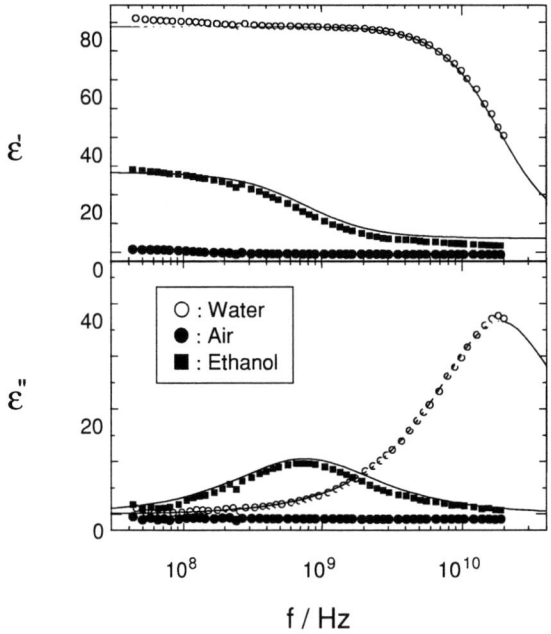

FIGURE 4. Microwave dielectric spectra of air, water and ethanol at room temperature. Solid lines were calculated by Debye equation [13] for the water's data, and Cole-Cole equation [14] for the ethanol's data.

RESEARCH OBJECTS

The intermolecular interactions within the solvent and between the solvent and the solute are the important factor to understand solution properties. The chemical reaction in the solution such as the variety of chemical and biological systems is quite different from that in the air because of the effects from the solvent. The dynamics in the aqueous solution is clearly related to the water structure. In polar liquids such as water and alcohol, the absorption peaks around $700 cm^{-1}$ and $200 cm^{-1}$ are remarkable because it arises from the vibration, which involves the hydrogen bonding atoms [1]. It is sensitive to the hydrogen-bonding configuration in the liquid. However, there are a few experimental spectra of the aqueous solution below $1000 cm^{-1}$ [7]. This is because water absorbs IR intensively and there are only few intense broadband light sources in the lower frequency infrared and far infrared regions.

The MIRRORCLE 20 is a particularly useful light source for the FTIR spectroscopy in the infrared and

far-infrared regions [3]. The constructing beam line will transmit strong synchrotron light. The critical wavelength is 10 μm ($1000 cm^{-1}$), and available wave number range is from $5000 cm^{-1}$ to $30 cm^{-1}$. Especially, high power is obtained in the range from $1000 cm^{-1}$ to $100 cm^{-1}$, which agrees to the intermolecular vibration band of the polar liquids. This feature enables MIRRORCLE 20 to best suit the spectroscopy research of the water-rich samples.

We start the research of the aqueous solutions of simple molecules such as organic solvents, sugars and ascorbic acid. The molecular dynamics in the mixtures differs from that in the pure solvent; for example, the liquid structure in the mixture has a local heterogeneity like concentration fluctuation. The presence of hydrogen bonding increases the complexity of the network structure even in the water mixed with simple organic solvent such as lower alcohol. The higher order association and the local concentration fluctuation have been observed as the function of composition, and a water-rich mixture has a different structure from an alcohol-rich mixture [15]. Our first object is to clarify the vibration mechanism in the infrared and far-infrared region and to obtain fundamental knowledge about the intermolecular interaction related to the liquid structure in the aqueous solutions.

We will expand our research toward the solution of biopolymers such as DNA, proteins, and phospholipid. The water to form network around the proteins was observed in the x-ray crystal structure analysis [16]. Interaction between biopolymer and water is considered important since water has a critical role in the stability and the functional expression [17].

We are also interested in the activation process of the hydrated water or the solute molecule by the monochromatic IR irradiation system. When the wave number of the specific normal mode is found by the FTIR spectroscopy, the mode is possible to be activated by the monochromatic IR irradiation. A schematic figure of the water network structure with the glucose is shown in Figure 5. Molecular orbital of water and glucose was calculated by WinMOPAC, and the molecules of ball and stick model were drown by RasMol. The molecules and the hydrogen bonds were manually arranged at appropriate places. When the vibrational mode of the solute is activated, temperature may increase locally. Hydrogen bonds around the solute are expected to be unstable.

To detect the effect by the selective activation, we construct the DRS system with IR radiation as shown in Figure 3. The peak frequency of the dielectric relaxation process in pure water is 19GHz at 25°C [18]. This peak is broadened by distribution of the cooperative molecular motion in the aqueous solution [19]. When the solute's vibrations excited by monochromatic IR disturb the water structure near the solute, changes of the peak position and shape will be observed. The experimental results may give us knowledge about the interaction between the solute and the solvent.

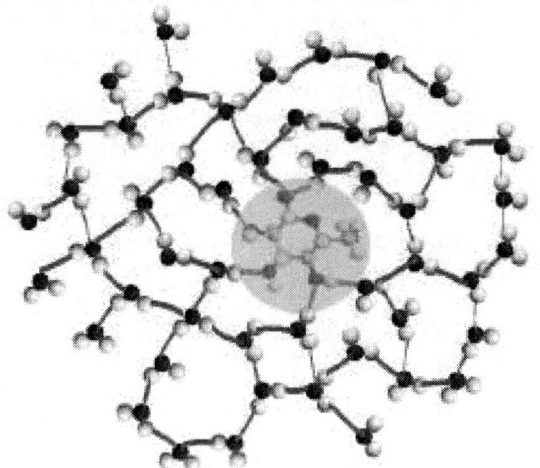

FIGURE 5. Schematic figure of the water network structure with glucose in the IR irradiation. The water and glucose molecules are drown by ball and stick model, and the hydrogen bonds are shown by the straight lines. The vibrational modes in the shaded area are activated by the irradiation of monochromatic IR, and the temperature is higher than that in other areas locally. Hydrogen bonds in this area are expected to be unstable.

This water analysis system can be utilized for a general purpose of water study not only in the basic research of chemical physics but also in the biological environments and any other water systems. The selective activation by the IR irradiation arises to increase local temperature and instability in the water structure, and these effects may affect kinetics of the chemical reactions. We would like to apply our knowledge to the related fields such as medical treatment and food science. We are also planning to develop the time resolved measurements of IR and dielectric spectroscopy in the irradiation system with monochromatic IR impulse, which is useful for the investigation of intra- or inter-molecule energy transfer in the aqueous solutions.

REFERENCES

1. Eisenberg, D., and Kauzmann, W., "The structure and properties of water", London: Clarendon Press, 1997, Chap. 4.

2. Hirata, F. (Ed.), "Molecular theory of solvation", Dordrecht: Kluwer Academic Publishers, 2003.

3. Yamada, H., *Adv. Colloid Interface Sci.* **71-72**, 371-392 (1997).

4. Yamada, H., *Journal of Synchrotron Radiation* 5, 1326-1331 Part 6 (1998).

5. Ladanyi, B. M., and Skaf, M. S., *J. Phys. Chem.* **100**, 1368-1380 (1996).

6. Gaiduk V. I., and Vij, J. K., *Phys. Chem. Chem. Phys.* **4**, 5289-5299 (2002).

7. Venables, D. S., and Shmuttenmaer, *J. Chem. Phys.* **113**, 11222-11236 (2000).

8. Wang, Z., Pakoulev, A., and Dlott, D.D., *Science* **296**, 2201-2203 (2002).

9. Moritsugu, K., Miyashita, O., and Kidera, A., *Phys. Rev. Lett.* **85**, 3970-3973 (2000).

10. Yuzawa, T., Kato, C., George M. W., and Hamaguchi, H., *Appl. Spectrosc.* **48**, 684-690 (1994).

11. Pal, S. K., Peon, J., and Zewail, A., *Proc. Natl. Acad. Sci. USA* **99**, 1763-1768 (2002).

12. Debye, P., "Polar molecules", New York: Chemical Catalog Co., 1929.

13. Cole, K. S., and Cole, R. H., *J. Chem. Phys.* **9**, 341 (1941).

14. Hill, N. E., Vaughan, W. E., Price, A. H., Mansel, D., "Dielectric properties and molecular behavior", London: Van Nostrand Reinhold Company, 1969, Chap. 5.

15. Nishikawa, K., Hayashi, H., and Iijima, T., *J. Phys. Chem.* **93**, 6559-6565 (1989).

16. Nakasako, M., *J. Mol. Biol.* **289**, 547-564 (1999).

17. Zhou, Y., Zhou, H., and Karplus, M., *J. Mol. Biol.* **326**, 593-606 (2003).

18. Barthel, J., Bachhuber, K., Buchner, R., Hetzenauer, H., *Chem. Phys. Lett.* **195**, 369-373 (1990).

19. Ryabov, Y. E., Feldman, Y., Shinyashiki, N., and Yagihara, S., *J. Chem. Phys.* **116**, 8610-8615 (2002).

Volumetric Properties of Hydrophobic Hydration under High Pressure

Seiji SAWAMURA

Department of Applied Chemistry, Ritsumeikan University, Kusatsu, Shiga, 525-8577, Japan
e-mail: sawamura@se.ritsumei.ac.jp

Abstract. In this article, partial molar volumes of hydrophobic compounds in water were reviewed. Distorted surfaces of the partial molar volumes for alkylbenzenes in water were drawn in a wide range of pressure, 0.10-400 MPa, and temperature, 273-323 K. They include a domain of negative compressibility of the partial molar volume in a low temperature and low pressure region. The negative compressibility was ascribed to a well-known property of bulk water; more compressible at lower temperature in the observed region. Aromatic ring which was included in naphthalene, anthracene, etc. contribute to the positive sign of the volume change for hydrophobic hydration though the negative sign has been usually observed for a methylene group of many hydrophobic compounds.

INTRODUCTION

Several interesting thermodynamic phenomena have been observed when a hydrophobic molecule is transferred from a nonpolar solvent to water phase: The partial molar entropy decreases and partial molar heat capacity increases. These phenomena have been attributed to ordering of water molecules around a hydrophobic solute by Frank and Evans [1], who called the ordering structure "iceberg" in 1945. Kauzmann [2] pointed out that this iceberg formation plays an important role in controlling the stability of biological macromolecules such as proteins and proposed the idea of a "hydrophobic bond" or "hydrophobic interaction" in 1959. It has increasingly become important in the new century of bioscience and biotecnology [3-5]

In this article, hydrophobic hydration and hydrophobic interaction in the stand point of volumetric property are reviewed including the progress since Kauzmann and our recent studies under high pressure are also included. Hydrophobic interaction defined by Kauzmann corresponds to a transfer of a hydrophobic molecule from the water phase to a hydrophobic solvent and hydrophobic hydration is thought to be the reverse process. Other definition such as Ben-Naim's one [6] is not used in this review.

VOLUME CHANGE FOR THE HYDROPHOBIC INTERACTION

Kauzmann pointed out an importance of increase in the volume change accompanying the formation of hydrophobic interaction as well as the large entropy reduction and heat capacity increment in his original article [2]. This was based on the fact that the partial molar volumes of methane and ethane in water in ambient condition are smaller than those in hexane by *ca.* 20 cm^3 mol^{-1}. Comparing this large volume difference with the volume change accompanying the denaturation of proteins, he tried to relate the protein denaturation to the hydrophobic interaction. But in the present time, the volume change accompanying the protein denaturation is known to be several tens cm^3 mol^{-1} [5,7]. The quantity corresponds to a transfer of only a few methane molecules into water phase. It is too small compared with exposure of hydrophobic residues of protein to water phase accompanying a drastic structure change of proteins in denaturation; much more hydrophobic residues should be exposed to water phase.

Kauzmann and collaborators in his laboratory started to measure the high-pressure solubility of liquid hydrocarbons in water in order to estimate the volume change accompanying the hydrophobic hydration of solutes except for methane and ethane. Volume change, ΔV_{sol}°, for the dissolution is estimated by pressure coefficient of the solubility, x, using eq. (1), which corresponds to a difference of the partial molar volume, V^{∞}, of a hydrophobic solute in water and molar volume, V^*, of the solute in the pure liquid state as shown in eq. (2).

$$(\partial \ln x / \partial p)_T = -\Delta V_{sol}^{\circ}/RT \qquad (1)$$
$$\Delta V_{sol}^{\circ} = V^{\infty} - V^* \qquad (2)$$

In these estimation, activity coefficient of a solute in water is thought to be unit and ΔV_{sol}° corresponds to the volume change accompanying the hydrophobic hydration, ΔV_{hy}°.

Kliman in 1969 [8] measured the solubility of 4-octanone as a model compound of hydrophobic solute in water under high pressure and estimated a large negative value; $\Delta V_{sol}^{\circ} = -30$ cm^3 mol^{-1} at 298. 2 K. This

CP716, *Portable Synchrotron Light Sources and Advanced Applications,*
edited by H. Yamada, N. Mochizuki-Oda, and M. Sasaki

value was reduced to -9.2 cm^3 mol^{-1} by remeasuring by Kim [9] in the same laboratory eleven years later. Generally the volume change accompanying the hydration of ketone, considering a difference between the partial molar volume of a ketone in water and nonpolar solvent such as CCl$_4$, is measured to be *ca.* -10 cm^3 mol^{-1} [10, 11], near to Kim's value rather than Kliman's one. In this negative value, volume reduction by hydrogen-bond formation between the solute and water is included besides the hydrophobic hydration. Zipp [12] in the same laboratory measured the solubility of propylbenzene in water under high pressure in 1973 and estimated ΔV_{sol}° = -5 cm^3 mol^{-1}. This value is also not so remarkable negative value compared with methane and ethane.

In these studies of high-pressure solubility of liquid hydrophobic compounds in water, another important property was observed; that is, the solubilities of all solutes measured by them showed a maximum at high pressures around 100 or 200 MPa. This means ΔV_{sol}° < 0 up to 100 or 200 MPa and then ΔV_{sol}° < 0 at the higher pressure following eq. (1) [8, 9, 12, 13]. The volume change accompanying hydrophobic hydration depends on pressure and reverses the sign from negative to positive with increasing pressure. Kauzmann, at first, cited the negative volume change of -20 cm^3 mol^{-1} accompanying the hydration of methane and ethane in order to attribute the protein denaturation by pressure around several hundreds MPa, which causes a negative volume change, to hydrophobic hydration. But in this stage, even sign of the volume change does not coincide each other.

On the other hand, with progress in the calorimetry, temperature dependence of thermodynamic properties such as Gibbs free energy for the protein denaturation and those for dissolution of liquid hydrocarbon in water have been measured and compared with each other. Because of the similarity of these temperature dependences of the thermodynamic properties, importance of hydrophobic interaction in protein stability or denaturation has seemed to be verified, and hydrophobic interaction has more and more attracted our attention [14, 15]. But such similarity has been discussed only temperature dependence at atmospheric pressure not the pressure dependence. We should understand these properties in more general situation including not only temperature axis but also pressure one as Kauzmann had pointed [16].

PARTIAL MOLAR VOLUME OF HYDROPHOBIC COMPOUNDS IN WATER

ΔV_{sol}° in eq. (1) is the volume change for dissolution and corresponds to a difference between the partial molar volumes of a solute in water and hydrophobic solvent (in some cases, the latter may be the molar volume of a liquid hydrocarbon). Though the volume change for hydrophobic interaction has been used in considering the denaturation of protein, the partial molar volume of hydrophobic compounds itself has recently become of interest in the view of hydrophobic hydration. With progress in experimental methods and techniques, *e. g.* vibrational densitometer [17], many data of the partial molar volumes have been accumulated in spite of the experimental difficulty because of the low solubility of hydrophobic compounds in water.

Partial molar volumes of organic compounds having a series of alkyl group in water are plotted in Fig. 1 as a function of the number, *n*, of methylene groups (-CH$_2$-) in the compounds. The value of 16.1*n* (cm^3 mol^{-1}) is subtracted from the partial molar volume in the ordinate. The values V^{∞}-16.1*n* show a linearity against *n* for each series of compound suggesting an volume additivity of methylene groups. The slopes of the linear lines in Fig. 1 are zero or negative for water and positive for organic (hydrophobic) solvents. The values are listed in Table 1. These results show that methylene group has an inherent partial molar volume of 16.4-16.7 cm^3 mol^{-1} in organic solvents and

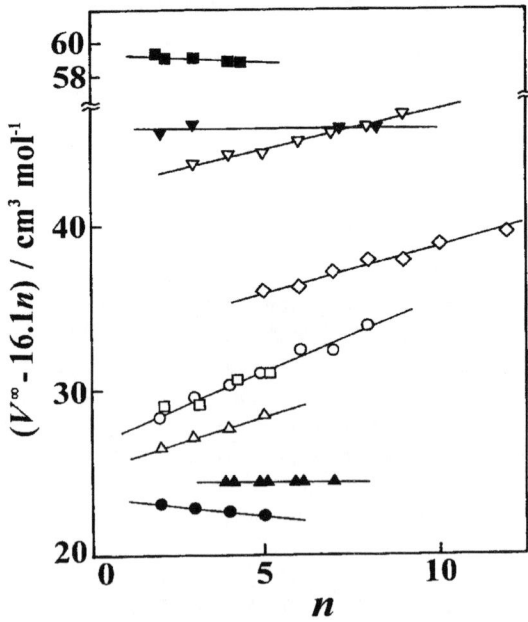

FIGURE. 1. (V^{∞}-16.1*n*)vs. *n* for several organic compounds at 0.10 MPa and 298.15 K [18].
○, H(CH$_2$)$_n$OH in CCl$_4$ [11]; □, H(CH$_2$)$_n$OH in benzene [19]; △, H(CH$_2$)$_n$OH in C$_2$H$_5$OH[19]; ▽, H(CH$_2$)$_n$Br in CCl$_4$ [20]; ◇, H(CH$_2$)$_n$H in CCl$_4$[21]; ●, H(CH$_2$)$_n$OH in H$_2$O[10]; ▲, HO(CH$_2$)$_n$OH in H$_2$O [22,23]; ■, H(CH$_2$)$_n$O-C$_2$H$_4$OH in H$_2$O [24,25]; ▼, H(CH$_2$)$_n$NH$_3$Br in H$_2$O[26].

Table 1 Partial molar volume (V^∞) of organic compounds having methylene group in water at 0.10 MPa and 298.15 K[18].

solute	solvent	n	$V^\infty/$ cm^3 mol^{-1}	ref
H(CH$_2$)$_n$OH	benzene	2-5	27.48+16.72n	19
	c-hexane	2-10	37.39+16.41n	19
	EtOH	2-5	25.33+16.62n	19
	neat	2-6	25.29+16.66n	27
	CCl$_4$	2-16	28.13+16.58n	11
H(CH$_2$)$_n$Br	CCl$_4$	3-9	42.35+16.48n	20
H(CH$_2$)$_n$H	CCl$_4$	5-32	32.33+16.65n	21
H(CH$_2$)$_n$OH	H$_2$O	2-6	23.44 + 15.69n	
				10,22,23,28-30
HO(CH$_2$)$_n$OH	H$_2$O	3-7	23.79+16.11n	22,23
H(CH$_2$)$_n$OEtOH	H$_2$O	2-4	59.33+16.11n	24,25
H(CH$_2$)$_n$NH$_3$Br	H$_2$O	2-8	45.84+16.00n	26

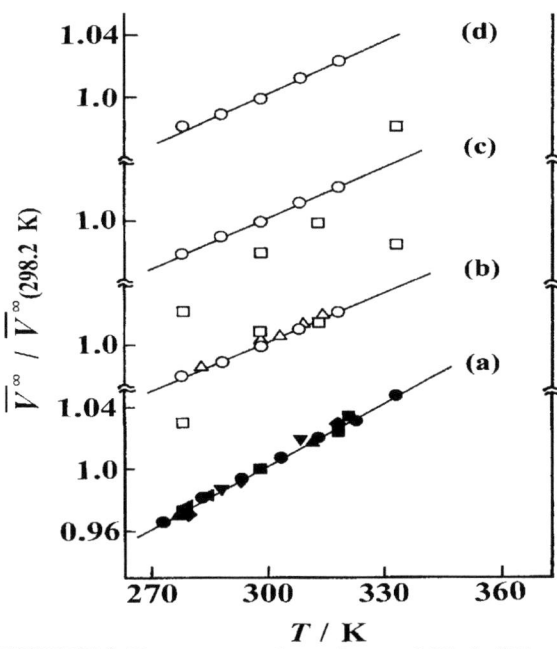

FIGURE 2 Temperature dependence of V^∞ of -CH$_2$- group (a), benzene (b), toluene (c), and ethylbenzene (d) in water at 0.10 MPa, normalized to 298.2 K. ●, H-(CH$_2$)$_n$OH [22,23,28-30]; ▲, H(CH$_2$)$_n$COONa[32]; ■, H-(CH$_2$)$_n$NH$_3$Cl[32]; ▼, H(CH$_2$)$_n$NH$_3$Br[26]; ◆, H(CH$_2$)$_n$O-(CH$_2$)$_2$OH[24,25]; ▲, HO(CH$_2$)$_n$OH[22, 23]; ○, Ref. 33; □, Ref. 34; △, Ref. 35.

EXPANSION COEFFICIENT OF THE PARTIAL MOLAR VOLUME OF HYDROPHOBIC COMPOUNDS IN WATER UNDER ATMOSPHERIC PRESSURE.

In Fig. 1, the partial molar volumes at only 298.15 K are shown. They should generally depend on temperature. In this section, the partial molar volumes of methylene groups (-CH$_2$-) in water are collected at several temperatures. They are plotted against temperature in Fig. 2, where the values are normalized at 298.15 K. Temperature dependence of the partial molar volumes of alkylbenzenes is also plotted in Fig. 2. All of the volumes linearly increase with increasing temperature. From the slopes, expansion coefficient, α^∞ [$\equiv (\partial V^\infty/\partial T)_p/V^\infty$], of the partial molar volume in water are estimated to be $(1.33 \pm 0.06) \times 10^{-3}$ K^{-1} for methylene group, 1.03×10^{-3} K^{-1} for benzene, 1.08×10^{-3} K^{-1} for toluene, and 1.12×10^{-3} K^{-1} for ethylbenzene at 298.15 K. These values do not so differ from the expansion coefficient, α [$\equiv (\partial V/\partial T)_p/V$], of pure liquid hydrocarbons such as toluene (ca. 1-2 $\times 10^{-3}$ K^{-1}) and are larger than the α of pure water (0.25×10^{-3} K^{-1}) by several times.

Compressibility of the partial molar volume of hydrophobic compounds in water under atmospheric pressure

Adiabatic compressibility, κ_S^∞ [$\equiv -(\partial V^\infty/\partial p)_S/V^\infty$], of the partial molar volume has been estimated from sonic measurements for several organic compounds that have alkyl substituents in water. However, there are few measurements of isothermal compressibility, κ_T^∞ [$\equiv -(\partial V^\infty/\partial p)_T/V^\infty$] of the partial molar volume [36]. In Fig. 3, $V^\infty\kappa_S^\infty$, i.e., $-(\partial V^\infty/\partial p)_S$, of a series of aliphatic compounds in water are plotted against the number of methylene groups, n. Linear relation between $V^\infty\kappa_S^\infty$ and n (>3) is observed at each temperature as well as the V^∞ does. It indicates the presence of an intrinsic value of $-(\partial V^\infty/\partial p)_S$, and hence κ_S^∞, for the methylene group at each temperature. The values are plotted in Fig. 4, including other series of aliphatic compounds. $V^\infty\kappa_S^\infty$ for phenyl group is estimated from the difference of $V^\infty\kappa_S^\infty$ between phenylalanine and alanine in water. It is worth noting that κ_S^∞ for both methylene and phenyl groups is negative at low temperature; that is, the volume adiabatically expands with increasing pressure and $V^\infty\kappa_S^\infty$ increases with increasing temperature changing the sign from negative to positive. A similar $V^\infty\kappa_S^\infty$ - T curve has been reported by Chalikian, et al. [42]. To relate these adiabatic

15.7-16.1 cm^3 mol^{-1} in water, resulting a difference of 0.3-1 cm^3 mol^{-1} corresponding to the volume change for hydrophobic hydration of methylene group. Similar volume differences of methylene group have been observed for a formation of micelle of aliphatic carboxylates [7, 31].

values to the high-pressure solubilities, κ_S^∞ should be translated to the isothermal condition. This can be achieved by using the following equation [43]:

$$\kappa_T^\infty = \kappa_S^\infty + \frac{TV_W\alpha_W}{C_{PW}}\left(2\alpha^\infty - \alpha_W\frac{V_WC_p^\infty}{V^\infty C_{PW}}\right) \quad (3)$$

where V_W, α_W, and C_{PW} are the molar volume, expansion coefficient, and isobaric heat capacity of water, respectively. C_p^∞ is the partial molar isobaric heat capacity of a solute in an infinitely diluted aqueous solution. C_p^∞ (-CH$_2$-) of the methylene group can be estimated from the C_p^∞ of a series of aliphatic compounds as well as the case of $V^\infty\kappa_S^\infty$ (-CH$_2$-) using the data of H(CH$_2$)$_n$OH [44-46], HO(CH$_2$)$_n$OH [45], H(CH$_2$)$_n$OEtOH [47], H(CH$_2$)$_n$H [48-51], H(CH$_2$)$_n$-C$_6$H$_5$ [52, 53], H(CH$_2$)$_n$NH$_3$Br [26], H(CH$_2$)$_n$COOH [54], H(CH$_2$)$_n$NH$_2$ [54], CH$_3$NHCO(CH$_2$)$_n$H [54], and H(CH$_2$)$_n$NHCOCH$_3$ [54]. The value is 88 ± 8 J mol^{-1} K^{-1}, independent of temperature between 273 and 333 K. The error of *ca.* 10 % in C_p^∞ (-CH$_2$-) is not significant in estimation of κ_T^∞ because the contribution of this term to κ_T^∞ in eq. (3) is small. The estimated values of κ_T^∞ are plotted as a solid line in Fig.

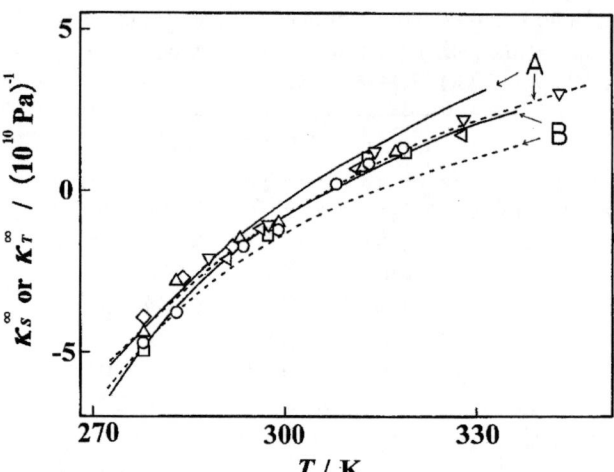

FIGURE 4 Adiabatic (κ_S^∞) and isothermal (κ_T^∞) compressibilities of the partial molar volume in water at 0.10 MPa. A, methylene group; B, phenyl group [38, 40]; ·····, κ_S^∞; ———, κ_T^∞. Plots are κ_S^∞ for methylene groups (◯, H(CH$_2$)$_n$OH[22, 37-39]; △, HO(CH$_2$)$_n$OH [22, 37, 39]; □, H(CH$_2$)$_n$OC$_2$H$_4$OH [24]; ◇, H(CH$_2$)$_n$-CH(OH)CH$_3$ [39]; ▽, H(CH$_2$)$_n$CH(NH$_2$)COOH [38, 40]; △, H$_2$N(CH$_2$)$_n$COOH[41]).

4. It can be seen that κ_T^∞ does not differ significantly from κ_S^∞ with negative values at low temperature and increases with increasing temperature. The C_p^∞ per gram of the phenyl group is similar to that of the methylene group. Therefore κ_T^∞ for the phenyl group is also temperature dependent, similar to that of the methylene group.

PARTIAL MOLAR VOLUMES OF ALKYLBENZENES IN WATER AS A FUNCTION OF PRESSURE AND TEMPERTATURE

In the above section, anomalous negative compressibility of the partial molar volume of hydrophobic groups has been cleared at ambient pressure. It is of interest to us how high the volume continues with such anomaly in the pressure. Then we have measured the high-pressure solubility of alkylbenzenes, *i.e.*, toluene, ethylbenzene, and propylbenzene, in water [13], following the method used by Kauzman's group. The volume change, ΔV_{sol}°, for the dissolution was estimated from the high-pressure solubility using eq. (1) and the partial molar volume of alkylbenzene in water was estimated from ΔV_{sol}° and V^* using eq.(2). The result for ethylbenzene in water is shown in Fig. 5 [13]. Negative compressibility is observed in the low temperature and low pressure region of the V^∞- p - T surface. In the

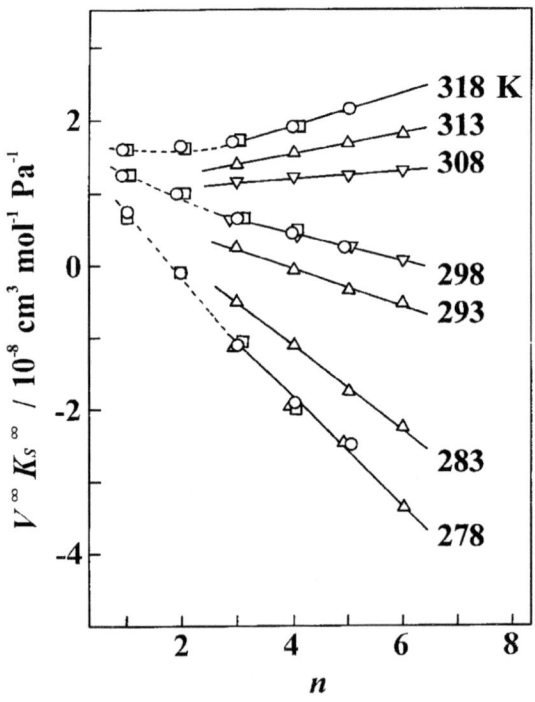

FIGUTRE 3 $V^\infty\kappa_S^\infty$ vs n for H(CH$_2$)$_n$OH in water at 0.10 MPa.◯,Ref. 37;□, Ref. 38;△, Ref. 22;▽, Ref. 39.

other region, usual "positive" compressibility is observed. Similar distorted surfaces of the partial molar volume including a negative compressibility region were observed for toluene and propylbenzene in water [13].

Kharakoz [55] has measured the adiabatic compressibility of proteins in water at pressures up to 20 MPa and presumed that the partial molar volume of unfolded proteins in water have a maximum at 70 MPa contrary to the native (folded) ones decreasing with increasing pressure without any maximum. This maximum is similar to that observed in isothermal curves in Fig. 5, suggesting that the hydrophobic hydration contributes to the volumetric property of proteins in the unfolded form in water.

Though, in section 4, the α^∞ of hydrophobic groups such as -CH_2- in water at atmospheric pressure was described not to appear to differ from α of general organic solvents; 1×10^{-3} - 2×10^{-3} K^{-1}, the α^∞ of ethylbenzene in water at atmospheric pressure is clearly large compared with that at high pressures when it is viewed in a wide pressure range as shown in Fig. 5.

Pressure coefficient of the V^∞ in Fig. 5 strongly depends on pressure causing the maximum. On the other hand, temperature coefficient of that dose not so depend on temperature, seemingly the V^∞ being not sensitive to the temperature. But the situation is not so. The molar volume of pure water decreases by 12 % with increasing pressure up to 400 MPa. It increases, on the contray, by only 1 % with increasing temperature from 273 K to 323 K. Such difference in the property of solvent water may contribute to the partial molar volume. It means that the pressure scale is much wider than the temperature when Fig. 5 is considered in the stand point of the volume.

Though negative compressibility is not generally possible for ordinary solids, liquids, and gases, the situation is different for partial molar volume because molecular interaction between the solute and solvent occurs. Considering that partial molar volume is divided in two contributions from solute and solvent, separately, the latter corresponds to the volume change accompanying the change of water molecules around a solute to hydration water molecules. It is counted to be $N(V_h - V_b)$, where V_b is the molar volumes of pure (bulk) water, and N is the number of water molecules in the domain of hydrophobic hydration around a solute molecule, in which the water has another molar volume of V_h as a hydration water. The other contribution is the exclusion volume, V_{ex}, due to a solute molecule. Then the partial molar volume is expressed by [56]

$$V^\infty = V_{ex} + N(V_h - V_b). \qquad (4)$$

The value of N has been not clear. Considering the hydration structure is similar to that of clathrate hydrate, the N for ethylbenzene is estimated to be *ca.* 36 from computer simulations [57,58] taking the number of water molecules in the first hydration shell as N. Using the value, volume of the hydration domain for one mole of ethylbenzene amounts to 648 cm^3 mol^{-1} (= NV_b = 18x36). On the other hand, V_{ex} of ethylbenzene is though to be not so differ from 123 cm^3 mol^{-1}, *i.e.* molar volume of the liquid solute. Then the NV_b is remarkably large compared with V_{ex} and NV_h would be also large as well as NV_b. Such difference between V_{ex} and NV_b (or NV_h) becomes much remarkable when the second and third hydration shell is taken as the hydrophobic domain.

The volumes of V_{ex}, V_h, and V_b should decrease with increasing pressure according to the thermodynamic principles. It means that compressibility of these volumes must not negative nor contribute to the negative compressibility of the partial molar volume shown in Fig. 5. Then the negative compressibility should be attributed to a contribution from $(V_h - V_b)$. Namely, $(V_h - V_b)$ increases with increasing pressure when hydration water is less compressible than the bulk water, resulting in a contribution of negative compressibility of V^∞. This contribution is remarkable at low temperature in Fig. 5. It may be attributed to large compressibility of bulk water at low temperature, suggesting that compressibility of hydration water does not become so large at low temperature compared with that of V_b. Large expansion coefficient of V^∞ at atmospheric pressure compared with that at high pressures in Fig. 5 may be attributed to an anomaly of water, that is, more compressible at lower temperature. In other words, this anomaly of water corresponds to the fact that expansion coefficient of (bulk) water becomes large with increasing pressure. When eq. (4) is differentiated with respect to temperature, we obtain eq. (5).

$(\partial V^\infty / \partial T)_p$
$$= (\partial V_{ex} / \partial T)_p + N(\partial V_h / \partial T)_p - N(\partial V_b / \partial T)_p \quad (5)$$

The third term in the right side of this equation becomes large with increasing pressure and should decrease $(\partial V^\infty / \partial T)_p$ or α_T^∞. In order to estimate the contribution, we consider the expansion coefficient of the partial molar volume of ethylbenzene in water at 0.10 MPa and 400 MPa. The difference defined as $\Delta(\partial V^\infty / \partial T)_p$ (≡$(\partial V^\infty / \partial T)_p$(at 400 MPa) - $(\partial V^\infty / \partial T)_p$(at 0.10 MPa)) can be estimated to be -0.086 cm^3 mol^{-1} K^{-1} from Fig. 5. On the other hand, reduction of expansion coefficient of V_b with increaisng pressure up to 400 MPa defined as $-N\Delta(\partial V_b / \partial T)_p$ (≡-N $(\partial V_b / \partial T)_p$(at 400 MPa) + $N(\partial V_b / \partial T)_p$(at 0.10 MPa)) can be estimated to be -0.090 cm^3 mol^{-1} using 0.26×10^{-3} K^{-1} and 18.1 cm^3 mol^{-1} for expansion coefficient and molar volume of V_b

at 0.10 MPa, and 0.45x10⁻³ K⁻¹ and 16.0 cm³ mol⁻¹ at 400 MPa [59], respectively, and $N = 36$. This value is an enough large quantity to contribute to reduction of expansion coefficient of the partial molar volume with increasing pressure up to 400 MPa. Though the first and second terms in the right hand of eq. (5), of course, may contribute to the expansion coefficient of the V^∞, we insist that the third term at least can not neglect in considering expansion coefficient of the V^∞.

Hydration structure around a hydrophobic solute has often been imagined as a structure similar to that of clathrate hydrate [2, 3]; a cage-like structure formed by hydrogen bonding between water molecules around a

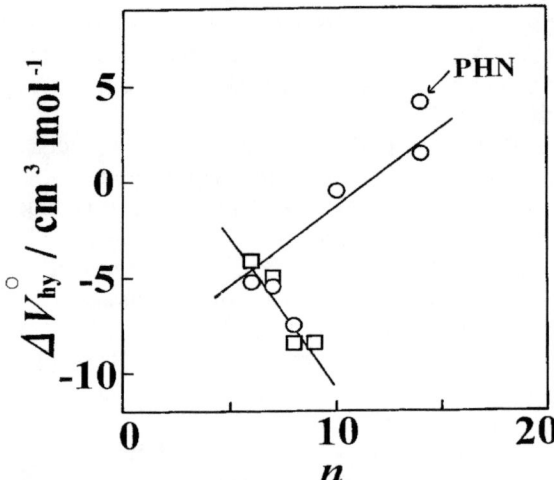

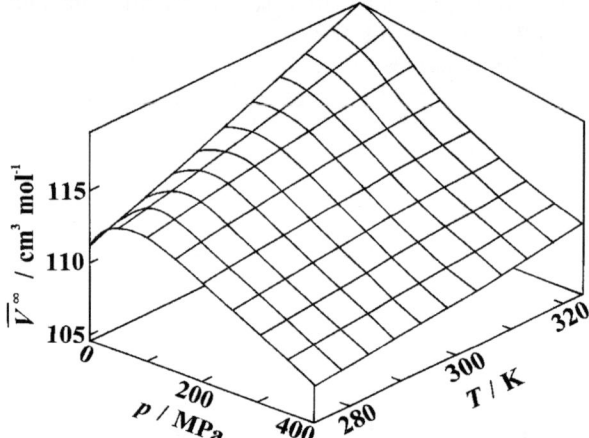

FIGURE 5 Partial molar volume of ethylbenzene in water [13].

FIGURE 7 Volume difference vs. n for several aromatic hydrocarbons at 298.15 K. n is the number of carbon atom in the solute. O, $\Delta V_{hy}^\circ (= V^\infty - V_{CT}^\infty)$; □, $\Delta V_{hy}^\circ (= V^\infty - V_{liq}^*)$. Solutes are benzene ($n$=6), toluene (7), ethylbenzene (8), n-propylbenzene (9), naphthalene (10), anthracene (14), phenanthrene (PHN) [61].

named the domain of hydrophobic hydration (hs), corresponds to a sum of V_{ex} and NV_h, that is, $V_{hs} = V_{ex} + NV_h$, and then, $V_{hs} = V^\infty + NV_b$ from eq. (4). The V_{hs} of ethylbenzene in water on using N=36 is compared with V_b [18] in Fig. 6, where both values are normalized at 400 MPa and 298.2 K. In computer simulation, the hydration domain has been thought to be much larger than N=36, including the second or third hydration shell. So V_{hs} may be thought to be the volume of only hydration water because the contribution of V_{ex} is negligibly small in V_{hs}. In this figure, V_{hs} decreases with increasing pressure and any negative compressibility can not be observed. And V_{hs} and V_b have a similar value each other at pressures higher than 200 MPa. On the contrary, the deviation between them can be caused with decreasing pressure from 200 MPa to zero. In other words, V_b seems to steeply increase when the pressure goes to zero. This phenomenon of bulk water may be ascribed to bulky and hydrogen-bonded open structure of (bulk) water at atmospheric pressure. Possibility of this bulky structure of water is supported by the fact that the difference between V_{hs} and V_b becomes large with decreasing temperature at atmospheric pressure, suggesting an enhancement of open structure by hydrogen bonding. From such consideration for the structures, the phenomena observed in Fig. 6 is understandable as follows. Bulk water has a bulky and open structure and the structure becomes remarkable at low temperature because of enhancement of hydrogen bonding between water molecules. Well-known

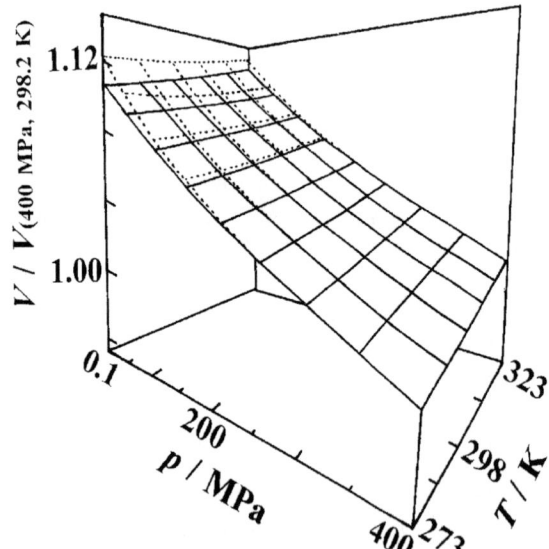

FIGURE 6 V_b and V_{hs} of ethylbenzene in water, normalised to 400 MPa and 298.2 K [18]. —, V_{hs}; ……, V_b.

minimum of V_b around 277 K at atmospheric pressure is ascribed to a balance between breaking of the structure and thermal expansion. This bulky structure is broken by pressure of 200 MPa. On the other hand, such bulky structure is not clear for the water in the hydration domain and the volume reduction of hydration water accompanying any breaking by pressure is thought to be small. Therefore compressibility of V_{hs} would not be so large as that of V_b at pressures below 200 MPa. At high pressures bove 200 MPa, compression phenomena of V_{hs} and V_b becomes similar each other because the bulky structure of bulk water has already broken by pressure. In these consideration, $N = 36$ for ethylbenzene is used though it is not clear. The conclusion does not change even if much large value for N is used. Only the difference between V_{hs} and V_b reduces.

HYDROPHOBIC HYDRATION OF AROMATIC RING

Recently we have measured the high-pressure solubility of aromatic compounds such as naphthalene, anthracene, and phenanthrene in water at 298.15 K and estimated the volume change, ΔV_{sol}°, accompanying the dissolution using eq. (1) [60, 61]. In this case, ΔV_{sol}° should be expressed as eq. (6) because the solute is in a solid state.

$$\Delta V_{sol}^{\circ} = V^{\infty} - V_{solid}^{*} \qquad (6)$$

where V_{solid}^{*} is a molar volume of a solid solute usually estimated from x-ray data of the crystal structure. Then the partial molar volume, V^{∞}, of these aromatic compounds in water can be estimated from the ΔV_{sol}° and V_{solid}^{*} though estimation of V^{∞} of such insoluble solutes is impossible from usual density measurement of the solution. Using the partial molar volume of these compounds in water and those in carbon tetrachloride, V_{CT}^{∞}, we can estimate the volume change, ΔV_{hy}°, accompanying the hydrophobic hydration defined as eq. (7), where CCl₄ is taken as a hydrophobic solvent. The results are shown

$$\Delta V_{hy}^{\circ} = V^{\infty} - V_{CT}^{\infty} \qquad (7)$$

in Fig. 7 [61], where the values are plotted to the number, n, of carbon atoms constituting the aromatic molecules. In the order of benzene, naphthalene, and anthracene, the ΔV_{hy}° linearly increases against n suggesting a volume additivity for aromatic carbon atoms. In comparison, ΔV_{hy}° for alkylbenzenes are also plotted in Fig. 7. It is clear in the figure that aromatic hydrocarbons contribute to a positive volume change for hydrophobic hydration contrary to methylene group in alkylbenzenes contributing to the negative one.

For a long time since Kauzmann has proposed the concept of hydrophobic bond, the volume change for a transfer of hydrophobic compounds or groups from an oil phase to water has been negative values at atmospheric pressure. The reason why they have been negative is that only aliphatic hydrocarbons or groups have been used as a model compound for hydrophobic interaction or hydration. Now we should consider various sorts of hydrophobic compounds.

As one of the reason why the signs of ΔV_{hy}° of aromatic ring and methylene groups are different each other, their molecular sizes should be considered [61]. The diameter of a chain of methylene groups is 0.55 nm as a cylinder. It is smaller than the inner diameter of the largest clathrate hydrate, 0.66 nm, which is usually cited as a model structure of the hydrophobic hydration. Alkyl chains of tertraalkylammonium salt is known to form a clathrate hydrate [63-65]. On the other hand, diameter of even short axes for aromatic ring such as naphthalene and anthracene is 0.74 nm as an ellipsoid. It exceeds 0.66 nm and seems to be negative for any accommodation of the compounds in the clathrate hydrate. Such difference may cause the difference in the sign for the volume change for hydrophobic hydration.

Another noticeable point in the high-pressure solubility of naphthalene, anthracene, etc. in water is the fact that the logarithm of the solubility linearly decreases with increasing pressure, observed at pressures up to 200 MPa at 298.2 K [60, 61]. The linearity means that the ΔV_{sol}° dose not depend on pressure from eq. (1) and then V^{∞} of naphthalene and anthracene in water is estimated to decrease with increasing pressure from eq. (6) because V_{solid}^{*} generally decreases with increasing pressure. This result is contrary to expansion of V^{∞} of ethylbenzene in water by pressure as shown in Fig. 5. The situation means that V_h takes the same property for compression as V_b and the hydration structure imagined from clathrate hydrate seems not to be formed for naphthalene and anthracene. Of course, such consideration is base on the high-pressure solubility observed only at room temperature of 298.2 K. Measurement in wide temperature range should be continued in order to compare the volumetric property with the methylene group.

CONCLUSION

In this article, volumetric properties of hydrophobic hydration were reviewed in a wide pressure and temperature range. One of the most typical conclusion is on the distorted surface of the partial molar volume of ethylbenzene in water in Fig. 5. The distortion is not originated from any distorted property of hydration water but from the well-known property of bulk water as shown in Fig. 6; the isothermal compressibility of bulk water is larger at lower temperature at atmospheric pressure and such temperature dependence reduces with increasing pressure.

On the other hand, such distortion of the partial molar volume seems not to be observed for naphthalene and anthracene. So far alkyl chains have been used exclusively as a model compound of hydrophobic hydration as shown in Figs. 1-4 and the volumetric property of them has been taken as the hydrophobic one. The matters which are not soluble in water or phobic for water have been called hydrophobic compounds. But even in the same hydrocarbons such as aliphatic and aromatic hydrocarbons, thermodynamic properties such as partial molar volume may be different each other.

Though we have measured high-pressure solubility of hydrocarbons in water to clarify the hydrophobic hydration which was found in anomaly of heat capacity and entropy change in the middle of the last century, it seems to make the situation complex more and more. But by considering the hydrophobic hydration in much more wide view, the understanding of it should deepen. As was pointed out by Kauzmann [16], it is important not only to see the figures of hydrophobic hydration in conditions where we are easy to see them, but also to try to see them in the place where no one has arrived. It is the problem in the new century.

REFERENCES

[1] Frank, H. S., Evans, M. W., *J. Chem. Phys.* **13**, 5 (1945).

[2] Kauzmann, W., *Adv. Protein Chem.*, **14**, 1 (1959).

[3] Suzuki, K., *Mizu Oyobi Suiyoueki* (*Water and Aqueous Solutions*), Japan: Kyoritu, 1980.

[4] Suzuki K., *Hyomen* (*Surface*) **34**, 43 (1996).

[5] Taniguchi, Y., *Zairyo* (*J. Soc. Mater. Sci., Jap.*) **41**, 839 (1992).

[6] Ben-Naim, A., *Hydrophobic Interaction*, New York: Plenum Press, 1980.

[7] Taniguchi, Y., Suzuki, K., *J. Phys. Chem.* **87**, 5185 (1983).

[8] Kliman, H., Ph. D. Dissertation, Dept. of Chem., Princeton Univ., 1969.

[9] Kim, K., Ph. D. Dissertation, Dept. of Chem., Princeton Univ., 1980.

[10] Edward, J. T., Farrell, P. G., Shahidi, F., *J. Chem. Soc., Faraday Trans.* 1 **73**, 705 (1977).

[11] Edward, J. T., Farrell, P. G., Shahidi, F., *Can. J. Chem.* **57**, 2585 (1979).

[12] Zipp, A., Ph. D. Dissertation, Dept. of Chem., Princeton Univ., 1973.

[13] Sawamura, S., Nagaoka, K., Machikawa, T., *J. Phys. Chem.* B **105**, 2429 (2001).

[14] Privalov, P. L., Gill, S. J., *Adv. Protein Chem.* **39**, 191 (1988).

[15] Privalov, P. L., Gill, S. J., *Pure & Appl. Chem.* **61**, 1097 (1989).

[16] Kauzmann, W., *Nature*, **375**, 763 (1987).

[17] Kratky, O., Leopold, H., Stabinger, H., *Z. Angew. Phys.* **27**, 273 (1969).

[18] Sawamura, S. *J. Mol. Liq.*, to be submitted.

[19] Edward, J. T., Farrell, P. G., Shahidi, F., *Can. J. Chem.* **57**, 2887 (1979).

[20] Shahidi, F., Farrell, P. G., Edward, J. T., *J. Phys. Chem.* **83**, 419 (1979).

[21] Edward, J. T., Farrell, P. G., Shahidi, F., *J. Phys. Chem.* **82**, 2310 (1978).

[22] Hoiland, H., *J. Solution Chem.* **9**, 857 (1980).

[23] Nakajima, T., Komatsu, T., Nakagawa, T., *Bull. Chem. Soc. Jpn.* **48**, 783 (1975).

[24] Harada, S., Nakajima, T., Komatsu, T., Nakagawa, T., *J. Solution Chem.* **7**, 463 (1978).

[25] Roux, G., Perron, G., Desnoyers, J. E., *J. Solution Chem.* **7**, 639 (1978).

[26] Leduc, P.-A., Fortier, J.-L., Desnoyers, J. E., *J. Phys. Chem.* **78**, 1217 (1974).

[27] Riddick, J. A., Bunger, W. B., *Organic Solvents*, 3rd Ed. New York, Wiley-Interscience, 1970.

[28] Alexander, D. M., *J. Chem. Eng. Data* **4**, 252 (1959).

[29] Sakurai, M., Nakamura, K., Nitta, K., *Bull. Chem. Soc. Jpn.* **67**, 1580 (1994).

[30] Cabani, S., Conti, G., Matteoli, E., *J. Solution Chem.* **5**, 751 (1976).

[31] Kaneshina, S., *Hyomen* (*Surface*) **22**, 248 (1984).

[32] Sakurai, M., *Bull. Chem. Soc. Jpn.* **46**, 159 (1973).

[33] Sakurai, M., *Bull. Chem. Soc. Jpn.* **63**, 1695 (1990).

[34] Makhatadze, G. I., Privalov, P. L., *J. Chem. Thermodyn.* **20**, 405 (1988).

[35] Shahidi, F., *J. Chem. Soc., Faraday Trans.* 1 **77**, 1511 (1981).

[36] Yayanos, A. A., *J. Phys. Chem.* **76**, 1783 (1972).

[37] Nakajima, T., Komatsu, T., Nakagawa, T., *Bull. Chem. Soc. Jpn.* **48**, 788 (1975).

[38] Kharakoz, D. P., *J. Phys. Chem.* **95**, 5634 (1991).

[39] Hoiland, H., Vikingstad, E., *Acta Chem. Scand.* A **30**, 692 (1976).

[40] Kikuchi, M., Sakurai, M., Nitta, K., *J. Chem. Eng. Data* **40**, 935 (1995).

[41] Chalikian, T. V., Sarvazyan, A. P., Breslauer, K. J., *J. Phys. Chem.* **97**, 13017 (1993).

[42] Chalikian, T. V., Sarvazyan, A. P., Breslauer, K. J., *Biophys. Chem.* **51**, 89 (1994).

[43] Desnoyers, J. E., Philip, P. R., *Can. J. Chem.* **50**, 1094 (1972).

[44] Arnett, E. M., Kover, W. B., Carter, J. V., *J. Am. Chem. Soc.* **91**, 4028 (1969).

[45] Jolicoeur, C., Lacroix G., *Can. J. Chem.* **54**, 624 (1996).

[46] Alexander, D. M., Hill, D. J. T., *Aust. J. Chem.* **23**, 347 (1969).

[47] Roux, G., Perron, G., Desnoyers, J. E., *J. Solution Chem.* **7**, 639 (1978).

[48] Olofsson, G., Oshodj, A. A., Qvarnstroem, E., Wadsoe, I., *J. Chem. Thermodyn.* **16**, 1041 (1984).

[49] Dec, S. F., Gill, S. J., *J. Solution. Chem.* **14**, 827 (1985).

[50] Naghibi, H., Dec, S. F., Gill, S. J., *J. Phys. Chem.* **90**, 4621 (1986).

[51] Naghibi, H., Dec, S. F., Gill, S. J., *J. Phys. Chem.* **91**, 245 (1987).

[52] Gill, S. J., Nichols, N. F., Wadsoe, I., *J. Chem. Thermodyn.* **8**, 445 (1976).

[53] Makhatadze, G. I., Privalov, P. L., *J. Chem. Thermodyn.* **20**, 405 (1988).

[54] Konicek, J., Wadsoe, I., *Acta Chem. Scand.* **25**, 1541 (1971).

[55] Kharakoz, D. P., *Biochemistry* **36**, 10276 (1997).

[56] Sawamura, S., Kitamura, K., Taniguchi, Y., *J. Phys. Chem.* **93**, 4931 (1989).

[57] Ravinshanker, G., Mehrotra, P. K., Mezei, M. Beveridge, D. L., *J. Am. Chem. Soc.* **106**, 410 (1984).

[58] Jorgensen, W. L., Gao, J., Ravimohan, C., *J. Phys. Chem.* **89**, 3470 (1985).

[59] Minassian, L. T., Pruzan, P., Soulard, A., *J. Chem. Phys.* **75**, 3064 (1981).

[60] Sawamura, S., Tsuchiya, M., Ishigami, T., Taniguchi, Y., Suzuki, K., *J. Solution Chem.* **22**, 727 (1993).

[61] Sawamura, S., *J. Solution Chem.* **29**, 369 (2000).

[62] Sloan, E. D., Jr., *Clathrate Hydrates of Natural Gases*, New York, Marcel Dekker, 1990, Chap.2.

[63] Nakayama, H., Yamanobe, M., Baba, K., *Bull. Chem. Soc. Jpn.* **64**, 3023 (1991).

[64] Feil, D., Jeffrey, G. A., *J. Chem. Phys.* **35**, 1863 (1961).

[65] McMullan, R. K., Bonamico, M., Jeffrey G. A., *J. Chem. Phys.* **39**, 3295 (1963).

Pressure Effect on the Helix-Coil Transition of an Ala-Rich Peptide in Aqueous Solution: A FT-IR Spectroscopic Study

Takahiro Takekiyo[*], Akira Okuno[*], Takashi Imai[†], Akio Shimizu[¶],
Minoru Kato[*], and Yoshihiro Taniguchi[*1]

[*]Department of Applied Chemistry, Ritsumeikan University, Kusatsu, Shiga, 525-8577, Japan.

[†]Research Organization of Science and Engineering, Ritsumeikan University, Kusatsu, Shiga, 525-8577, Japan.

[¶]Department of Environmental Engineering for Symbiosis, Soka University, Hachioji, Tokyo, 192-8577, Japan

Abstract. We investigated the pressure effect on the helix-coil transition of Ala-rich peptides in aqueous solutions by FT-IR spectroscopy. The population of the helix conformer increased with increasing pressure. The differences in the partial molar volume between the α-helix and random coil conformers of Ala-rich peptides are determined from the pressure dependence of the absorbance. The partial molar volume of the helix conformer is smaller than that of the coil conformer. The present result is different from the previous study (H. Noguchi, *Prog. Polym. Sci. Jpn.*, 1975, 8, 191-199.), which reported that the partial molar volume of the helix form of polypeptide is larger than that of the coil form.

INTRODUCTION

Thermodynamics of the conformational change of a protein in aqueous solution depends not only on the intramolecular interaction of the protein but also on the solute-solvent interaction. The hydration of proteins is an important factor for the structural stability of proteins. The study of the hydration of model peptides can be a significant step toward understanding the structural stability of proteins. The helix-coil transition of oligopepitdes in aqueous solution has been extensively studied by theoretical and experimental methods as models of the fold-unfold transition of proteins [1]. However, there are few studies of the pressure effect on the transition of the model peptides.

It is reported that the volume changes associated with the helix formation of poly-L-glutamic acid (PGA) are $+1.1 \ cm^3/mol$ by dilatometry [2] and $-0.3 \pm 1.3 \ cm^3/mol$ by densimetry [3]. However, it is difficult to exactly estimate the volume change of PGA for the helix formation by using these methods, because the observed volume change involves the dissociation volume change of the COOH group of the side-chain.

In this study, we focus on the pressure effect on the helix-coil transition of an Ala-rich oligopeptide (AK16:YGAAKAAAAKAAAAKA-NH$_2$) in aqueous solution. The Ala-rich peptide is designed to avoid complication arising from interactions among side-chains and interactions between side-chains and helix-macrodipole such as PGA [4]. They are based on the sequence $K(AAAAK)_n$, and they undergo a monomeric helix-coil transition in aqueous solution [5]. The Y residue enables accurate determination of the

---

1) To whom correspondence should be addressed.

Phone: +21-77-561-2785. Fax: +81-77-561-2659

E-mail: taniguti@se.ritsumei.ac.jp

CP716, *Portable Synchrotron Light Sources and Advanced Applications,*
edited by H. Yamada, N. Mochizuki-Oda, and M. Sasaki
© 2004 American Institute of Physics 0-7354-0195-0/04/$22.00

peptide concentration by Tyr absorbance, and the G residue provides a flexible linker. The A residue has the highest helix propensity of all amino acids. The K residue dissolves the peptide to water.

FT-IR spectroscopy is one of the most powerful techniques for investigating the conformational change of biomolecules in solutions. Compared with NMR method, FT-IR spectroscopic experiment has a much shorter time resolution (on the order of 10-100 fs). Therefore, all existing conformer populations are observed simultaneously [6].

In this article, we investigated the pressure effect on the helix-coil transition of the Ala-rich peptide in aqueous solution by pressure tuning FT-IR spectroscopy, and determined the volume change from the helix to the coil conformations. Besides, we discuss the influence of the difference in the PMV for the acetylation of the N-terminal of AK16.

EXPERIMENTAL AND METHOD

Peptide synthesis. AK16 and acetylated-AK16 (Ac-AK16) were synthesized by the solid-phase method using the Millipore 9050 peptide synthesizer. In the process of the peptide synthesis, TFA is used to disrupt the bond between the Fmoc group and the peptide. These peptides were purified by high performance liquid chromatography (HPLC), and the identity was confirmed by the time of flight mass spectrometry (TOF-MS). The concentration of sample was 2.0 weight % (pD 9.0) for 20 mM Tris buffer solution. The influence for the secondary structures due to the peptide-peptide interaction is negligible under this condition. [1] All exchange backbone amide protons in AK16 and Ac-AK16 were deuterated by incubating the solution at 298 K. The pD was read directly from a pH meter and no adjustments were made for isotope effects. All sample solutions were prepared immediately prior to the infrared measurement.

FT-IR spectroscopy. The FT-IR spectra were measured using a FT-IR-680 plus spectrometer (JASCO, Tokyo) equipped with a liquid nitrogen cooling DLA-MCT detector. Each spectrum was obtained by co-adding 256 scans at a spectral resolution of 2.0 cm^{-1}. The infrared beam was condensed by a zinc selenide lens system onto the sample in the diamond anvil cell. The completion of hydrogen-deuterium exchange was confirmed by no further changes in the amide II band. This amide band in the frequency region around 1550 cm^{-1} shifts to around 1450 cm^{-1} as a result of deuteration of the backbone amide protons. For the pressure experiments, the sample solutions were placed together with a small amount of powdered α-quartz in a 1.0 mm diameter hole of a 0.05 mm thick Hasteloy C-276 gasket mounted on a diamond anvil cell. The α-quartz was used as an internal pressure calibrant.

RESULTS AND DISCUSSION

The amide I' mode consists of the C=O stretching, C-N stretching, and C-C-N deformation modes, and appears in the region from 1620 to 1690 cm^{-1}. This mode has been known to be highly sensitive to the secondary structures of polypeptides and proteins so that it has served as an indicator of the α-helix and/or β-sheet conformations. Therefore, this mode has been the marker band for detecting the secondary structures of polypeptides and proteins [7].

In a previous study by CD spectroscopy, Clarke *et al.* [8] reported that the helix and coil conformers of AK16 peptides exist in aqueous solutions. Figures 1 and 2 show (a) the original FT-IR spectra and (b) the difference spectra of the amide I' region of AK16 and

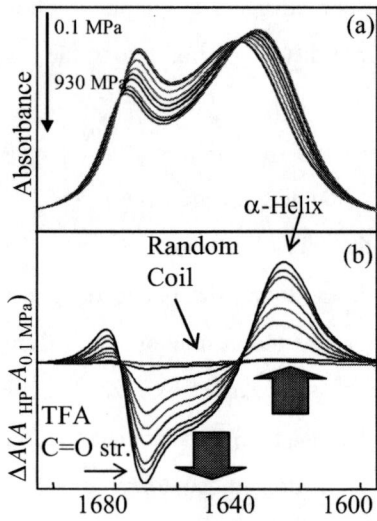

FIGURE 1. (a) Original FT-IR spectra and (b) difference spectra of the amide I' region of AK-16 in the aqueous solution (pD 9.0) at various pressures.

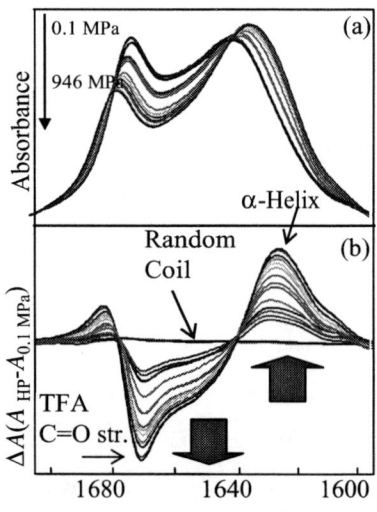

FIGURE 2. (a) Original FT-IR spectra and (b) difference spectra of the amide I' region of Ac-AK-16 in the aqueous solution (pD 9.0) at various pressures.

Ac-AK16 in aqueous solutions at various pressures.

The peaks centered at around 1635, 1656, and 1672 cm^{-1} are assigned to the solvent exposed α-helix and random coil conformations, and the asymmetric C=O stretching mode of trifluoroacetic acid (TFA),

TABLE 1. Volume (cm³/mol) difference between the helix and coil conformers of AK16 and Ac-AK16 in aqueous solution.

Peptide	$\Delta V^{coil \rightarrow helix}$	$\Delta V^{coil \rightarrow helix}$ / residue
AK16	-23.8 ± 1.1	-1.4 ± 0.1
Ac-AK16	-17.1 ± 0.3	-1.0 ± 0.0

respectively [9-11]. Figures 1 and 2 show that the peaks of the helix conformer of AK16 increases with increasing pressure, while that of the coil conformer decreases. We determine the volume difference between the helix and coil conformers from the pressure dependence of the absorbance. Assuming that the ratio of the absorption coefficients of the conformers is independent of pressure, the volume difference between the two conformers (ΔV) is given by

$$\Delta V^{coil \rightarrow helix} = -RT \left(\frac{\partial \ln(A_{helix} / A_{coil})}{\partial p} \right)_T, \quad (1)$$

where R, T, and p indicate the gas constant, temperature, and pressure, respectively. A_{helix} and A_{coil} indicate the absorbance of the helix and coil conformations, respectively. The volume change ($\Delta V^{coil \rightarrow helix}$) by the helix formation of AK16 and Ac-AK16 are -23.8 ± 1.1 cm³/mol (-1.4 ± 1.1 cm³/mol per residue) and -17.1 ± 0.3 cm³/mol (-1.0 ± 0.0 cm³/mol per residue), respectively. Table 1 lists the volume differences. The PMV of the helix conformer is smaller than that of the coil conformer in both cases. Our result is different from the previous study of PGA by Noguchi [2], which reported that the PMV of the helix form is larger than that of the coil form ($\Delta V^{coil \rightarrow helix} = +1.1$ cm³/mol per residue).

Next we discuss why the PMV of the helix conformer is smaller than that of the coil conformer. Here, we decompose the PMV difference into two contributions. $\Delta V^{coil \rightarrow helix}$ is given by

$$\Delta V_{obs}^{coil \to helix} = \Delta V_{M}^{coil \to helix} + \Delta V_{hyd}^{coil \to helix}, \quad (2)$$

where $\Delta V_{M}^{coil \to helix}$ is the difference of the molecular volumes, which are composed of the van der Waals volume difference (ΔV_w) and the void volume difference (ΔV_v). The void volume is the structural voids within the solvent inaccessible core of the solute molecule. $\Delta V_{hyd}^{coil \to helix}$ is the hydration volume difference, which results from the intermolecular interaction between the peptide and water molecules. From a theoretical study for the volume changes associated with the helix-coil transition of oligopepitdes, ΔV_w between the helix and coil conformers is negligible. However, the helix conformer has larger void spaces than the coil conformer [12]. Therefore, V_M of the helix conformer is larger than that of the coil conformer ($\Delta V_{M}^{coil \to helix} > 0$). From $\Delta V_{obs}^{coil \to helix} < 0$ and $\Delta V_{M}^{coil \to helix} > 0$, the contribution of $\Delta V_{hyd}^{coil \to helix}$ is negative. Therefore, the PMV of the helix conformer is smaller than that of the coil conformer due to the contribution of the hydration volume rather than that of the molecular volume.

Finally, we discuss the influence of the acetylation of the N-terminal of AK16 on the difference in the PMV. In previous studies [13-14], the acetylation of the N-terminal of AK16 increases the helix conformer due to the disappearance of the charge of the N-terminal NH_2 groups by the N-capping effect. From Table 1, the absolute value of ΔV of AK16 is smaller than that of Ac-AK16. ΔV by the acetylation of the N-terminal of AK16 is $+ 6.7$ cm^3/mol. This result could be explained by the increase of the void in the helix by the N-capping effect.

REFERENCES

1. A. Chakrabartty and R. L. Baldwin, *Adv. Protein Chem.*, 1995, 46, 141-176.

2. H. Noguchi, *Prog. Polym. Sci. Jpn.*, 1975, 8, 191-199.

3. G. D. Noudeth, N. Taulier, T. V. Chalikian, *Biopolymers*, 2003, 70, 563-574.

4. A. Chakrabartty, T. Kortemme, R. L. Baldwin, *Protein Sci.*, 1994, 3, 843-852.

5. S. Marqusee, S and R. L. Baldwin, *Proc. Natl. Acad. Sci. USA*, 1987, 84, 8898-8902.

6. H. H. Mantsch, A. Perczel, M. Hollosi, G. D. Fasman, *Biopolymers*, 1993, 33, 201- 207.

7. J. Bandeker, *Biochimica Biophysica Acta*, 1992, 1120, 123-139.

8. D. T. Clarke, A. J. Doig, B. J. Stapley, G. R. Jones, *Proc. Natl. Acad. Sci. USA*, 1999, 96, 7232-7237.

9. S. M. Decatur and J. Antonic, *J. Am. Chem. Soc.*, 1999, 121, 11914-11915.

10. R. A. G. D. Silva, J. Y. Nguyen, S. M. Decatur, *Biochemistry*, 2002, 41, 15296-15303.

11. S. Williams, T. P. Causgrove, R. Gilmansihn, K. S. Fang, R. H. Callender, W. H. Woodruff, R. B. Dyer, *Biochemistry*, 1996, 35, 691-697.

12. T. Imai, Y. Harano, A. Kovalenko, F. Hirata, *Biopolymers*, 2001, 59, 512-519.

13. J. S. Richardson and D. C. Richardson, *Science*, 1988, 240, 1648-1652.

14. A. Chakrabartty, A. J. . Diog, R. L. Baldwin, *Proc. Natl. Acad. Sci. USA*, 1993, 90, 11332-11336.

Protein Dynamics Research Station on IR Beam Line of MIRRORCLE-20

T. Kikuzawa [a], H. Yamada [a, b], N. Mochizuki-Oda [b], N. Miura [b], A. Moon [a]

[a] Faculty of Science & Engineering, Ritsumeikan University
[b] Synchrotron Light Life Science Center, Ritsumeikan University

Abstract. The purpose of our study is to examine the molecular dynamics under synchrotron radiation with wavelength in far infrared region that has not yet been applied. Yamada and collaborators have succeeded in developing tabletop synchrotron light sources having 15 cm orbit radius for brilliant IR radiation named "MIRROCLE-20". The IR radiation from the entire electron orbit of exact circular is collected by circular mirror and quasi-ellipsoidal mirror. The IR synchrotron light spectrum is centered at 10 μm wavelength and covers from few to 100 μm. The unique beam line is the monochromatic IR irradiation system, which was unable before due to the limited IR irradiation system. We are constructing the dispersive type monochrometer with gratings that reveals 1 to 1/2 % bands monochromatic IR-rays. Specimens are placed along 15 cm long focal plane with 30 chambers, where monochromatic IR-rays are applied through slits with 1 mm width. We have a plan to detect chemical products due to the high flux IR irradiation with specific wavelength to the proteins in water. Molecular dynamics and enzymatic reaction under IR or FIR radiation have not yet been fully understood. Generally speaking, intramolecular vibration mode will be spread and finally transfer into thermal energy. Our research object is to clarify the relaxation process of the vibration energy and to control the chemical reactions through excitation of certain vibration mode of the target proteins with monochromatic IR rays. The structural change in the proteins will be detected by its chemical products as well as change in absorption and emission spectra. Sample in our mind is for instance Horseradish peroxidase (HRP).

INTRODUCTION

Organisms receive various environmental information including photon, sounds, heats, and chemicals, and respond in consequence of the integration of these signals. Each process of signal transduction occurs in the specific organs and cells via receptor proteins. Based on the homology search of amino acid sequence with known proteins, many of these new proteins are supposed to be functional in cells. However, actual behavior of soluble proteins in cytoplasm is not clearly understood. For example, enzyme protein is considered to have three-dimensional structure and fluctuates in the solvent or cytoplasm. So it can get special function. The purpose of our research is to clarify the dynamics of protein in solvent or under the condition similar to cytoplasm by using high brilliance infrared (IR) lights. By searching the infrared light with wavelength that can activate the target molecule, we are seeking to control protein function selectively [1].

SYNCHROTRON RADIATION

In our laboratory we can use high brilliance IR from tabletop synchrotron "MIRRORCLE-20" [2]. One of the features of synchrotron radiation (SR) is that its spectrum wavelength covers from x-ray to far infrared region. So IR light with specific wavelength is available. Other sources of black body such as CO_2 laser is difficult to change lasing frequency continuously. Yamada developed synchrotron light sources with 0.15 m obit radius [3]. SR light can be collected from all electron trajectories [4]. Figure 1 shows calculated 20 MeV SR spectra and black body spectra by using equation (1) below in ref. [5]. It shows advantage of SR light particularly in far infrared region.

$$N = 5.494 \times 10^{22} \times kA\Omega / \lambda^3$$
$$\times \left[\exp(1.43 \times 10^4 / \lambda T) - 1\right]^{1} \tag{1}$$

CP716, *Portable Synchrotron Light Sources and Advanced Applications,*
edited by H. Yamada, N. Mochizuki-Oda, and M. Sasaki
© 2004 American Institute of Physics 0-7354-0195-0/04/$22.00

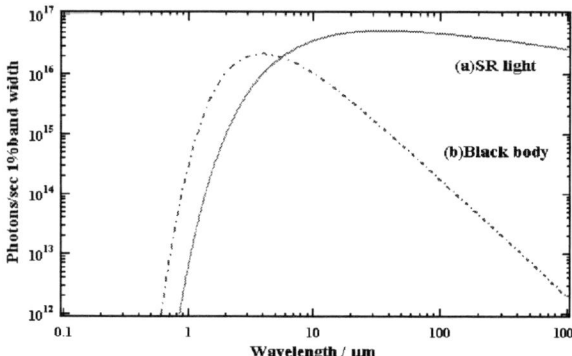

FIGURE 1. Comparison of synchrotron radiation and black body radiation is made. (a) SR lights collected from whole arc of electron orbit at ring current=1 A. (b) Black body flux is calculated from equation (1) at k=1 % band width, A=0.5 cm^2 source area, Ω=0.075 acceptance solid angle, and T=1300 K absolute temperature.

Figure 2 shows angular distribution of SR light calculated using Schwinger's equation [6]. As SR light is highly brilliant, so we can focus it easily.

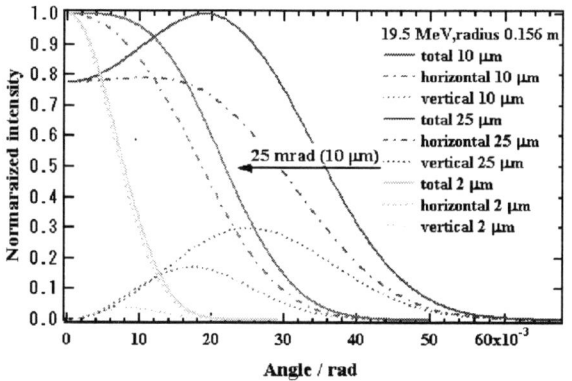

FIGURE 2. Angular spread and polarization of SR lights for wavelengths of 2, 10, 25 μm.

MONOCHROMATIC IR RAY IRRADIATION SYSTEM

We have designed dispersive type monochromatic IR irradiation system (Fig. 3) ranging wavelength of 2-25 μm. The grating efficiency is reported as shown in Figure 4. This system is able to irradiate samples with wide range of wavelengths in same time. Specimens are placed on a long focal plane (150 mm). Bands of monochromatic IR rays are made to vary by changing gratings (three sets of 20, 30, 40 line/mm, see table 1). IR rays from SR are supplied through 1 mm slits to each sample cells. Temperature is controlled between 273 K and 318 K by water bath system. The power of

IR is detected by thermo pail with its lower detection limit of 10 μW/cm^2.

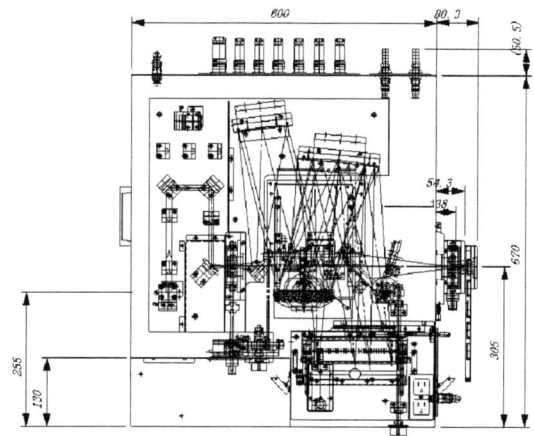

FIGURE 3. Designed dispersive-type grating monochromater

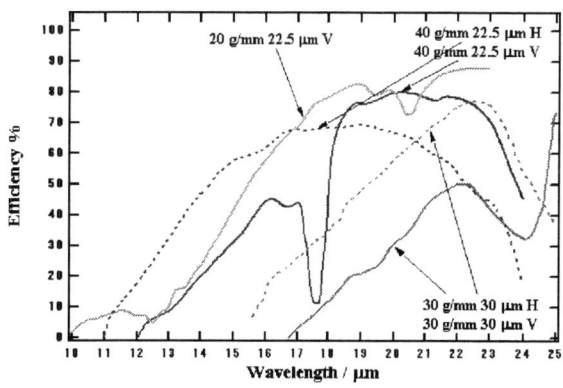

FIGURE 4. Diffraction efficiency of brazed gratings of 4.8, 22.5, 30 μm (each has 20, 30, 40 grating/mm, respectively). The solid curves show the vertically polarized (V) light. The horizontal (H) components are shown by dashed curves sited from ref. [7].

TABLE 1. Gratings parameter.

Brazing wavelength	dλ/dl	Δλ/λ (slit 1 mm)
22.5 μm	0.1225 μm/mm	2.6 %(at 4.8 μm)
4.8, 22.5 μm	0.0604 μm/mm	0.27 %(at 22.5 μm)
30 μm	0.0811 μm/mm	0.27 %(at 30 μm)

STOPPED-FLOW ASSEMBLY

To detect enzymatic reactions, stopped-flow methods must have advantages for observing rapid chemical reaction. By mixing two samples we start the reaction, and time-resolved signal will be detected within the ms of order. This method requires only a few amounts of samples. Stopped-flow methods follow three processes. (1) Start reaction by mixing substrates and enzyme rapidly. (2) Stop the flow. (3) Observe chemical reaction products occurred in a sample cell by measuring proper signals. Figure 5 shows stopped-flow assembly developed by Inada [8].

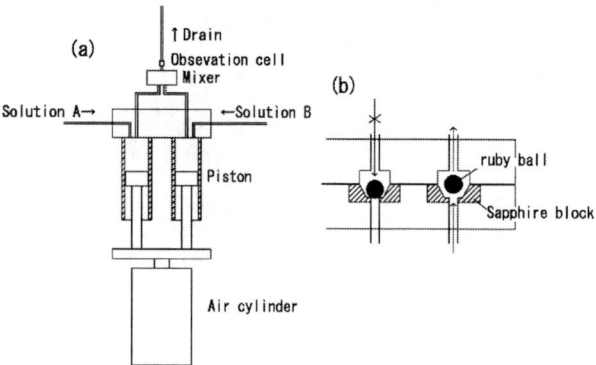

FIGURE 5. (a) Stopped-flow assembly. (b) Mechanism changing the flow direction by using ruby ball.

By this system automatic repetition is achieved, and the stopped-flow unit is reasonably compact. The dead time of the mixing procedure will be 3.6 ms. There are many detection techniques including pH monitor, conductimetry, Raman scattering, ESR, EXAFS, etc. We will measure transmitted UV, VIS, and IR lights, through CaF_2 window equipped to the cell in our stopped-flow unit. We will measure time dependence absorption changing in UV, VIS radiation. Time resolution will be at least ms order. The design of optical system for measuring UV, VIS absorption under IR irradiation is shown in Fig. 6. Components listed in Table 2 will be incorporated in this system. The UV and VIS light is detected by photo diode or photo multiplier, and intensified by using high-speed current amplifier. An oscilloscope or an A/D converter digitizes the signal outputs from the detectors. Optical fibers and mixing solution tube can be connected to each sample cells. Additionally, UV, VIS emission, X-ray absorption and IR absorption are also measurable with this system.

TABLE 2. Components of measurements system.

Components	Feature
Light source	IR (2-25 μm) from SR
	Halogen lamp 100 W
Observation cell	CaF_2 with Stopped-flow assembly
Spectrometer (SHIMAZU SPG-100S)	F3 dλ/dl 5.7 nm/mm
	Slit width 5 μm
Photo diode (HAMAMATSU S1227)	Sensitivity 34 mA/W (633 nm)
	Rise time 0.5 μs
Photo multiplier (HAMAMATSU R2949)	Sensitivity 68 mA/W (400 nm)
	Rise time 2.2 ns
CCD (HAMAMATSU C4880)	Sensitivity 70 mA/W (400 nm)
	Reading frame time 10 msec
A/D converter	Resolution 16 bit
	Time resolution 10 μsec
Oscilloscope (LECROY WR6050)	Vertical Resolution 8 bit
	Memory 68 Mbyte
	Time resolution 20 ps

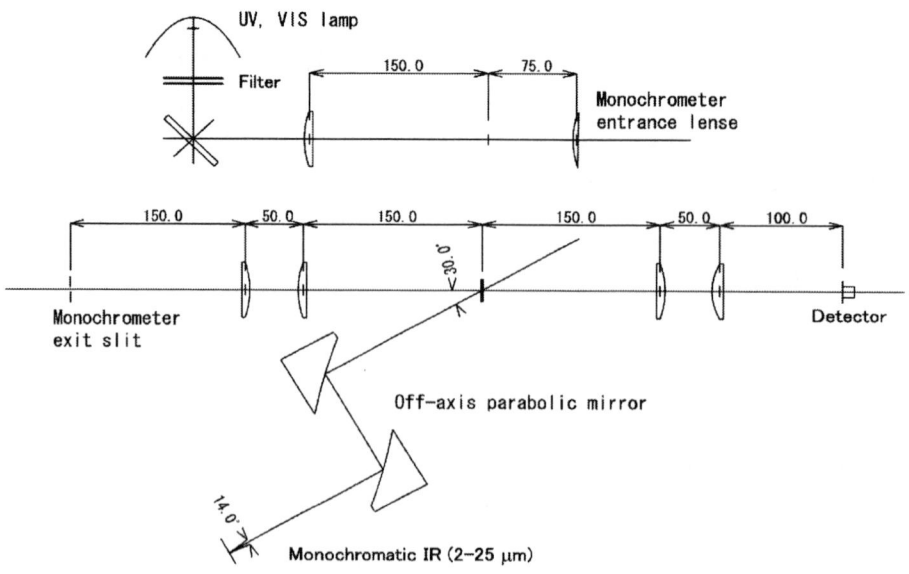

FIGURE 6. Optical system for UV, VIS absorption under IR irradiate condition

SAMPLES

Establishing the noble method for controlling the protein reaction by far-infrared ray irradiation is our target. For the first experiment we select well-known sample. One of the samples in our mind is horseradish peroxidase (HRP), which oxidizes substrates in the presence of hydrogen peroxide. HRP works as a catalase. Important functions of HRP are (1) Decomposition of hydrogen peroxide and reactive oxygen species. (2) Activation of biosynthesis of various hormones. These reactions are written as follows [9].

1. Complex I formation:

$$Fe(III) + H_2O_2 \rightarrow Fe(IIII) = O^{+} + H_2O \qquad (2)$$

2. Formation of Complex II from Complex I:

$$Fe(IIII) = O^{+} + AH_2 \rightarrow Fe(III) + AH^{\cdot} \qquad (3)$$

3. Reduction of Complex II to resting state:

$$Fe(IIII) = O + AH_2 \rightarrow Fe(III) + AH^{\cdot} + H_2O \qquad (4)$$

AH_2 is the electron donor. This enzymatic reaction by stopped-flow methods is observed by Chance [10]. He measured the rate constant by detecting the absorption changes. In this reaction, intermediate complexes exist which have Fe=O stretching vibration bond in equation (2)-(4). So we are planning to detect time dependence absorption changes under conditions in which Fe=O stretching mode is excited. And we also try to know about thermal denaturation mechanism of HRP. To detect small changes in absorption spectra, two-dimensional correlation FTIR spectroscopy (2D-FTIR) by using temperature perturbation [11] is helpful. Time-dependent variations in infrared spectra can be induced by an external perturbation, such as mechanical, thermal, chemical, electrical, or acoustic stimulations. A correlation analysis of these fluctuations generates two-dimensional map that increases the spectral resolution by spreading peaks along the second dimension and that reveals the order of actual sequence of process.

By using MIRRORCLE-20 1mW IR power per 1 % band is available; it will be possible to examine the new method to study the protein dynamics and control the protein reaction.

REFERENCES

1. H. Yamada, *Adv. Colloid Interface Sci.*, **71-72**, 371-392 (1997).

2. H. Yamada, *Journal of Synchrotron Radiation*, (1998) pp.1326-1331.

3. H. Yamada, *Nucl. Instr. And Meth.* B 199 (2003) 509.

4. H. Yamada, Japan. *J. Appl. Phys.* 28, L1655-L1668 (1989).

5. T. Nanba, *Rev. Sci. Instrum.*, **60 (7)**, 1680 (1989).

6. S. Krinsky, M. L. Perlman and R. E. Watson, *Handbook on Synchrotron Radiation*, ed. E.E. Koch, (North-Holland) Vol. 1(1983) 65.

7. http://www.spectra-physics.com/gratings

8. Y. Inada, H. Hayashi, and S. Funahashi, M. Nomura, *Rev. Sci. Instrum.*, **68 (8)**, 2973-2977 (1997).

9. K. Fukuyama, N. Kunishima, F. Amada, T. Kubota, and H. Matsubara, *J. Biol. Chem.*, **270**, 21884-21892 (1995).

10. B. Chance, *J. Biol. Chem.*, **151**, 553-577 (1943).

11. I. Noda , *Appl. Spectrosc.*, **44**, 550 (1990).

Effects of Near-Infrared Laser on Neural Cell Activity

Noriko Mochizuki-Oda[1], Yosky Kataoka[2], Hisao Yamada[2] and Kunio Awazu[3]

1.Synchrotron Light Life Science Center, 21st Century COE Program, Ritsumeikan University, Kusatsu, Shiga 525-8577, 2.Dept. Anatomy and Cell Science, Kansai Medical University, Moriguchi, Osaka 570-8506, 3. Institute of Free Electron Laser, Graduate School of Engineering, Osaka University, Hirakata, Osaka 573-0128, Japan

Abstract. Near-infrared laser has been used to relieve patients from various kinds of pain caused by postherpetic neuralgesia, myofascial dysfunction, surgical and traumatic wound, cancer, and rheumatoid arthritis. Clinically, He-Ne (λ=632.8 nm, 780 nm) and Ga-Al-As (805 ± 25 nm) lasers are used to irradiate trigger points or nerve ganglion. However the precise mechanisms of such biological actions of the laser have not yet been resolved. Since laser therapy is often effective to suppress the pain caused by hyperactive excitation of sensory neurons, interactions with laser light and neural cells are suggested. As neural excitation requires large amount of energy liberated from adenosine triphosphate (ATP), we examined the effect of 830-nm laser irradiation on the energy metabolism of the rat central nervous system and isolated mitochondria from brain. The diode laser was applied for 15 min with irradiance of 4.8 W/cm^2 on a 2 mm-diameter spot at the brain surface. Tissue ATP content of the irradiated area in the cerebral cortex was 19 % higher than that of the non-treated area (opposite side of the cortex), whereas the ADP content showed no significant difference. Irradiation at another wavelength (652 nm) had no effect on either ATP or ADP contents. The temperature of the brain tissue was increased 4.5 – 5.0 °C during the irradiation of both 830-nm and 652-nm laser light. Direct irradiation of the mitochondrial suspension did not show any wavelength-dependent acceleration of respiration rate nor ATP synthesis. These results suggest that the increase in tissue ATP content did not result from the thermal effect, but from specific effect of the laser operated at 830 nm. Electrophysiological studies showed the hyperpolarization of membrane potential of isolated neurons and decrease in membrane resistance with irradiation of the laser, suggesting an activation of potassium channels. Intracellular ATP is reported to regulate some kinds of potassium channels. Possible mechanisms of laser effect on neural activity through interaction between ATP and potassium channels will be discussed.

INTRODUCTION

Near-infrared laser has been used to relieve patients from various kinds of pain caused by postherpetic neuralgesia, myofascial dysfunction, surgical and traumatic wound, cancer and rheumatoid arthritis[1]. Clinically, He-Ne (λ=632.8 nm, 780 nm) and Ga-Al-As (805 ± 25 nm) lasers are used to irradiate trigger points or nerve ganglion. However the precise mechanisms of such biological actions of the laser have not yet been resolved. Near infrared light has relatively high penetrability of biological tissues, and its light scattering is somewhat less than that of visible light. Since lasers are often effective to suppress the pain caused by hyperactive excitation of sensory neurons, interactions with laser light and component of neural cells are suggested (Fig. 1).

However, the precise mechanisms underlying such a beneficial action of lasers on cells have not yet been defined. Several hypotheses have been proposed from in vivo and in vitro experiments: (1) an increase in peripheral and cerebral microcirculation [2,3], (2)

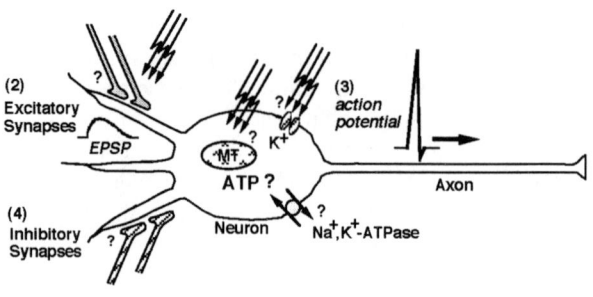

FIGURE 1. Hypothesis of effects of near-infrared laser on neuron. (2) ～(4) correspond the parenthesis in the text, respectively.

inhibition of release of algesic substances at nerve endings [4], (3) change in conductivity of primary afferent neurons [5,6], and (4) activation of inhibitory

CP716, *Portable Synchrotron Light Sources and Advanced Applications,*
edited by H. Yamada, N. Mochizuki-Oda, and M. Sasaki
© 2004 American Institute of Physics 0-7354-0195-0/04/$22.00

neural inputs [7]. These hypotheses, especially (2) to (4), concern the cellular physiology of neurons.

MATERIALS AND METHODS

Wistar male rats (8 week-old) were anesthetized with halothane. Two holes (2.0 mm diameter) were drilled in the skull over the parietal cortex on both sides (2 mm posterior and 1.5 mm lateral to the bregma). Continuous laser light (4.8 W/cm^2) was applied through the arachnoid membrane to one region for 15 min via a glass fiber positioned at 3 mm above the hole (Fig. 2). Another region was not treated with the laser. Irradiation of 830 nm was performed by means of a diode laser. The tissue temperature was monitored by a thermocouple needle-probe of 0.4 mm diameter inserted into the edge of the irradiated area to a depth of ca. 1.26 mm and at an angle of 57 degrees to the surface of the brain [8].

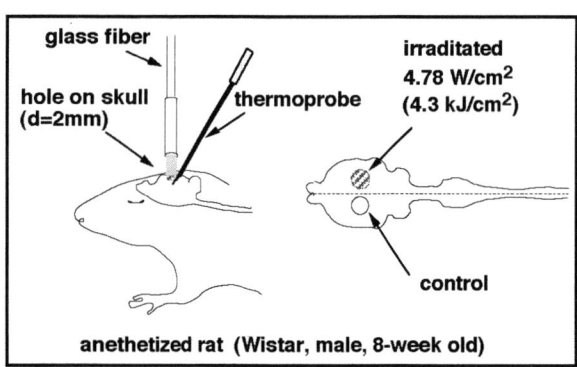

FIGURE 2. Irradiation of rat brain (parietal cortex) by laser light.

After the laser irradiation, the brain tissue was homogenized immediately in a solution composed of 0.5 M HClO$_4$, 1 mM EDTA, and 0.02 % methyl red. The content of total ATP was assayed by the luciferase-luciferin method. ADP was converted to ATP by pyruvate kinase in the solution containing 750 u/ml pyruvate kinase, 50 mM phosphoenol pyruvate, 35 mM KNO$_3$, 6 mM MgSO$_4$, 100 mM TES (pH 7.4) for 40 min at 20 °C; and the pyruvate kinase was then denatured at 100 °C for 10 min. The amount of ADP was obtained by subtraction of the amount of ATP determined by the luciferase-luciferin method without the conversion step from that with the step.

Rat brain mitochondria was suspended in the reaction mixture, 300 mM manitol, 20 mM Tris-HCl, 10 mM

KCl, 4 mM MgSO$_4$, 0.2 mM EDTA (pH7.5). The rate of oxygen consumption was monitored by oxygen electrode.

RESULTS AND DISCUSSION

The parietal cortex containing gray and white matter (1 mm diameter and thickness of ca. 2 mm) was obtained by use of a punch. The ATP content of non-irradiated cortical tissue (cortical area opposite to the laser-irradiated site) was 17.63 ± 4.39 nmol/mg protein (n = 10,). This value corresponded to values reported by other investigators [9]. On the other hand, the ATP content of the section irradiated by 830-nm with 4.8 W/cm^2 for 15 min was 20.36 ± 3.52 nmol/mg protein (n = 10). Though the contents varied among the animals, the ATP ratio (irradiated/control) showed an increase in ATP by irradiation with 830-nm laser. The paired t-test revealed that the increase was significant (p-value less than 0.05). On the other hand, irradiation with the 652-nm laser resulted in no significant change in the content of ATP. The mean ratio was 1.19 in the case of 830 nm and 1.00 in that of 652 nm. The amount of ADP was larger than that of ATP in both control and irradiated tissues. However, the laser light of either wavelength did not affect the ADP content, and thus the ratio was near unity. These results suggest that the ATP/ADP ratio was increased by the irradiation only at 830 nm laser light, specifically.

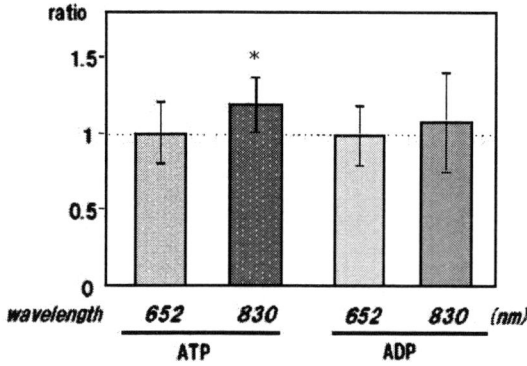

FIGURE 3. Effect of laser irradiation on ATP and ADP contents. Ordinate shows the relative concentration of nucleotides in the tissues with or without irradiation. * p<0.05 (paired t-test)

No detectable ablation, coagulation or large bleeding in the arachnoid was observed during irradiation by either 830-nm or 652-nm laser. The needle-type thermo-probe inserted into the illuminated area was shaded at its portion outside the irradiated tissue by planer aluminum-foil so as not to receive direct and scattered light on its heat-sensitive component. Therefore, the record revealed the temperature of the tissue itself. The brain temperature before irradiation was 30.0 ± 1.89 °C (n=9) under halothane anesthesia, suggesting the effect of lowering of the cerebral circulation. The temperature during the irradiation was 34.4 ± 3.56 °C (n = 6) for 830-nm laser and 34.7 ± 0.49 °C (n = 3) for 652-nm laser, respectively (Fig.4). There was no significant difference between the increases in the temperature with the wavelength.

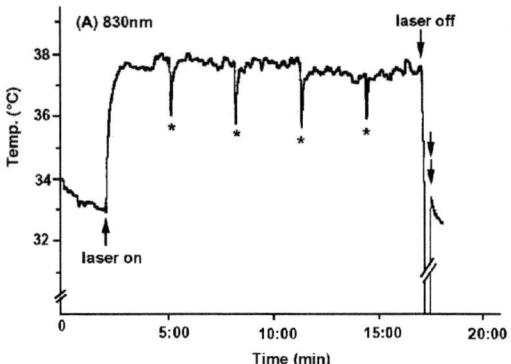

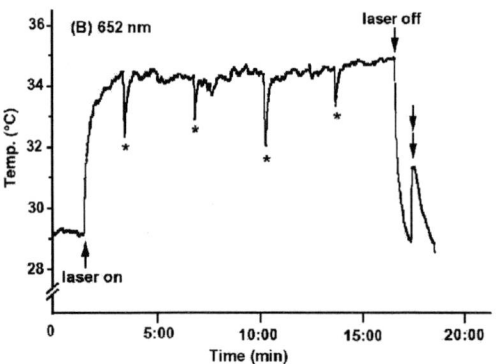

FIGURE 4. Representative examples of the change in the tissue temperature during laser irradiation are shown. Diode laser was switched off automatically every 180 (A) or 200 (B) sec, and was restarted within 2 sec. The start temperatures presented as 0 in the ordinate were 30.4 (A) and 28.5 °C. (B), respectively. At the point of the double arrows, the inserted needle-probe was pulled out from the tissue. Asterisks indicate the times of the auto power-off and restart of the diode laser.

These findings suggest that the increment of ATP caused by 830-nm laser light cannot be attributed to any thermal effect. The absorption by soft tissue is low between 700 – 1000 nm, so these wavelengths are called the 'window' of the spectrum. As the penetration depth of the light in this window was shown to reach 1 - 5 mm for soft tissue [10], the effect of the 830-nm light might have extended throughout the whole tissue obtained with the punch. The tissue temperature measured by the needle-type probe reflected the mean value from the surface of the brain to a depth of approx. 1.26 mm. Provided that light at the 830-nm wavelength penetrates deeper than that at 652 nm, the volume of the region illuminated by the infrared laser might be larger than that by the visible one. The difference in the mass of the area affected by the laser might be one reason for the wavelength specificity.

Although the measurement of the temperature at a given depth of the brain is not possible yet, it is necessary to consider the distribution of tissue temperatures under the condition of laser light irradiation. The cellular concentration of ATP is determined by the balance of its production and consumption rate. Passarella et al. [11] reported an increase in the proton electrochemical potential gradient and ATP synthesis in rat liver mitochondria irradiated by a He-Ne laser (632.8 nm). Therefore, the sensitivity of mitochondrial electron transport activity to infrared laser irradiation was examined by using mitochondrial suspension. However, neither the rate of oxygen consumption nor ATP synthesis of mitochondria isolated from rat brain showed wavelength specific change with irradiation of 830-nm or 652-nm laser.

Another possible mechanism of the laser-induced increase in ATP content might involve ATP consumption via Na^+, K^-, Ca^{2+}, and H^+-ATPase. Kudoh et al. reported activation of the Na^+, K^+-ATPase following diode laser ($\lambda = 830$ nm) irradiation of the rat saphenous nerve [12]. Any direct effect of 830 nm laser on isolated mitochondria and cellular ATPase activity remains to be clarified.

The role of ATP in the neural transmission is not clear now, however, several lines of experiments suggested the regulatory mechanism of ion channels by intracellular ATP. As shown in Fig. 1, generation of the action potential in neuron is caused by transient opening of cation channels such like Na^+ channels and subsequent long-lasting opening of K^+ channels. Membrane current carried by Na^+ flows from outside of the cells to inside inducing depolarization, whereas K^+ current flows to opposite direction, resulting in re-polarization of the cellular membrane potential. If intracellular ATP activate some K^+ channels, the membrane potential becomes deeper, thus the frequency of action potential decreases. Kataoka et al. reported that in rat brain slices, field excitatory postsynaptic potential (EPSP) that reflects the number of neurons excited by test pulses, decreased during the irradiation with 830-nm laser [13]. Furthermore, in isolated primary cultured neurons, patch-clamp study showed the hyperpolarization of membrane potential and simultaneous decrease in membrane resistance [14]. These results suggest that the hyperpolarization was caused by the increase in the opening probability of K^+ channels. Among the various types of K^+ channels, some are regulated by intracellular ATP [15, 16]. Thus indirect or direct regulation of K^+ channels by near-infrared laser may be one of the possible mechanisms of the suppression of neuropathic pain.

The present experimental results and others strongly suggest the wavelength-specific interaction of laser and cellular components. A more precise search of the action spectrum may give useful information about the target molecule(s) involved in the neuronal mechanisms of pain relief. The tabletop synchrotron light source, "MIRRORCLE 20", developed by Yamada and collaborators generates infrared rays much brighter than any others SR sources [17]. It provides dispersive infrared irradiation system by using a grating monochromer. The monochromatic beam lines with 0.5 % band width may be a powerful tool to determine the basic properties of the interaction between infrared light and biological tissues or molecules.

REFERENCES

1. A. Honmura, A. Ishii, M. Yanase, J. Obata, and E. Haruki, Lasers Surg. Med. **13**, 463-469 (1993)
2. Z. F. Gourgoulitatos, A. J. Welch, K.R. Diller, and S.J. Aggarwal: Laser Surg. Med., **10**, 524-532 (1990)
3. M. Schaffer, H. Bonel, R. Sroka, P.M. Schaffer, M. Busch, M. Reiser, and E. Duhmke: J. Photochem. Photobiol. B: Biol., **54**, 55-60 (2000)
4. A. Honmura, M. Yanase, J. Obata, and J. Haruki: Laser Surg., **12**, 441-449 (1992)
5. K. Jimbo, K. Noda, K. Suzuki, and K. Yoda: Neurosci. Lett. **240**, 93-96 (1998)
6. A. Miura, and M. Kawatani: Pain Res., **11**, 175-183 (1996)
7. M. Shimoyama, K. Iijima, T. Mizuguchi, M. Shimoyama and Y. Hori: Pain Res., **6**, 111-115 (1991)
8. N. Mochizuki-Oda, Y. Kataoka, Y. Cui, H. Yamada, M. Heya and K. Awazu: Neurosci. Lett. **323**, 207-210 (2002)
9. J.M. Gray, and K.M. Raley-Susman: Brain Res., **798**, 223-231 (1998)
10. M. Tamura, O. Hazeki, S. Nioka, and B. Chance: Ann. Rev. Physiol., **51**, 813-834 (1989)
11. S. Passarella, F. Casamassima, S. Molinari, D. Pastore, E. Quagliariello, I.M. Catalano and A. Cingolani: FEBS lett. **175**, 95-99 (1984)
12. C. Kudoh, K. Inomata, K. Okajima, M. Motegi and T. Ohshiro: Laaser Therapy, **1**, 63-67 (1989)
13. Y. Kataoka, Y. Cui, Y. Maegawa, T. Ito and Y. Watanabe: Jpn. J. Physiol. **50** (suppl) S166 (2000)
14. Y. Tamura, Y. Kataoka, Y.Cui, N. Oda-Mochizuki and H. Yamada: Neurosci. Res. **25**, S151 (2000)
15. L. Aguilar-Bryan, C.G. Nicols, S.W. Wechsler, J.P. Clement, A.E. Boid, G. Gonzalez, H Herrera-Sosa, K Nguy, J. Bryan and D.A. Nicols: Science **268**, 423-426 (1995)
16. H.-L. Zhang and T.B. Bolton: J. Pharmacol. **118**, 105-112 (1996)
17. H. Yamada: Nuclear Instruments and Methods in Physics Research, **B199**, 509-516 (2002)

Study on C8-substituent effects of synthetic bacteriochlorophyll-*d/f* analogues

Shin-ichi Sasaki and Hitoshi Tamiaki

Department of Bioscience and Biotechnology, Faculty of Science and Engineering,
Ritsumeikan University, Kusatsu, Shiga 525-8577, Japan

Abstract. Bacteriochlorophyll(=BChl)s-*c/d/e* are known to be major pigments constructing the main light-harvesting antennae of photosynthetic green bacteria. In this study, methyl bacteriopheophorbide-*d* derivatives having a series of alkyl or alkenyl substituents at the 8-position and their zinc complexes were synthesized, and their visible absorption spectra and self-aggregation in a solution were examined. It was shown that the C8-substituents conjugating with the chlorin π-system cause red-shifts of the Q_x peaks in their monomeric forms, but do not strongly affect the supramolecular structure of the self-aggregates along the Q_y axis. Furthermore, 3^1-epimerically pure Zn-MBPhe-*f* analogues possessing the C8-formyl group were also synthesized, and their spectral characteristics were compared to those of the C7-formyl compounds. It was found that the C8-formyl group cause drastic spectral changes both in monomeric and oligomeric forms compared to the spectra of Zn-MBPhe-*f* having the C7-formyl group. For example, self-aggregates of the former showed more red-shifted Q_y peak in 6%THF-H_2O than the latter, and larger diastereomeric control in the oligomeric Q_y peaks was observed for C8-formyl derivatives than for C7-formyl compounds.

Chlorosome, an extramembrane antenna of green photosynthetic bacteria, is a unique light-harvesting apparatus. It is known that bacteriochlorophylls-(BChls)-*c*, -*d*, and -*e* in chlorosome self-aggregate to form antennas without any assistance from proteins.[1] As a model compound of bacteriochlorophyll(BChl)-*d*, we synthesized methyl bacteriopheophorbide-*d* derivatives **1-5** (MBPhes-*d*) having a series of substituents at the 8-position and their zinc complexes (Zn-**1-5**).[2]

Absorption spectra of **1-5** in CH_2Cl_2 showed that the C8-substituents conjugating directly with the chlorin ring shifted the Soret and Q_x peaks to longer wavelengths while retaining the position of the Q_y band at around 650 nm as shown in Table 1. The observed Q_x peaks in wavelength were situated in the order of ethyl (**1**) ≈ phenylethyl (**2**) < vinyl (**3**) ≈ *cis*-styryl (**4**) < *trans*-styryl (**5**) as the C8-substituent reflecting the degree of the conjugation with the chlorin π-system.

Absorption spectra of epimerically pure zinc chlorins Zn-**1-5** were measured in CH_2Cl_2. In this

solution, little difference was observed between 3^1R and 3^1S epimers. The absorption maxima of the Soret and Q_x peaks of each monomer are red-shifted in the order of Zn-**1** (Soret, Q_x: 422, 513 nm) ≈ Zn-**2** (423, 515) < Zn-**3** (430, 519) ≈ Zn-**5** (430, 519) < Zn-**4** (440, 524), while their Q_y peaks are almost the same at around 648 nm. This tendency is consistent with those of the free-base **1-5**, reflecting the effect of the C8-substituents.

In nonpolar organic solvent such as 1%CH_2Cl_2-cyclohexane, zinc chlorins Zn-**1-5** self-aggregated to form oligomers, which was confirmed by the large red-shift and broadening of Q_y peak. The observed S-shape CD peaks at the region of red-shifted Q_y band also suggest the formation of the well-ordered aggregates. It should be pointed out that 3^1R epimers of Zn-MBPhes-*d* tend to form aggregates with more red-shifteds Q_y peaks than those of 3^1S epimers, although the differences are dependent on the C8-substituents. Among the examined compounds, self-aggregates of Zn-**4R/S** with *trans*-styryl group showed the largest diastereomeric control in the Q_y band. The

CP716, *Portable Synchrotron Light Sources and Advanced Applications*,
edited by H. Yamada, N. Mochizuki-Oda, and M. Sasaki
© 2004 American Institute of Physics 0-7354-0195-0/04/$22.00

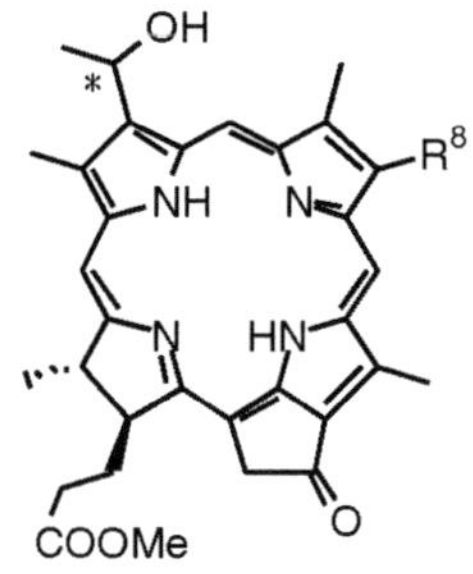

FIGURE 1. Molecular structures of synthetic methyl bacteriopheophorbides-*d* 1-5.

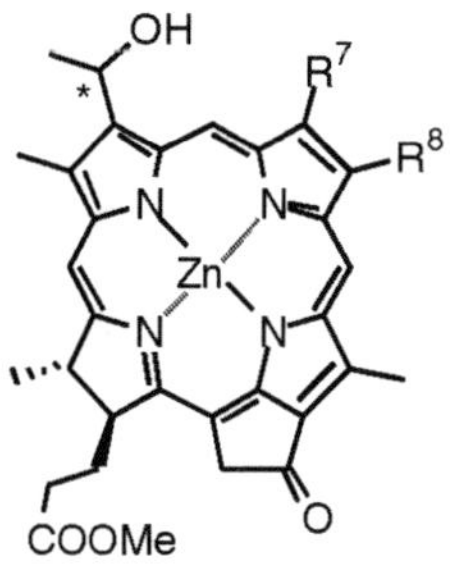

1: R^8 = Et

2: R^8 = CH$_2$CH$_2$Ph

3: R^8 = CH=CH$_2$

4: R^8 = *trans*-CH=CHPh

5: R^8 = *cis*-CH=CHPh

FIGURE 2. Molecular structures of zinc methyl bacterio-phoephorbides-*f* Zn-6 and its C8-formyl analogue Zn-7.

Zn-6: R^7=CHO, R^8=Et

Zn-7: R^7=Me, R^8=CHO

Table 1. Absorption maxima (nm) of methyl bacteriopheophorbide-*d* 1-5 in CH$_2$Cl$_2$.

Comound	Soret	Q_x	Q_y
1	409	504, 535	660
2	411	504, 536	660
3	418	509, 541	660
4	423	512	663
5	419	510, 541	662

Table 2. Absorption maxima (nm) of epimeric zinc chlorins 6 and 7.

	THF		6%THF-H$_2$O
	Soret	Q_y	Q_y
Zn-**6R**(3^{1}R)	448	628	671
Zn-**6S**(3^{1}S)	448	628	671
Zn-**7R**(3^{1}R)	448	650	702
Zn-**7S**(3^{1}S)	448	650	744

red-shift in the Q_y peaks accompanied by the formation of aggregates were 63 and 46 nm for Zn-**4R** and Zn-**4S**, respectively, and the difference of the Q_y peaks in the aggregation form is 340 cm^{-1}. Since the Q_y peak of Zn-**4R** (713 nm) was located at a more red-shifted region (*ca.* 10 nm) than the other Zn-**R** aggregates, the *trans*-styryl group might support formation of the tighter self-aggregates through any additional intermolecular interaction.

On the other hand, 3^1-epimerically pure C8-formyl analogues of Zn-MBPhe-*f* Zn-**7** were synthesized, and their spectroscopic and self-aggregation properties were compared to those of Zn-MBPhe-*f* Zn-**6** possessing the C7-formyl group as a model compound of bacteriochlorophyll-*f*.[3] Movement of a formyl group from the 7- to 8-position caused large spectral changes both in monomeric and oligomeric forms.

Monomeric Zn-**7** in THF showed a more red-shifted Q_y peak (650 nm) than Zn-**6** (628 nm), but both afforded the same Soret maximum at 448 nm. Self-aggregates of Zn-**7** in 6%(v/v)THF-H$_2$O gave a more red-shifted Q_y peak accompanied by a large diastereomeric control in the oligomeric Q_y peaks than Zn-**6** did.

When excited in the Soret region, the self-aggregate of Zn-**7S** in 6%THF-water showed intense fluorescence emission at a peak maximum of 749 nm. Although BChl-*e* (7-CHO) mainly absorbs longer wavelength lights around 500 nm than BChls-*c/d* (7-CH$_3$) do, the emission peak of BChl-*e* was located at a shorter wavelength region (~730 nm) due to its blue-shifted Q_y peak. Therefore, the self-aggregate of Zn-**3S** in the aqueous medium is quite interesting in view of both its longer wavelength Soret absorption (up to ca. 530 nm) and emission bands (~750 nm), compared with those of BChls-*c/d/e*. Since the self-aggregate can effectively absorb the most intense region of the solar spectrum and the excited singlet energy can closely fit that of energy-acceptor in natural chlorosomes, baseplate (795-nm maximum), Zn-**7S** self-aggregates are promising for construction of an efficient artificial antenna system.

REFERENCES

1. Saga, Y., Oh-oka, H., Hayashi, T., and Tamiaki, H., *Anal. Sci.* **19**, 1575-1579 (2003).
2. Sasaki, S., Omoda, M., and Tamiaki, H., *J. Photochem. Photobiol. A: Chem.* in press.
3. Sasaki, S., and Tamiaki, H., *Bull. Chem. Soc. Jpn.*, in press.

Diastereoselective Self-Assembly of Synthetic Bacteriochlorin Mimicking the Molecular Structure of Chlorosomal Bacteriochlorophylls

Michio Kunieda, Tadashi Mizoguchi and Hitoshi Tamiaki

Department of Bioscience and Biotechnology, Faculty of Science and Engineering, Ritsumeikan University, Kusatsu, Shiga 525-8577, Japan.

Abstract. We prepared self-aggregative bacteriochlorin (**1**) by modifying naturally occurring bacteriochlorophyll-*a*. These synthetic 3^1-epimers were separated and the absolute configuration at the 3^1-position was determined. Epimerically pure 3^1R- and 3^1S-**1** were mildly oxidized to afford the corresponding chlorins, 3^1R- and 3^1S-**2**, respectively. A dichloromethane solution of 3^1R- or 3^1S-**1** was diluted with 1000 fold volume of cyclohexane to give self-aggregation species absorbing lights at a near-infrared region (< 910 nm). The resulting Q_y maximum in 3^1R-**1** was 860 nm and red-shifted by 2170 cm^{-1} from the monomeric one, whereas epimeric 3^1S-**1** showed a less red-shifted peak at 798 nm. In contrast, self-aggregation of 7,8-dehydro-compound **2** showed much smaller diastereomeric control. Such visible spectra indicated that the self-aggregates of 3^1R-**1** gave different supramolecular structures from those of 3^1S-**1**. Molecular modeling calculation suggested that 3^1R-**1** self-aggregated in the same manner as 3^1R-**2** did to give large oligomer but 3^1S-**1** self-aggregated in a less-ordered orientation than 3^1S-**2** did.

INTRODUCTION

In an extramembranous antenna system of photosynthetic green bacteria, self-aggregates of bacteriochlorophyll(BChl)s-*c*, *d* and *e* consisting of a chlorin π-conjugate without any assistance of proteins are available, so called chlorosome.[1] The chlorosomal BChl molecules have a 3^1-hydroxy group, a magnesium atom as the central metal and a 13-keto-carbonyl group in a straight line. In chlorosomes, self-aggregation of natural BChls is achieved by systematic intermolecular interactions among the above moieties. In the proposed models of the self-aggregates,[2] the coordination of 3^1-hydroxy group to magnesium atom of another chlorin builds up a column of many chlorins and then the hydrogen-bonds between 13-carbonyl groups of one column and 3^1-hydroxy groups of another combine numerous columns to grow large self-aggregates. The large self-aggregates are also stabilized with the π-π interaction among the neighboring chlorins. The resulting supramolecular structures were dependent on the 3^1-stereochemistry.[2]

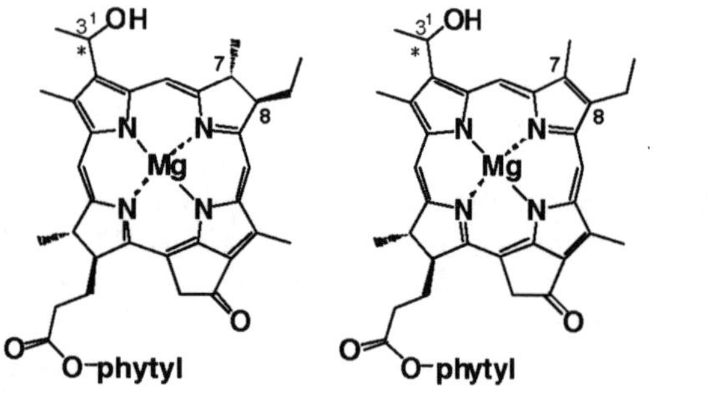

FIGURE 1. Molecular structures of self-aggregative bacteriochlorin **1** and chlorin **2**.

We have preliminarily reported self-aggregation of zinc methyl 3-(1-hydroxyethyl)bacteriopyropheophorbide-*a* as an artificial chlorosome model possessing a broad Q_y band in the near infrared region (< 900 nm).[3] Neither separation of the 3^1-epimers nor the diastereomeric self-aggregation has yet been investigated and the observed broad Q_y band might be ascribable to the diastereomeric mixture of samples.

CP716, *Portable Synchrotron Light Sources and Advanced Applications,*
edited by H. Yamada, N. Mochizuki-Oda, and M. Sasaki
© 2004 American Institute of Physics 0-7354-0195-0/04/$22.00

Here we report preparation of novel magnesium bacteriochlorins **1** possessing substituents requisite for chlorosomal BChls by modification of BChl-*a*, and also their 3^1-diastereoselective self-aggregation in non-polar organic solvents by measurements of ultraviolet-visible-near infrared (UV-VIS-NIR) and circular dichroism (CD) spectra. As reference compounds, epimerically pure chlorins **2** corresponding to the 7,8-dehydrogenated derivatives of **1** were also prepared.

RESULTS AND DISCUSSION

Figure 1 shows the molecular structures of 3-deacetyl-3-(1-hydroxyethyl)bacteriopyrochlorophyll-*a* (**1**) and [E,M]BChl-d_P (**2**). Model **1** is only differentiated from **2** in the π-conjugate skeleton; the former is a bacteriochlorin and the latter is a chlorin. Epimerically pure (bacterio)chlorins $3^1R/S$-**1** and $3^1R/S$-**2** were prepared according to previous report.[4,5]

When a dichloromethane solution of 3^1R-**1** or 3^1S-**1** was diluted with 1000 fold volume of cyclohexane, significant spectral changes were observed compared to the monomeric spectra in THF. In UV-VIS-NIR spectra of bacteriochlorins **1**, a spectral differences were controlled by the 3^1-stereochemistry. In the non-polar medium, 3^1R-**1** afforded a largely red-shifted Q_y band (860 nm), while 3^1S-**1** did less red-shifted one (798 nm). Positive and negative CD signals were observed around the red-shifted Q_y bands for 3^1R-**1** and 3^1S-**1**, respectively. These CD spectra suggest a formation of chlorosome-like self-aggregates. At the Q_x and Soret band regions, large differences were observed in CD spectra of 3^1R-**1** and 3^1S-**1** self-aggregates. The self-aggregate of 3^1R-**1** had strong CD signals which arose from the Soret and Q_x bands, whereas those of 3^1S-**1** had less intense CD signals at the regions. The enhanced CD intensities in 3^1R-**1** self-aggregates suggest that the Q_x dipole moment of 3^1R-**1** oriented linearly in the supramolecular structures. In contrast, in the 3^1S-**1** self-aggregates, the weak signals at the Soret and Q_x regions indicated that 3^1S-**1** self-aggregation had composite Q_x dipole moments less-ordered. The above UV-VIS-NIR and CD spectral results clearly propose that 3^1R-**1** self-aggregates afford large oligomers extending to both Q_y and Q_x axes in the well-ordered orientations, and that 3^1S-**1** self-aggregates afford small oligomers in a less ordered manners. In contrast, UV-VIS-NIR and CD spectra of 3^1R-**2** and 3^1S-**2** self-aggregates showed no significant 3^1-diastereomeric control (721 and 719 nm for Q_y maxima of 3^1R-**2** and 3^1S-**2** self-aggregates, respectively).

The above experimental results were successively explained by use of molecular modeling calculation. Trans-hydrogenation of the 7,8-double bonds in the energy-minimized molecular model of 3^1R-**2** dodecamer[2] was regarded as a calculated model of 3^1R-**1** dodecamer. In the model, a pair of 7,8-substituent groups in a 3^1R-**1** molecule stand clear without any disturbance to afford a tightly packing supramolecule in much the same way as 3^1R-**2** did. The molecular model of the 3^1S-**1** self-aggregate was calculated in a similar manner and showed that 7,8-substituent groups of 3^1S-**1** molecules got close to each other in the dodecamer and were unable to aggregate in the same manner with 3^1S-**2** self-aggregate. Therefore, 3^1-stereochemical control on the self-aggregates was enhanced by the trans-7,8-reduction.

ACKNOWLEDGMENTS

We thank Dr. Yoshitaka Saga of Ritsumeikan University and Dr. Shiki Yagai of Chiba University for their helpful discussions. This work was partially supported by Grants-in-Aid for Scientific Research (No. 15033271) on Priority Areas (417) from the Ministry of Education, Culture, Sports, Science and Technology (MEXT) of Japanese Government and for Scientific Research (B) (No. 15350107) from Japan Society for the Promotion of Science (JSPS). M.K. is grateful for Research Fellowship of JSPS for young scientists.

REFERENCES

1. Tamiaki, H., *Cood. Chem. Rev.* **148**, 183–197 (1996).

2. Yagai, S., Miyatake, T., Shimono, Y., and Tamiaki, H., *Photochem. Photobiol.* **73**, 153–163 (2001).

3. Tamiaki, H., Kubota, T., and Tanikaga, R., *Chem. Lett.* 639–640 (1996).

4. Kunieda, M., Mizoguchi, T., and Tamiaki, H., *J. Photosci.* **9**, 353-355 (2002).

5. Kunieda, M., Mizoguchi, T., and Tamiaki, H., *Photochem. Photobiol.* **79**, in press (2004).

Alterations of Bacteriochlorophyll *d* to *c* in Chlorosomes Seemed to Be Induced *in vitro* by Reverse Mutations of the Inactivated *bchU* Gene in a Photosynthetic Green Sulfur Bacterium *Chlorobium vibrioforme* NCIB8327

Jiro Harada[1], Hirozo Oh-oka[2], Yoshitaka Saga[1] and Hitoshi Tamiaki[1]

[1]*Department of Bioscience and Biotechnology, Faculty of Science and Engineering, Ritsumeikan University, Kusatsu, Shiga 525-8577, Japan*

[2]*Department of Biology, Graduate School of Science, Osaka University, Toyonaka, Osaka 560-0043, Japan*

Chlorosomes are unique light-harvesting antennas in photosynthetic green sulfur bacteria. The major chlorophyllous pigments are bacteriochlorophyll (=BChl)s *c*, *d* and *e*, which are only found in chlorosomes. Photosynthetic green sulfur bacteria usually contain only one kind of these three BChl species. However, it has recently been clarified that some strains in a photosynthetic green sulfur bacterium, *Chlorobium* (*Chl.*) *vibrioforme*, exhibited different profiles of chlorosomal pigment compositions. We isolated two strains containing exclusively BChl *c* or BChl *d* from *Chl. vibrioforme* NCIB 8327 (called C sub-strain or D sub-strain, respectively) [1]. We also reported that *Chl. vibrioforme* DSM 263 (called DSM 263 strain) possessed both BChls *c* and *d* [2]. As our recent investigation of the 16S rRNA indicated that all of these strains seemed to be derived from the same clone, we considered that these strains would have various activity of Bchl *c* 20-methyltransferase on the level of the corresponding gene *bchU* expression and/or the enzyme itself. Nearly around the same time, Bryant and his colleagues have demonstrated that the *bchU* gene in *Chl. vibrioforme* NCIB 8327 strain, which mainly contained BChl *d*, was flame-shifted by one base insertion resulting in the inactivation of its gene [3].

In this study, we analyzed the DNA sequences of the *bchU* gene in these strains. The obtained results suggested that the DSM 263 strain and the C sub-strain were derived from the D sub-strain by the reverse mutations in the inactivated *bchU* gene during relatively short-time cultivation just as in laboratories. We also investigated the oxygen sensitivity of C and D sub-strains, respectively. The relatively higher tolerance of the C sub-strain against oxygen compared with that of the D sub-strain would serve as a selective pressure toward reverse mutations of the less-functional *bchU* gene *in vitro*.

REFERENCES

1. Y. Saga, H. Oh-oka, T. Hayashi and H. Tamiaki, *Anal. Sci.*, **19**, 1575-1579 (2003).
2. Y. Saga, H. Higuichi, J. Harada, H. Oh-oka and H. Tamiaki, IWGHB 2003, Kisarazu, Japan (2003).
3. D. A. Bryant, N. U. Frigaard, A. G. M. Chew, J. A. Maresca and H. Li, ISPP 2003, Tokyo, Japan (2003).

CP716, *Portable Synchrotron Light Sources and Advanced Applications*, edited by H. Yamada, N. Mochizuki-Oda, and M. Sasaki

Microbial community in biofilm formed on reed surface

Suguru Okunishi[*], Yuko Kawasaki[†], Aya Takeda[†], Shin Yoda[†]
and Hisao Morisaki[†]

[*]Center for Promotion of the COE Program, Ritsumeikan University, 1-1-1 Noji-higashi, Kusatsu 525-8577, Japan
[†]Faculty of Science Engineering Ritsumeikan University, 1-1-1 Noji-higashi, Kusatsu 525-8577, Japan

Abstract. Biofilms formed on the submerged part of reed and stone surface contained various types of microorganisms, such as eukaryotes, heterotrophic and autotrophic bacteria. Culturable bacteria from biofilms of reed and stone surface were classified into wide range of phylogenetic groups. It was also found that light irradiation affect the mineralization activity of these biofilms.

INTRODUCTION

Many microorganisms are in an attached state on surfaces rather than in a free-living state. Attached microorganisms are mixed populations of various types of eukaryotes and prokaryotes forming a microbial community 'biofilm'. The microbial communities are different in function and activity from cells of single species. We have found some bacterial strains relating with photosynthetic bacteria in biofilm on the surface of reed and stone. These bacteria may have a role as an energy source to sustain the microbial community in the biofilm.

The aim of the present study is to clarify the relation between the community structure and functions of the microbes from the aspect of energy flow through the biofilm.

MATERIALS AND METHODS

Biofilms formed on reed and stone surfaces were collected from northern and southern basin of Lake Biwa. The subsamples were diluted with sterilized water and poured on nutrient broth (NB) and diluted nutrient broth (DNB) agar plates (1). These plates were incubated at 20 °C. Sequences of 16S rDNA of the isolates from these media were determined by using a BigDye Terminator v3.1 Cycle Sequencing Kit (Applied Biosystems) with an ABI 3100-Avant Genetic Analyzer automated DNA sequencer (ABI-PRISM). Phylogenetic tree was conducted using CLUSTAL W version 1.7 (2).

The mineralization activity of glucose by the biofilm and a decomposed biofilm was determined by using double-vial method (3), which enables to count the evolved $^{14}CO_2$ as consequence of mineralization of glucose labeled uniformly with ^{14}C. The mineralization activity was measured under two different conditions, with and without light irradiation (ca. 100 μ Einstein s^{-1} m^{-2}).

RESULT AND DISCUSSION

Almost a half of the isolates from the biofilm on the reed surfaces in northern and southern basin of Lake Biwa were categorized into alpha-proteobacteria phylogenetically. Some strains similar to photosynthetic bacteria were also found in the biofilm. Isolates from biofilm formed on stone surface were dominated as well by the alpha-proteobacteria.

It was clearly shown that the evolution of CO_2 from biofilm was reduced by light irradiation when the microbes were packed together in the biofilm. Photosynthetic microorganisms may utilize $^{14}CO_2$ evolved by other microbes in biofilm resulting in the decrease in CO_2 evolution. When the biofilm structure was decomposed, CO_2 evolution was largely changed. The structure of biofilm seems to be important to sustain the interrelating function of microorganisms.

CP716, *Portable Synchrotron Light Sources and Advanced Applications*,
edited by H. Yamada, N. Mochizuki-Oda, and M. Sasaki
© 2004 American Institute of Physics 0-7354-0195-0/04/$22.00

CONCLUSION

Our data of phylogenetic analysis for culturable isolates indicated that various microorganisms exist in biofilm formed on surface of reed and stone. Furthermore we obtained a clue to clarify the metabolic interrelation among the microbes in the biofilm through the effect of light irradiation on mineralization activity of biofilm.

REFERENCES

1. Suguru Okunishi, Ki-ichiro Yokota and Hisao Morisaki, *Microbes and Environments*, **15**, 1-10 (2000).

2. Thomopson, J. D., D. G. Higgins and T. J. Gibson, *Nucleic Acids Res.*, **22**, 4673-4680 (1994).

3. Vicky, L. Mckinley, Thomas W. Federle and J. Robie Vestal, *Appl.Environ. Microbiol* **45**, 255-259 (1983).

Three Group-I introns in 18S rDNA of Endosymbiotic Algae of *Paramecium bursaria* from Japan

Ryo Hoshina, Shin-ichiro Kamako and Nobutaka Imamura

Department of BioScience and Biotechnology, Ritsumeikan University, Noji Higashi 1-1-1 Kusatsu, Japan

Abstract. In the nuclear encoded small subunit ribosomal DNA (18S rDNA) of symbiotic alga of *Paramecium bursaria* (F36 collected in Japan) possesses three intron-like insertions (Hoshina et al., unpubl. data, 2003). The present study confirmed these exact lengths and insertion sites by reverse transcription-PCR. Two of them were inserted at *Escherichia coli* 16S rRNA genic position 943 and 1512 that are frequent intron insertion positions, but another insertion position (nearly 1370) was the first finding. Their secondary structures suggested they belong to Group-I intron; one belongs to subgroup IE, others belong to subgroup IC1. Similarity search indicated these introns are ancestral ones.

INTRODUCTION

The represent green ciliate, *Paramecium bursaria* Ehrenberg (Peniculida, Nassophorea), possesses several hundreds of endosymbiotic coccoid green algae. The trials for the identification of this alga have been carried out by many authors on the basis of its morphology and physiology, e. g. [1]; [2]; [3]. For this subject, we have studied several algal strains isolated from Japanese *P. bursaria* based on 18S rRNA gene, which resulted they phylogenetically close to *Chlorella* Beijerinck, Trebouxiophyceae (Hoshina et al., unpubl. data, 2003). In this gene, however, three intron-like alignments were contained. In the present study, we confirm their lengths and insertion positions, construct their secondary structure models, and consider their features.

MATERIALS AND METHODS

We used the total RNA obtained from algal symbiont isolate of *P. bursaria* strain, F36. Three introns were confirmed exact insertion positions and lengths by reverse transcription-PCR (RT-PCR) method using OneStep RT-PCR Kit (QIAGEN). Each intron was retrieved its homologous sequences from

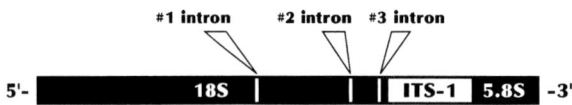

FIGIRE 1. Part of nuclear ribosomal DNA region indicating the insertion sites of the introns.

TABLE 1. FASTA similarity search results. The best three sequences are listed for each intron.

#1 intron Group IE, position 943*

Organism	GenBank	Intron type	Similarity	Overlapped	position	Phylogeny
Chlorella sp. AN 1-3	AY195964	IE	71.4%	287 nt	18S r (651*)	Trebouxiophyceae, Chlorophyta
Paecilomyces tenuipes	AB044642	IE	69.0%	261 nt	28S r	Trichocomaceae, Ascomycota
Beauveria bassiana	AF430703	IE	63.1%	312 nt	28S r	Clavicipitaceae, Ascomycota

#2 intron Group IC1, position nearly 1370*

Organism	GenBank	Intron type	Similarity	Overlapped	position	Phylogeny
Trebouxia gelatinosa	AJ249568	IC1	72.3%	260 nt	18S r (1512*)	Trebouxiophyceae, Chlorophyta
Trebouxia arboricola	Z68705	IC1	70.4%	260 nt	18S r (1512*)	Trebouxiophyceae, Chlorophyta
Trebouxia usneae	AJ249573	IC1	69.7%	274 nt	18S r (1512*)	Trebouxiophyceae, Chlorophyta

#3 intron Group IC1, position 1512*

Organism	GenBank	Intron type	Similarity	Overlapped	position	Phylogeny
Koliella spiculiformis	AF278744	IC1	69.3%	267 nt	18S r (943*)	Klebsormidiaceae, Embryophyta
Trebouxia jamesii	AJ249571	IC1	83.3%	142 nt	18S r (1512*)	Trebouxiophyceae, Chlorophyta
Trebouxia arboricola	Z68705	IC1	79.4%	160 nt	18S r (1512*)	Trebouxiophyceae, Chlorophyta

*Position numbering in 18S rRNA are based on *Escherichia coli* 16S rRNA.

CP716, *Portable Synchrotron Light Sources and Advanced Applications*,
edited by H. Yamada, N. Mochizuki-Oda, and M. Sasaki
© 2004 American Institute of Physics 0-7354-0195-0/04/$22.00

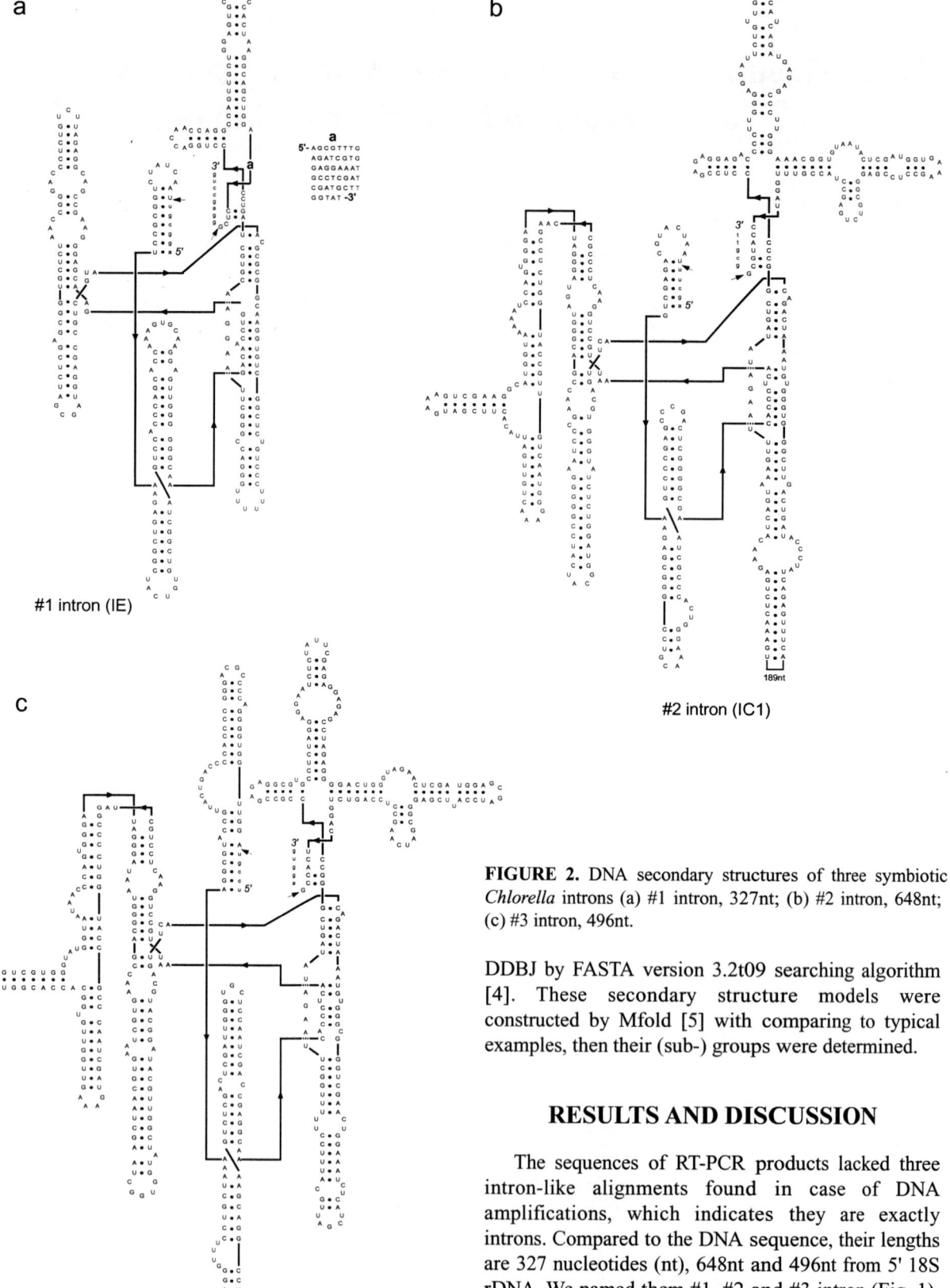

a 5'- AGCGTTTG
AGATCGTG
GAGGAAAT
GCCTCGAT
CGATGCTT
GGTAT -3'

#1 intron (IE)

#2 intron (IC1)

#3 intron (IC1)

FIGURE 2. DNA secondary structures of three symbiotic *Chlorella* introns (a) #1 intron, 327nt; (b) #2 intron, 648nt; (c) #3 intron, 496nt.

DDBJ by FASTA version 3.2t09 searching algorithm [4]. These secondary structure models were constructed by Mfold [5] with comparing to typical examples, then their (sub-) groups were determined.

RESULTS AND DISCUSSION

The sequences of RT-PCR products lacked three intron-like alignments found in case of DNA amplifications, which indicates they are exactly introns. Compared to the DNA sequence, their lengths are 327 nucleotides (nt), 648nt and 496nt from 5' 18S rDNA. We named them #1, #2 and #3 intron (Fig. 1). Similarity searches of the DDBJ sequence database with these intron sequences showed that #1 intron had moderate similarity with *Chlorella* (Trebouxiophyceae), but also similar to fungal sequences. #2 and #3 introns also had moderately similar sequences, however, these were small overlapped (Table 1). Of all similar sequences are classified as Group-I intron at the comparative RNA web (CRW) site [6], and our introns also have a feature of Group-I intron, that is, the last exon base U and the last intron base G [7]. To determine the types of introns, we tried to construct their secondary structures. We referred to above similarity search results and the typical structures (concretely we referred to the models of *Metarhizium anisopliae* var. *anisopliae* [in CRW] for #1 intron and of *Trebouxia* spp. [8] for #2 and #3 introns). The cores were manually arranged, then the stems which were basically utilized the results of Mfold prediction were added. The completed secondary structures were shown in Fig. 2. Conserved cores and stem-loop constructions are typical for Group-I intron, and they belong to subgroup IE (#1 intron) and IC1 (#2 and #3 introns). It is well known that Group-I intron inserts similar positions with other Group-I introns, and CRW site gives us the information of published data. Two of our introns, #1 and #3 introns were identified that these inserted in such spots, 943 and 1512 (numbering in *Escherichia coli* 16S rRNA). Oppositely, the identical insertion position with #2 intron was not found in CRW. The insertion position of this intron may be a new site; it inserts near position 1370 in *E. coli* 16S rRNA.

Although the functions of group-I introns are not well understood, they have been identified in organellar and nuclear genomes of diverse organisms. Similar sequences have been found among phylogenetically sporadic organisms, which strongly suggests they are mobile genetic elements capable of lateral transmission between evolutionarily distinct lineages [9]. In the study of lichen (it is composed of fungi and *Trebouxia* spp.), Bhattacharya et al. [10] and Friedl et al. [8] suggested a possibility that introns are laterally transferred in their thallus by cell-to-cell contact. The organism used in this study is an alga endosymbiotically included in the cell of the ciliate, which may be the perfect situation for lateral transmission. Similarity search (Table 1) suggested these introns were inherited from the pre-species before divergence to *Trebouxia* and *Chlorella*. It can be a new evidence for the hypothesis of lateral transmission when similar sequences were found in the genome of *P. bursaria*.

REFERENCES

1. W. Reisser, *Br. Phycol. J.* **19**, 309-318 (1984).
2. W. Reisser, S. Vietze and M. Widowski, *Symbiosis* **6**, 253-270 (1988).
3. A. E. Douglas and V. A. R. Huss, *Arch. Microbiol.* **145**, 80-84 (1986).
4. W. R. Pearson and D. J. Lipman, *PNAS* **85**, 2444-2448 (1988).
5. M. Zuker, *Nucleic Acids Res.* **31**, 3406-3415 (2003).
6. J. J. Cannone, S. Subramanian, M. N. Schnare, J. R. Collett, L. M. D'Souza, Y. Du, B. Feng, N. Lin, L. V. Madabusi, K. M. Muller, N. Pande, Z. Shang, N. Yu and R. R. Gutell, *BMC Bioinformatics* **3 (2)**, 1-31 (2002).
7. A. Mavridou, J. Cannone and M. A. Typas, *Fungal Genet. Biol.* **31**, 79-90 (2000).
8. T. Friedl, A. Besendahl, P. Pfeiffer and D. Bhattacharya, *Mol. Phylogenet. Evol.* **14**, 342-352 (2000).
9. B. Dujon, *Gene* **82**, 91-114 (1989).
10. D. Bhattacharya, T. Friedl and S. Damberger, *Mol. Biol. Evol.* **13**, 978-989 (1996).

International Symposium on Portable Synchrotron Light Sources and Advanced Applications

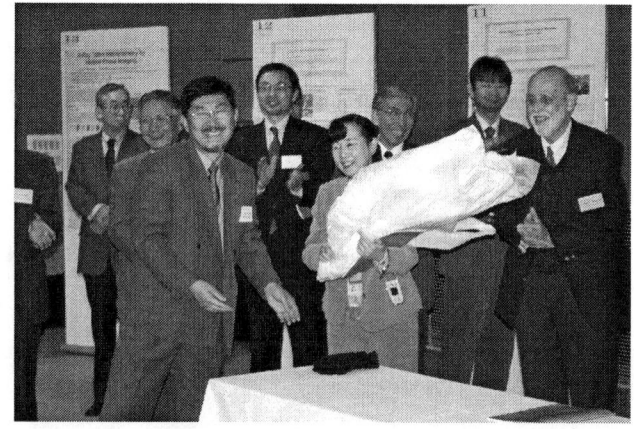

208

Symposium Program

January 13, Tuesday

Opening remarks: Sadao Kawamura (Vice President, *Ritsumeikan Universiy*)

Hironari Yamada / Junichi Chikawa / Yoshihiro Taniguchi

(Symposium Co-chairs)

Session 1: Light Sources and Instruments

1. 9:20-10:00 (Chairperson: Makoto Inoue, *Ritsumeikan University*)

Ultrabright Multikilovolt Coherent Tunable X-Ray Source at □2.71-2.93Å

Charles K. Rhodes

University of Illinois, USA

2. 10:00-10:40 (Inoue)

R&D Results on Laser-Compton Photon Beam Generation

Junji Urakawa

KEK, Japan

3. 10:50-11:30 (Chairperson: Jun-ichi Chikawa, *CAST*)

Features of the Portable Synchrotrons Named MIRRORCLE

Hironari Yamada

Ritsumeikan University, Japan

4. 11:30-12:10 (Chikawa)

Advances and Challenges of Synchrotron Medical Applications

Dean Chapman

Canadian Light Source, Canada

5. 13:10-13:50 (Chairperson: Hiroshi Kihara, *Kansai Medical University*)

The X-Ray Microscopy and Micro-Spectroscopy Facility at the ESRF

Jean Susini

European Synchrotron Radiation Facility, France

6. 13:50-14:10 (Kihara)

Large-Area Phase-Contrast X-ray Imaging System Based on a Two-Crystal X-ray

Interferometer

Akio Yoneyama, Tohoru Takeda, Yoshinori Tsuchiya, Jin Wu, Thet Lwin,
Kazuyuki Hyodo
Hitachi Ltd., Japan

7. 14:10-14:40 (Kihara)
Application of Infrared Synchrotron Radiation to Various Fields of Science
Takao Nanba
Kobe University, Japan

Session 2: Application to Materials Science
8. 14:50-15:30 (Chairperson: Yoshihiro Taniguchi, *Ritsumeikan University*)
Application of Synchrotron Radiation in the Study of the Structure and Dynamics
of Amorphous Ices and Clathrate Hydrates
John S. Tse
National Research Council of Canada, Canada

9. 15:30-16:00 (Taniguchi)
UV Photo-Oxidation Technology of Silicon: Present Status and Future Prospect
H. Oyanagi, A. Fukano
National Institute of Advanced Industrial Science and Technology, JAPAN

10. 16:00-16:30 (Taniguchi)
Hard X-ray Synchrotron Spectromicroscopy of Inhomogeneous Materials; XAFS
Analysis of Major and Minor Elements in Some Iron Meteorites
Ronald G. Cavell
University of Alberta, Canada

Poster Session: 16:30-17:30, ROHM PLAZA 3F
Open Laboratory: 17:30-18:30, MIRRORCLE 6X and MIRRORCLE 20
Banquet: 18:30-20:30, ROHM PLAZA 3F

January 14, Wednesday

Session 3: Application to Chemistry and Biophysics

11. 9:00-9:40 (Chairperson: Hitoshi Tamiaki, *Ritsumeikan University*)

Raman Determination of H-Bond Energies and H-Bond Pair Volume for Water

George E. Walrafen

University of Kansas, USA

12. 9:40-10:10 (Tamiaki)

High Pressure FTIR/Raman Studies on Molecular Conformation of Proteins and Model Peptides

Yoshihiro Taniguchi

Ritsumeikan University, Japan

13. 10:10-10:50 (Tamiaki)

Molecular Mechanism of Light-Driven Proton Pump of Bacteriorhodpsin

Mikio Kataoka

Nara Institute of Science and Technology (NAIST), Japan

14. 11:00-11:40 (Chairperson: Mikio Kataoka, *NAIST*)

Protein Crystallography: A "Must" Technology for Drug Design

Takao Matsuzaki

Mitsubishi Chemical Co. & Zoegene Co., Japan

15. 11:40-12:10 (Kataoka)

Dynamical Observations of Individual Protein Molecules using X-rays

Yuji C. Sasaki

Japan Science and Technology Corporation (JST/CREST Sasaki-team), Japan

Session 4: Imaging and Medical Application

16. 13:00-13:30 (Chairperson: Jean Susini, *European Synchrotron Radiation Facility*)

Micro XAFS study on Vanadium in Ascidians Alive Blood Cells Investigated by Fluorescence Scanning X-Ray Microscopy at ID21 at ESRF

K. Takemoto, T. Ueki, B. Fayard, A. Yamamoto, H. Sasaki, M. Salome, J. Susini, H. Michibata, H. Kihara

Kansai Medical University, Japan

17. 13:30-14:00 (Susini)

THz-Wave Parametric Source and its Imaging Application

<u>Kodo Kawase</u>

RIKEN, Japan

18. 14:00-14:30 (Chairperson: Masako Miyazaki, *Alberta University*)

The Biological Effects on Cancer Cells by Synchrotron Radiation Generated from MIRRORCLE-6X

<u>Teruki Teshima</u>, Toshiyuki Ogata, Atsuko Kawaguchi, Yuko Suzumoto, Daisuke Hasegawa, Noriko Mochizuki-Oda, Hironari Yamada

Osaka University, Japan

19. 14:30-14:55 (Miyazaki)

Homeostasis and cancer Symptom in Elemental Concentration Profiles of Hair Observed by Fluorescent X-ray Analysis with Synchrotron Radiation

<u>Jun-ichi Chikawa</u>

Center for Advanced Science and Technology, Japan

20. 15:05-15:35 (Chairperson: Masahiro Hiraoka, *Kyoto University*)

Infrared Free Electron Laser Induced Angioplasty for Arteriosclerotic Region of Blood Vessels

<u>Kunio Awazu</u>, Yuko Fukami

Osaka University, Japan

21. 15:35-16:15 (Hiraoka)

Pressure-Turning Infrared Spectroscopy: Applications to Biomedical Research and Diagnosis

<u>Patrick Wong</u>

University of Ottawa, Canada

22. 16:25-16:50 (Chairperson: Patrick Wong, *University of Ottawa*)

Mapping of Atheroma by a Synchrotron FT-IR Microscope

<u>Norio Miyoshi</u>, Tetsushi Yamada, Takao Nanba

University of Fukui, Japan

23. 16:50-17:30 (Wong)

Hyperthermia in Cancer Therapy: Current Status and Perspectives

Masahiro Hiraoka

Kyoto University, Japan

Concluding remarks: 17:30-17:40 Jun-ichi Chikawa / Yoshihiro Taniguchi

17:40-17:45 Hironari Yamada

Poster Session

P1 Quantum Mechanical Approach to the Meaning of Existence, Will and Life
<u>Hironari Yamada</u>
Ritsumeikan University, Japan

P2 The Portable Synchrotron MIRRORCLE-6X
<u>D. Hasegawa</u>, H. Yamada, A. I. Kleev, N. Toyosugi, Y. Kitazawa, T. Hayashi, T. Yamada, I. Tohyama, Y. D. Ro
Ritsumeikan University, Japan

P3 Photon Strage Ring
<u>Andrey I. Kleev</u>, Hironari Yamada
Ritsumeikan University, Japan

P4 Development of Low Energy and High Brilliance X-ray Source Using Portable Synchrotron MIRRORCLE
<u>T. Toyosugi</u>, Y. Narazaki, Y. Okazaki, T. Takashima, S. Imai, H. Yamada
Ritsumeikan University, Japan

P5 The Development of Hard X-ray Microscope with MIRRORCLE 6X
<u>T. Hirai</u>, T. Tokunaga, H. Yamada, M. Sasaki, D. Hasegawa
Ritsumeikan University, Japan

P6 Novel Edge-Enhanced X-ray Imaging by MIRRORCLE
<u>Toru Hirai</u>, Yukifumi Sonoda, Hironari Yamada, Shinya Maki, Takato Tokunaga, Daisuke Hasegawa, Norio Toyosugi, Masao Matsumoto, Kazuyuki Hyodo
Ritsumeikan University, Japan

P7 X-Ray Fluorescence Analysis of Heavy Elements with the Portable Synchrotron MIRRORCLE
<u>H. Saisho</u>, J. Hirano, T. Hirai, H. Yamada   *Ritsumeikan University, Japan*

P8 Problem of Radiation Safety in the Diagnosis Using MIRRORCLE

Y. Suetsugu, H. Yamada, D. Hasegawa, T. Hirai, Y. Sugihara

Ritsumeikan University, Japan

P9 Protein Crystallography Beam Line for MIRRORCLE

M. Sasaki, T. Hirai, H. Yamada

Ritsumeikan University, Japan

P10 High-Energy X-ray Microprobe by Multilayer Zone Plate and Microscopy

S. Tamura, M. Yasumoto, N. Kamijo, Y. Suzuki, M. Awaji, A. Takeuchi, H. Takano, K. Uesugi *National Institute of Advanced Industrial Science and Technology (AIST), Japan*

P11 Recent Progress with X-Ray Microscopy at Ritsumeikan Synchrotron Radiation Center

K. Takemoto, M. Kimura, K. Kojima, T. Matsumoto, B. Niemann, M. Hettwer, D. Rudolph, E. Anderson, D. Attwood, D. P. Kern, H. Iwasaki, H. Kihara

Kansai Medical University, Japan

P12 The X-ray microscopy project at Saga LS

M. Yasumoto, E. Ishiguro, K. Takemoto, T. Tomimasu, H. Kihara, N. Kamijo, Y. Chikaura

National Institute of Advanced Industrial Science and Technology (AIST)

P13 Possibility of Medical Phase Imaging Using X-Ray Talbot Interferometer

A. Momose, S. Kawamoto, I. Koyama

University of Tokyo, Japan

P14 Development of Dual-Energy X-ray CT using Synchrotron Radiation

M. Torikoshi, T. Tsunoo, M. Sasaki, M. Endo, Y. Noda, Y. Ohno, T. Kohno, M. Natsuhori, T. Kakizaki, N. Yamada, K. Hyodo, K. Uesugi, N. Yagi

National Institute of Radiological Science, Japan

P15 Analysis of Pheochromocytoma (PC12) Membrane Potential under the Exposure to Millimeter-wave Radiation

Maya Mizuno, Akihiko Hirata, Kodo Kawase, Chiko Otani, Tadao Nagatsuma

RIKEN, Japan

P16 FIR Beam Line for MIRRORCLE-20

A. Moon, Y. Nakamura, T. Toma, H. Yamada

Ritsumeikan University, Japan

P17 Mapping of Atheroma by a Synchrotron FT-IR Microscope

Norio Miyoshi, Tetsushi Yamada, Takao Nanba

University of Fukui, Japan

P18 Dynamical Study of Water Structure by Infrared Synchrotron Lights

Nobuhiro Miura, Hironari Yamada, Ahsa Moon, Kishi Nishikawa

Ritsumeikan University, Japan

P19 Withdrawn

P20 Volumetric Properties of Hydrophobic Hydration under High Pressure

Seiji Sawamura

Ritsumeikan University, Japan

P21 Pressure Effect on the Helix-Coil Transition of an Ala-Rich Peptides in Aqueous Solution: A FT-IR Spectroscopic Study

Takahiro Takekiyo, Akira Okuno, Takashi Imai, Akio Shimizu, Minoru Kato, Yoshihiro Taniguchi

Ritsumeikan University, Japan

P22 Protein Dynamics Research Station on IR Beam Line of MIRRORCLE-20

T. Kikuzawa, H. Yamada, N. Mochizuki-Oda, N. Miura, A. Moon

Ritsumeikan University, Japan

P23 Effects on Near-Infrared Laser on Neural Cell Activity

Noriko Mochizuki-Oda, Yosky Kataoka, Hisao Yamada, Kunio Awazu

Ritsumeikan University, Japan

P24 Study on C8-Substituent effects of Synthetic Bacteriochlorophyll-*d/f* Analogues
Shin-ichi Sasaki, Hitoshi Tamiaki
Ritsumeikan University, Japan

P25 Diastereoselective Self-Asembly of Synthetic Bacteriochlorin Mimicking the Molecular Structure of Chlorosomal Bacteriochlorophylls
Michio Kunieda, Tadashi Mizoguchi, Hitoshi Tamiaki
Ritsumeikan University, Japan

P26 Alterations of Bacteriochlorophyll *d* to *c* in Chlorosomes Seemed to be Induced *in vitro* by Reverse Mutations of the Inactivated *bchU* gene in a Photosynthetic Green Surfur Bacterium *Chlorobium vibrioforme* NCIB8327
Jiro Harada, Hirozo Oh-oka, Yoshitaka Saga, Hitoshi Tamiaki
Ritsumeikan University, Japan

P27 Microbial Community in Biofilm Formed on Reed Surface
Suguru Okunishi, Yuko Kawasaki, Aya Takeda, Shin Yoda, Hisao Morisaki
Ritsumeikan University, Japan

P28 The Origin of Endosymbiotic Algae of *Paramecium Bursaria* from Japan
Ryo Hoshina, Shin-ichiro Kamako, Nobutaka Imamura
Ritsumeikan University, Japan

LIST OF PARTICIPANTS

	Name	Institute/Company	Country
1	K. Awazu	Osaka Univ.	Japan
2	P.A.Cavell	Univ. Alberta	Canada
3	R. Cavell	Univ. Alberta	Canada
4	D. Chapman	Canadian Light Source	Canada
5	J. Chikawa	Hyogo Prefectural Center for Advanced Science and Technology	Japan
6	J. Der	Univ. Alberta	Canada
7	N. Dunwald	Univ. Alberta	Canada
8	K. Furuya	Nichicon Corporation	Japan
9	J. Harada	Ritsumeikan Univ.	Japan
10	J. Harada	Rigaku Corporation	Japan
11	D. Hasegawa	Ritsumeikan Univ.	Japan
12	T. Hirai	Ritsumeikan Univ.	Japan
13	A. Hirao	Ritsumeikan Univ.	Japan
14	M. Hiraoka	Kyoto Univ.	Japan
15	R. Hoshina	Ritsumeikan Univ.	Japan
16	Y. Ikeuchi	Ritsumeikan Univ.	Japan
17	S. Imai	Ritsumeikan Univ.	Japan
18	M. Inoue	Ritsumeikan Univ.	Japan
19	K. Kasesawa	Ritsumeikan Univ.	Japan
20	M. Kataoka	Nara Institute of Science and Technology	Japan
21	H. Katayama	TOYAMA Co.,Ltd	Japan
22	S. Katayama	Nitto Denko Corporation	Japan
23	M. Kato	Ritsumeikan Univ.	Japan
24	A. Kawaguch	Osaka Univ.	Japan
25	K. Kawase	RIKEN	Japan
26	H. Kihara	Kansai Medical Univ.	Japan
27	T. Kikuzawa	Ritsumeikan Univ.	Japan
28	Y. Kitazawa	Photon Production Laboratory, Lt'd	Japan
29	A. Kleev	Ritsumeikan Univ.	Japan
30	M. Kunieda	Ritsumeikan Univ.	Japan
31	H. Maekawa	Institute for Environmental Science	Japan

32 E. Mantey	Univ. Alberta	Canada
33 M. Matsumoto	Osaka Univ.	Japan
34 T. Matsuzaki	Mitsubishi Chemical Co. & Zoegene Co.	Japan
35 N. Miura	Ritsumeikan Univ.	Japan
37 M. Miyazaki	Univ. Alberta	Japan
38 T. Miyazaki	Spring-8/JASRI JST/CREST	Japan
36 N. Miyoshi	Fukui Univ.	Japan
39 M. Mizuno	RIKEN	Japan
40 N. Mochizuki-Oda	Ritsumeikan Univ.	Japan
41 A. Momose	Univ. Tokyo	Japan
42 A. Moon	Ritsumeikan Univ.	Japan
43 M . Mori	Ritsumeikan Univ.	Japan
44 O. Nakamatsu	APOLLOMEC CO.,LTD	Japan
45 T. Nanba	Kobe Univ.	Japan
46 H. Noguchi	Ritsumeikan Univ.	Japan
47 T. Ogata	Osaka Univ.	Japan
48 H. Ohyanagi	National Institute of Advanced Industrial Science and Technology	Japan
49 S. Okunishi	Ritsumeikan Univ.	Japan
50 A. Okuno	Ritsumeikan Univ.	Japan
51 T. Oochi	NTT Advanced Technology Corporation	Japan
52 H. Ooi	NEC TOKIN Corporation	Japan
53 C. Rhodes	Univ. Illinois	USA
54 H. Saisho	Ritsumeikan Univ.	Japan
55 A. Sakumoto	Yoshizawa-LA CO., LTD	Japan
56 M. Sasaki	Ritsumeikan Univ.	Japan
57 S. Sasaki	Ritsumeikan Univ.	Japan
58 Y. C. Sasaki	Japan Science and Technology Corporation, Spring-8	Japan
59 T. Sawa	Ritsumeikan Univ.	Japan
60 S. Sawamura	Ritsumeikan Univ.	Japan
61 Y. Suetsugu	Ritsumeikan Univ.	Japan
62 J. Susini	European Synchrotron Radiation Facility	France
63 Y. Suzuki	Osaka Univ.	Japan
64 T. Takahashi	Bunkoh-Keiki Co., LTD	Japan
65 T. Takekiyo	Ritsumeikan Univ.	Japan

66 K. Takemoto	Kansai Medical Univ.	Japan
67 H. Tamiaki	Ritsumeikan Univ.	Japan
68 S. Tamura	National Institute of Advanced Industrial Science and Technology, Kansai	Japan
69 Y. Taniguchi	Ritsumeikan Univ.	Japan
70 T. Tesima	Osaka Univ.	Japan
71 M. Torikoshi	National Institute of Radiological Science	Japan
72 N. Toyosugi	Ritsumeikan Univ.	Japan
73 J. S. Tse	National Research Council of Canada	Canada
74 T. Tsunoo	National Institute of Radiological Science	Japan
75 J. Urakawa	KEK	Japan
76 G. E. Walrafen	Univ. Kansas	USA
77 P. Wong	Otawa Univ.	Canada
78 H. Yamada	Ritsumeikan Univ.	Japan
79 R. Yamada	Photon Production Laboratory, Lt'd	Japan
80 A. Yoneyama	Hitachi Co. Ltd	Japan

W

Walrafen, G. E., 45
Webb, M. A., 36
Wong, P. T. T., 89
Wu, J., 22

Y

Yagi, N., 160
Yamada, H., 12, 73, 109, 116, 120, 124, 128,
  132, 135, 139, 141, 167, 171, 188, 192

Yamada, N., 160
Yamada, T., 98, 116
Yamamoto, A., 65
Yamashita, H., 128
Yasumoto, M., 144, 152
Yoda, S., 201
Yoneyama, A., 22

Z

Zhang, P., 5